Lecture Notes in Earth Sciences   127

Editors:

J. Reitner, Göttingen
M. H. Trauth, Potsdam
K. Stüwe, Graz
D. Yuen, USA

Founding Editors:

G. M. Friedman, Brooklyn and Troy
A. Seilacher, Tübingen and Yale

Ram S. Sharma

Cratons and Fold Belts of India

Springer

Prof. Dr. Ram S. Sharma
University of Rajasthan
Dept. Geology
Jaipur-302004
India
sharma.r.sw@gmail.com

ISSN 0930-0317
ISBN 978-3-642-01458-1   e-ISBN 978-3-642-01459-8
DOI 10.1007/978-3-642-01459-8
Springer Heidelberg Dordrecht London New York

Library of Congress Control Number: 2009926971

© Springer-Verlag Berlin Heidelberg 2009
This work is subject to copyright. All rights are reserved, whether the whole or part of the material is concerned, specifically the rights of translation, reprinting, reuse of illustrations, recitation, broadcasting, reproduction on microfilm or in any other way, and storage in data banks. Duplication of this publication or parts thereof is permitted only under the provisions of the German Copyright Law of September 9, 1965, in its current version, and permission for use must always be obtained from Springer. Violations are liable to prosecution under the German Copyright Law.
The use of general descriptive names, registered names, trademarks, etc. in this publication does not imply, even in the absence of a specific statement, that such names are exempt from the relevant protective laws and regulations and therefore free for general use.

Cover design: Bauer, Thomas

Printed on acid-free paper

Springer is part of Springer Science+Business Media (www.springer.com)

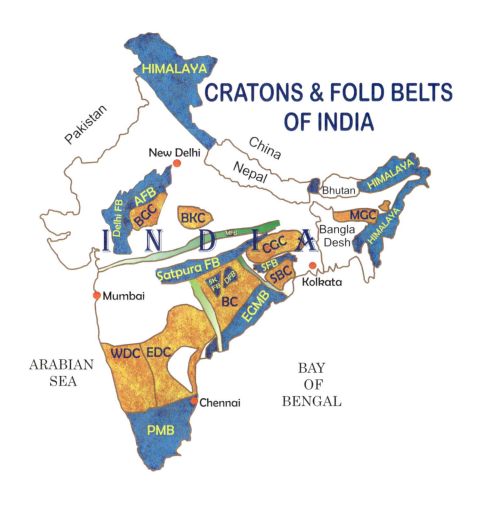

Cover Page Abbreviations

Blue Areas Are Fold Belts

AFB	Aravalli Fold Belt
Delhi FB	Delhi Fold Belt
DFB	Dongargarh Fold Belt
EGMB	Eastern Ghats Mobile Belt
MFB	Mahakoshal Fold Belt
PMB	Pandyan Mobile Belt
Satpura FB	Satpura Fold Belt
SFB	Singhbhum Fold Belt
SKFB	Sakoli Fold Belt

Orange Areas Are Cratons

BC	Bastar Craton
BKC	Bundelkhand Craton (massif)
BGC	Banded Gneissic Complex
CGC	Chhotanagpur Granite-Gneiss Complex
EDC	Eastern Dharwar Craton
MGC	Meghalaya Craton
SBC	Singhbhum Craton
WDC	Western Dharwar Craton

A hundred times a day I remind myself that my inner and outer lives are based on the labors of other people, living and dead, and that I must exert myself in order to give in the same measure as I have received and am still receiving.
Albert Einstein

Dedicated to the memory of my sister Rajbai and her mother-in-law Bhoribai whose benevolence and guardianship helped me attain what I could.

Foreword

Professor Ram Swaroop Sharma whom I affectionately address as *Bhratashri* has asked me to write a foreword for his book on "Cratons and Fold Belts of India". Bhratashri's wish is always my command and this foreword is written with both delight and humility.

Formation of rocks to make up the continental crust and the assembly of such rocks to form a contiguous and large physiographic feature called landmass is a first order geological problem. Several other geological problems such as the formation of various landforms (mountains, plateaus and plains), various mineral deposits, development/maintenance/evolution of hydrologic cycle, climate and life, and rivers and their basins are integral parts of this first order problem. Just as the fundamental thing called water is taken for granted, so also the existence of land on earth. Although land may not be a critical environmental resource, the existence of land is critical to the physics and chemistry of the life system on earth. This basic truth, perhaps realized by Vedic men, is dawned on to the contemporary scientific community only towards the end of the twentieth century, thanks to the developing environmental problems.

Our knowledge of the earth, its materials and processes, is very rapidly increasing due to advancement in technical tools of investigations and in theoretical aspects of physics, chemistry and biology. The advent of plate tectonic theory after the Second World War and the realization of environmental consequences of consumption of earth materials and ecosystem services contributed very significantly to our understanding of earth processes. Studies of the Precambrian Geology of India have made very good use of all these development as well as International scientific cooperation. Although our knowledge of the formation of India as a landmass is better today than ever before, it is still far from what is required to sustain human "civilization."

Prof. Sharma's book on "Cratons and Fold Belts of India" is an excellent exposition of the current status of our knowledge of the geology of Archaean cratonic parts of the Indian landmass and of the geology and processes of formation of Proterozoic fold belts on the cratonic basements (cratonic, of course, not during Proterozoic!). The book also provides, for an uninitiated but interested, in the first three chapters, concepts of terrain and terrain making processes, major cratonic terrains of India and processes of formation of ensialic fold belts. For each of the Proterozoic fold belts in India, possible models of evolution are discussed providing an excellent account for anyone interested in further research. The book is also an exposition of the volume of research that has been done in India in the past five decades as well as the volume of knowledge of Prof. Sharma on Indian Geology. Future generation of students of Geosciences will sure appreciate the mind and service of Professor Ram Sharma.

New Delhi, India   V. Rajamani

Preface

The fold belts, also known as mobile belts, whether the young Tertiary mountain of the Himalaya or the old Proterozoic fold belts, occur as linear belts of deformed and metamorphosed rocks within different continents of the Earth. To establish the continuity of these fold belts and to attempt correlations with other coeval fold belts elsewhere as well as to model their evolutionary mechanism is rather a difficult task. This is because of the subsequent lithospheric plate movements, which brought about assembly/reassembly of the continents during various geological periods. In the plate tectonic evolution of the fold belts, the Himalaya is the best example to understand the events of plate collision and to recognize colliding plates, subduction zones, arc, etc as the elements of the plate tectonics process. Undoubtedly, the evolution of the Himalaya would help us in organizing our observations and developing our skills to apply plate tectonics theory for the evolution of the Proterozoic fold belts of India. By studying the geological characteristics of the fold belts we can possibly decipher the tectonic history of the Earth preserved in these fold belts and nowhere else. Proterozoic fold belts are the main repositories of information about early plate tectonic interactions. The Proterozoic fold belts of India are significant in this respect, as they would help in tracing the past history of the cratonic regions of the Indian shield and in attempting global correlations. This needs a thorough investigation of these fold belts in regard to their geological/geochronological characteristics and evolutionary processes. The available geological information on these fold belts is, however, scattered and the different evolutionary models need a serious scrutiny towards a subjective overview, along with the role played in the process by the cratonic areas of the Indian shield. In the present book, we attempt to understand the nature, composition and antiquity of the cratonic regions of the Indian shield and their relationship with the fold belts in the neighbourhood, including the Himalaya, so as to arrive at a "definite point of view" concerning their origin. All published geodata, old and new, have been evaluated and interpreted in a logical way. A mere accumulation of facts would make science dull unless the facts are given a meaningful interpretation, either to strengthen a theory or to propose a new model replacing the existing paradigms. The author has adopted this approach while handling the available geological database.

The basic plan of the book is as follows: The first chapter begins with the fundamental concepts about Precambrian terrains and Precambrian tectonics. It is followed by a historical development of the theme of mountain building processes with a detailed account of plate tectonics and its features. In the second chapter, the cratonic areas of the Indian shield are described, because not only the evolution of the fold belts involved the process of rifting and collision of the cratonic areas, but also because the fold belts are either located at the continental margins or are sandwiched between two cratonic regions. In Chap. 3, evolution of the Himalaya is taken up as an illustrative example of the development of an orogenic belt in the plate tectonic setting. The tectono-thermal evolution of this Tertiary belt will help provide an understanding of the Proterozoic fold belts and their associated plate tectonic environments. In the subsequent chapters, each of the Proterozoic fold belts is described, incorporating its geological characteristics, the agreed-on observations and facts on the geological database pertaining to the deformation, metamorphism, and geochronology, with discussion of the evolutionary models proposed by different workers. Each of the proposed models has been evaluated in terms of these agreed observations and the geological database and a more plausible model is given by the author at the close of each chapter. Geophysical data inputs are, however, often missing in this modeling exercise, primarily for lack of author's own understanding of geophysics and for not having appreciation of relating the present subsurface features revealed by geophysics with the features of the geological past. The last chapter, Chap. 8, includes the Pandyan mobile belt of the southern India granulite terrain. This chapter is presented in somewhat greater detail for its different granulite blocks. Throughout this book, I have tried to keep the approach well-motivated and to strive for simplicity of presentation. As an interpreter of geosciences, care has been taken to clarify the idea of the science of the fold belts, to compare and contrast the different evolutionary models, some of which are only of historical value and even fit for rejection. By such an approach the younger generation will be benefited as well as encouraged.

The reader is advised to go back to the original literature whenever possible and see what that original writer had in mind. I have tried to aid such students with limited but essential references so that the reader should not miss the special points of the view that the authors put forward. When reporting words someone has used, I followed standard usage and used double quotation marks. This is to indicate the reader that the word or phrase is being used in some special way, with a special meaning or just to indicate that the discussion is calling the customary usage of the word in quotation.

Jaipur, India   Ram S. Sharma

Acknowledgements

While writing this book, I received encouragements from several friends and colleagues. Dr. K.R. Gupta of Department of Science and Technology (DST), New Delhi has been a constant reminder to this writer of the "academic responsibility" toward the younger generations of geoscientists of the country. It was not, however, until 2002 that the task was seriously begun. In this endeavor I got fullest support from my wife, Kanchan, and I took full liberty of her gentle nature, spending long hours with the task of writing chapters and typing or retyping them until the manuscript took the "final" version. A part of the book was revised when the author was in the USA with his sons, Professor Anurag Sharma at RPI in Troy and Dr. Prashant Sharma in Chicago. During writing this book, the author reckoned early formative days of his student career, which he spent under the guardianship of his elder sister, Rajbai, in her temple at Jaipur. The author is highly indebted to her for inculcating in him the pleasure of doing things for others. While writing this book, I received constant encouragements from friends and colleagues, whose regular enquiries about the progress of the book kept my morale all-times high. I express my appreciation to them, especially Dr. K.K. Sharma, Dr. Mallickarjun Joshi, Dr. Harel Thomas, Dr. Manoj Pandit and Professors A.K. Jain and V. Rajamani, Professor B.S. Paliwal and Professor C. Leelanandam. Thanks are also due to Dr. T.R.K. Chetty and Dr. Y.J. Bhaskar Rao of NGRI, Hyderabad for reading Chapters 7 & 8 of the manuscript. Special thanks are due to Dr. Tej Bahadur, senior research fellow in the department of geology of the University of Rajasthan at Jaipur, who prepared computer-aided sketches for this book without which it would not have been possible to bring the book in its present form. Dr. Rajiv Nigam is thanked for friendly suggestions pertaining to the figures accompanying the text. The editorial staff of Springer-earth sciences specially Drs. Chris Bendall, Petra van Steenbergen and Janet Sterritt, have done a marvellous job in bringing out this publication in a short time of seven months. Finally, this book could not have been published without the project grants from Indian National Science Academy, New Delhi.

Contents

1 Precambrian Terrains and Mountain Building Processes ... 1
   1.1 Introduction ... 1
   1.2 Archaean Terrains ... 3
   1.3 Proterozoic Terrains ... 7
   1.4 Mountain Building Processes ... 9
      1.4.1 Geosynclinal Theory ... 9
      1.4.2 Plate Tectonics Theory ... 12
   1.5 Continental Rifting and Sedimentary Basins ... 22
      1.5.1 Rifting and Triple Junctions ... 24
      1.5.2 Pull-Apart Basin ... 28
      1.5.3 Foreland Basin ... 30
   1.6 Orogeny and Plate Tectonics ... 30
   1.7 The Wilson Cycle ... 35
   1.8 Suspect Terrains ... 38
   References ... 39

2 Cratons of the Indian Shield ... 41
   2.1 Introduction ... 41
   2.2 Dharwar Craton ... 43
      2.2.1 Introduction ... 43
      2.2.2 Evolution of Dharwar Schist Belts ... 50
      2.2.3 Geodynamic Models for the Evolution of Dharwar Craton ... 52
   2.3 Bastar Craton ... 58
      2.3.1 Introduction ... 58
   2.4 Singhbhum Craton ... 65
      2.4.1 Geodynamic Evolution ... 72
   2.5 Chhotanagpur Granite-Gneiss Complex ... 72
      2.5.1 Archaean Status of CGGC Questioned ... 81
   2.6 Rajasthan Craton ... 83
      2.6.1 Banded Gneissic Complex and Berach Granite ... 84
      2.6.2 BGC: Antiquity Challenged and Nomenclature Changed ... 85
      2.6.3 Geological Setting ... 87

	2.6.4		Evolution of the Banded Gneissic Complex	92
	2.6.5		Bundelkhand Granite Massif (BKC)	94
	2.7	Meghalaya Craton	98
	References	102

3 The Young and Old Fold Belts ... 117
- 3.1 Introduction ... 117
- 3.2 Himalaya as an Example of the Young Fold Belt ... 118
  - 3.2.1 Geological Setting ... 118
  - 3.2.2 Evolution of Himalaya ... 122
- 3.3 Old Fold Belts ... 134
- 3.4 Proterozoic Fold Belts of India ... 136
- 3.5 Ensialic Orogenesis Model ... 136
- References ... 138

4 Aravalli Mountain Belt ... 143
- 4.1 Introduction: Geological Setting ... 143
- 4.2 Aravalli Fold Belt ... 143
  - 4.2.1 Aravalli Stratigraphy: Controversies ... 146
  - 4.2.2 Geochronology: Time of Aravalli Orogeny ... 148
- 4.3 Delhi Fold Belt ... 149
  - 4.3.1 Delhi Stratigraphy: Controversies ... 150
  - 4.3.2 Geochronology: Time of Delhi Orogeny ... 151
- 4.4 Deformation of the Proterozoic Fold Belts ... 153
- 4.5 Geohistory of Granulites ... 154
- 4.6 Agreed Observations and Facts ... 157
- 4.7 Evolutionary Models and Discussion ... 159
- 4.8 Geophysical Database ... 170
- 4.9 Conclusions ... 171
- References ... 172

5 Central Indian Fold Belts ... 177
- 5.1 Introduction ... 177
- 5.2 Mahakoshal Fold Belt ... 179
  - 5.2.1 Introduction: Geological Setting ... 179
  - 5.2.2 Agreed Observations and Facts ... 180
  - 5.2.3 Evolution of the Mahakoshal Fold Belt (MFB) ... 181
- 5.3 Satpura Fold Belt ... 182
  - 5.3.1 Introduction: Geological Setting ... 182
- 5.4 Agreed Observations and Facts ... 185
  - 5.4.1 Evolution of the Satpura Fold Belt (SFB) ... 186
- 5.5 Geological Significance of the Central Indian Tectonic Zone ... 189
- 5.6 Dongargarh Fold Belt ... 191
  - 5.6.1 Introduction: Geological Setting ... 191
  - 5.6.2 Agreed Observations and Facts ... 194
  - 5.6.3 Evolution of the Dongargarh Fold Belt (DFB) ... 194

	5.7	Sakoli Fold Belt	195
		5.7.1 Introduction: Geological Setting	195
		5.7.2 Agreed Observations and Facts	196
		5.7.3 Evolution of the Sakoli Fold Belt (SKFB)	197
	5.8	Discussion of Evolutionary Models of the Central Indian Fold Belts	198
	References		204
6	**Singhbhum Fold Belt**		209
	6.1	Introduction: Geological Setting	209
	6.2	Deformation	213
	6.3	Metamorphism	214
	6.4	Geochronolgy	215
	6.5	Agreed Observations and Facts	217
	6.6	Evolution of the Singhbhum Fold Belt (SMB)	218
	6.7	Evolutionary Models and Discussion	220
	References		226
7	**Eastern Ghats Mobile Belt**		231
	7.1	Introduction: Geological Setting	231
	7.2	Deformation	234
	7.3	Geochronology	235
	7.4	Petrological Characteristics	236
	7.5	Tectono-Thermal Evolution	236
	7.6	Northern Segment	237
		7.6.1 Southern Segment	239
	7.7	Is Godavari Graben a Major Terrain Boundary?	241
	7.8	Agreed Observations and Facts	242
	7.9	Evolutionary Models and Discussion	243
	7.10	Newer Divisions of the Eastern Ghats Mobile Belt	248
		7.10.1 Rengali Province	253
		7.10.2 Jeypore Province	253
		7.10.3 Krishna Province	253
		7.10.4 Eastern Ghats Province	254
	References		257
8	**Pandyan Mobile Belt**		263
	8.1	Introduction: Geological Setting	263
	8.2	Geological Attributes of the Granulite Blocks	265
	8.3	Pandyan Mobile Belt	272
	8.4	Agreed Observations and Facts	276
	8.5	Evolutionary Models and Discussion	276
	References		284
Postscript			291
Appendix			293

About the Book 295
About the Author 297
Index 299

List of Figures

1.1	Tectonic units of the continents: the Shields and Orogenic belts (after Umbgrove, 1947)	3
1.2	Non-uniformitarian evolutionary model of Greenstone belt (after Windley, 1984). (**a**) Synformal structure; (**b**) crustal thickening; (**c**) uplift of deformed greenstones	5
1.3	Plate tectonic model for the formation of Greenstone belt exemplified by Rocas Verdes Complex, South Chile (after Tarney et al., 1976). (**a**) Extensional phase and basin formation; (**b**) volcanic phase and emplacement of basalt; (**c**) continental sediments and arc lavas as supracrustals; (**d**) deformation due to basin closure and intrusion of synorogenic tonalite; (**e**) emplacement of late potassic Granite	6
1.4	Plate tectonic model for the evolution of greenstone-gneiss-granulite in the Archaean (redrawn from Windley, 1984). (**a**) Several small continental plates with ensialic back-arc basins. The Andean-type margins are invaded by mantle-derived tonalitic batholiths and plutons; (**b**) aggregations of the continents to produce a large continental plate consisting of greenstone and granulite belts	7
1.5	Geosynclinal model for the development of an orogenic belt	10
1.6	Depth zones and distribution of deformed and metamorphosed rocks in a crustal section	11
1.7	Block diagram showing the different plate margins. Mantle material rises below the Constructive plate margins (ocean ridges) and plate material descends into the mantle at Convergent (destructive) plate margins (ocean trenches). At Conservative plate margins (transform faults) plates slide past each other, without getting created or destroyed	13
1.8	Representation of development of a ridge-to-ridge transform (adapted from Verhoogen et al., 1970, Fig. 3.75). (**a**) Oceanic ridge (*double line*) along which new crustal material added is displaced by a sinistral strike-slip fault (*single line*);	

	(**b**) addition of magma between line A and A and slip on the fault; (**c**) addition of material between lines B and B; scar of previously active sections of the fault extend laterally into the surrounding crust. Note that the sense of slip on the transform fault is opposite of that inferred from the displacement of the ridge by strike-slip faulting	15
1.9	Physiographic-cum-geologic units in a convergence of the Ocean-Ocean plates (see text for details). e = eclogite facies; gl = glaucophane schist facies; pp = prehnite-pumpellyite facies; z = zeolite facies	17
1.10	Different mechanisms for the formation of Back-arc basin (after Moores and Twiss, 1995). (**a**) Forceful injection of a diapir from the subducting slab and extension of the overlying crust; (**b**) Convection currents driven by the drag of the downgoing slab resulting into extension of oceanic crust behind the island arc; (**c**) Retreat of overriding plate by global plate motion changes compression to extension of the oceanic crust behind the arc; (**d**) Back-arc basin formed by change of a Transform fault to a subduction zone	21
1.11	Marginal Basin (Back-arc basin) formed due to secondary spreading (see text)	21
1.12	Ductile stretching as a possible mechanism for rift basin formation (see text)	23
1.13	Stages in the evolution of a Rift (see text for details)	23
1.14	Development of an Extensional basin in a continental lithosphere. (**a**) Rising convection currents generate tensional stresses, causing stretching of the crust; (**b**) deposition of sediments and basaltic lava in the extensional basin	24
1.15	Evolution of triple junction from one time to later time (after Kearey and Vine, 1996); see also text	25
1.16	RTF triple junction and velocity vectors as indicators of stability/instability of Plate boundaries (after Kearey and Vine, 1996); see text	26
1.17	Development of an Aulacogen from a RRR triple junction; see text for details	27
1.18	Pull-apart basins due to strike-slip motion and extension. (**a**) Basin at a fault curvature during transtensional regime, (**b**) sharp pull-apart basin by strike-slip motion, (**c**) basins formed as a result of fault termination	29
1.19	Stages in the formation of Pull-apart Basins (see text for details)	29
1.20	Foreland basin formation as a flexural depression of the continental crust	30
1.21	Different sedimentary deposits in shelf, slope and deep-sea conditions within a geosyncline	31

List of Figures   xxiii

1.22	Schematic cross section of converging ocean-continent plates, showing island arc and the distribution of metamorphic facies. High P/T facies series develops in the subduction zone while low P/T facies series is formed in the arc. *Dotted line* is trace of an erosion plane along which a progressive sequence of regional metamorphism is encountered in eroded fold belts	32
1.23	Schematic map showing the development and location of plate tectonic elements and the distribution of metamorphic facies at a convergent continent-ocean plate boundary	34
1.24	Schematic illustration of the Wilson cycle (after Dewey and Bird, 1970)	37
2.1	Outline map of the Indian shield showing the distribution of cratons, including Chhotanagpur Granite Gneiss Complex, Rifts and Proterozoic fold belts. Megahalaya craton (formerly Shillong Plateau) is not outlined here but is shown in Fig. 2.10. The fold belts are: 1 = Aravalli Mt. belt, 2 = Mahakoshal fold belt, 3 = Satpura fold belt, 4 = Singhbhum fold belt, 5 = Sakoli fold belt, 6 = Dongargarh fold belt, 7 = Eastern Ghats mobile belt, 8 = Pandyan mobile belt. Abbreviations: Opx-in = orthopyroxene-in isograd; AKS = Achankovil shear zone; PC-SZ = Palghat Cauvery Shear Zone; SGT = Southern Granulite Terrain; SONA Zone = Son-Narmada lineament	42
2.2	Simplified geological map of the southern Indian shield (after GSI and ISRO, 1994) showing the Dharwar (-Karnataka) craton and its Schist belts, Southern Granulite Terrain (SGT) and shear zones. Metamorphic isograds between greenschist facies and amphibolite facies and between amphibolite and granulite facies (i.e. Opx-in isograd) are after Pichamuthu (1965). Greenstone (Schist) belts in the Dharwar craton (Eastern and Western blocks) are: 1 = Shimoga, 2 = Bababudan, 3 = Western Ghats, 4 = Chitradurga, 5 = Sandur, 6 = Kolar, 7 = Ramagiri, 8 = Hutti. Locality: BR = Biligiri Rangan Hills, Ch = Chitradurga, Cg = Coorg, Dh = Dharwar, Jh = Javanahalli, Kn = Kunigal, S = Shimoga. Shear zones: Bh = Bhavani; Ch-Sz = Chitradurga Shear Zone, M = Moyar; M-Bh = Moyar-Bhavani	44
2.3	Geological sketch map of central part of the Kolar Schist Belt (after Hanson et al., 1988)	52
2.4	Block diagram showing the crustal evolution in the Kolar area (after Hanson et al., 1988). See text for details	56
2.5	Central Indian fold belts and cratons. (**a**) Location of central Indian fold belts. (**b**) Geological setting of Bastar Craton in relation to adjacent cratons and Central Indian Tectonic	

	Zone (CITZ). Abbreviations: BC = Bastar craton, CGGC = Chhotangapur Granite Gneiss Complex, CIS = Central Indian shear zone, DC = Dharwar craton, SBC = Singhbhum craton, SONA ZONE = Son-Narmada Lineament zone bounded by Son-Narmada North Fault (SNNF) and Son-Narmada South Fault (SNSF). (**c**) Simplified geological map (*bottom sketch*) shows the CITZ sandwiched between Bastar craton in the south and Bundelkhand craton in the north. The four localities of granulites described in the Satpura fold belt are: BBG = Bhandara-Balaghat granulite, BRG = Bilaspur-Raipur granulite, MG = Makrohar granulite, and RKG = Ramakona-Katangi granulite	61
2.6	Location and Geological map of Singhbhum (-Orissa) craton comprising Archaean rocks of Older Metamorphic Group (1) and Older Metamorphic Tonalite Gneiss (2), Singhbhum Granite Group (SBG) with three phases (I, II, & III) of emplacement, and Iron-Ore Group (IOG) made up of: 1 – lavas and ultramafics, 2 – shale-tuff and phyllite, 3 – BHJ, BHQ, sandstone and conglomerate. Abbreviations: C = Chakradharpur, D = Daiteri, K = Koira, SSZ = Singhbhum shear zone. (1) = Singhbhum Granite, (2) = Bonai Granite, 3 = Mayurbhanj Granite	66
2.7	Location and distribution of Chhotanagpur Granite Gneiss Complex (CGGC). The five geological belts (I–V) outlined by zigzag line in the CGGC terrain are drawn after Mahadevan (2002). *Dotted lines* are boundaries of the States	75
2.8	Simplified geological map of Rajasthan craton (after Heron, 1953 and GSI, 1969), made up of Banded Gneissic Complex (BGC), Berach Granite and other Archaean granitoids. Granulite outcrops are in the BGC terrain and in the metasediments of the Delhi Supergroup (see text for details). Blank area occupied by Proterozoic fold belts and sand cover. Abbreviations: BL = Bhilwara, BW = Beawar, N = Nathdwara, M = Mangalwar. Inset shows the location of BBC (Banded gneissic complex-Berach Granite) and BKC (Bundelkhand) cratonic blocks that together constitute what is here termed the Rajasthan (-Bundelkhand) Craton, abbreviated RC	83
2.9	Location and geological map of the Bundelkhand craton/massif (after Basu, 1986). BB = Bijawar basin, GB = Gwalior basin, SB = Sonrai basin	95
2.10	Location of Meghalaya in the northeastern states of India (**a** and **b**) and Geological map (**c**) of the Shillong-Mikir Hills Plateau, herein called the Meghalaya craton, with location of Garo Hills (G), Khasi-Jantia Hills (K-J), Mikir	

	Hills (renamed as Kabri Hills by some authors). Other abbreviations: DF = Dauki Fault, DT = Disang thrust, MBT = Main Boundary thrust, MT = Mizu thrust, NT = Naga thrust, SPB = Schuppen Belt (Zone of imbrication)	99
3.1	Generalized geological map of the Himalaya showing the principal geological divisions (litho-tectonic units), based on the compilation map of GSI (1962), Gansser (1964), and Valdiya (1977). Bottom figure is a crustal section showing the relationship of different shear zones. Abbreviations: HHC = Higher Himalayan Crystallines, ITSZ = Indus-Tsangpo shear zone, MBT = Main Boundary Thrust, MCT = Main Central Thrust, MFT (MBF) = Main Frontal thrust (Main Boundary fault)	119
3.2	Sequence of events in the evolution of the Himalaya (redrawn after Allegre et al., 1984). Abbreviatons: BNS = Bongong-Nujiang suture, ITSZ = Indus-Tsangpo suture zone, KS = Kokoxili suture, MCT = Main Central Thrust, MBT = Main Boundary Thrust	123
3.3	A plausible model (after Sharma, 2005) for the inverted metamorphism in the Himalayan metamorphic belt (HMB), shown in stages: (**a**) Indian-Asian plate collision, but Indian plate was too buoyant to be carried down further, (**b**) Plastic deformation and recrystallization of the leading edge of the Indian plate, being at greater depth and hence at higher P & T, (**c**) Northward push of the subducted Indian plate resulted into southward thrusting of the plastically deformed and recrystallized rocks over the Precambrian rocks (Lesser Himalayan/basement) and overlying sediments. Emplacement of these rocks reduced drag load at the "tip" of the subducted continental crust, retreating and raising up the Indian plate, and this flip also caused southern convexity of the thrust belt all along the Himalayan length, (**d**) Continued northward push of the Indian plate resulted into renewed southward thrusting of still higher grade Higher Himalayan Crystalline (HHC) rocks during Oligocene-Miocene. Upthrusting of HHC caused decompression melting (later solidified as the Leucogranite), occasional folding of isograds and parallelism of "S" and "C" fabrics in the HHC rocks. Bottom figure shows thrusting in 3-dimension, differentiating MCT into I and II	128
3.4	Northward drift of India with respect to Asia from 71 Ma to the present, determined from magnetic lineations in the Indian and Atlantic oceans (redrawn from Molnar and Tapponier, 1975)	134

4.1	Generalized geological map of the Proterozoic fold belts (Aravalli and Delhi) and cratonic rocks (BGC) in Rajasthan (after GSI, 1969, 1993). Inset (**a**) is the location map and (**b**) shows the major Crustal Terranes in Rajasthan (after Sinha-Roy, 2000). Abbreviations: A = Alwar; AM = Ajmer; ANT = Antalia; B = Bhilwara; BN = Bhinder; BS = Banswara; BY = Bayana; CH = Chittaurgarh; DS = Dausa; GBF = Great Boundary Fault; J = Jahazpur; K = Kankroli; KHT = Khetri; N = Nathdwara; PHL = Phulad; S = Sirohi; SL = Salumbar; SM = Sandmata; SN = Sendra; SR = Sarara	144
4.2	Diagrammatic representation of the sequence of deformational phases and fold styles in the Precambrian rocks of Rajasthan, NW India (based on Naha and others, 1967, 1984; Mukhopadhyay and Dasgupta, 1978; Roy, 1988; Srivastava et al., 1995)	155
4.3	Evolutionary model of the Aravalli fold belt and related belts shown by cartoons (redrawn after Sinha-Roy, 1988). Abbreviations: AG = Anjana Granite; DG = Derwal Granite; J = Jahazpur belt; PBDB = Pur-Banera-Dariba-Bhinder belt; RU = Rakhabdev ultramafics. SDB = South Delhi Ocean Basin. Toothed line is a major dislocation	159
4.4	Shows relation of the Aravalli subduction and opening of (**a**) North Delhi Rifts and (**b**) South Delhi Ocean Basin (SDB), illustrating diachronous nature of the North and South Delhi fold belts (after Sinha-Roy, 1988). Abbreviations: AB = Aravalli Fold Belt; AFB = Alwar-Bayana Fold Belts; ICS = Intracontinental Subduction; RS = Rakhabdev Suture; SDF = South Delhi Ocean Opening (Basin); SJDT = Sambhar-Jaipur-Dausa Transcurrent Fault	160
4.5	Evolutionary model of Delhi fold belt (after Sinha-Roy, 1988). Abbreviations: AFB = Aravalli Fold Belt; AS = Arc Sequence; DMB = Delhi Marginal Basin; RS = Rakhabdev Suture; SG = Sendra Granite; SD = South Delhi Rift; SJDT = Sambhar-Jaipur-Dausa Transcurrent/Tansform Fault; SDFB = South Delhi Fold Belt; VA = Volcanic Arc. See text for details	161
4.6	Evolutionary model of Aravalli-Delhi fold belts (after Sugden et al., 1990). (**a**) Rifting of the Archaean lithosphere of Rajasthan, (**b**) generation of volcanic arc by eastward subduction of an oceanic crust, or (**b'**) volcanic arc generated by westward subduction of continental lithosphere, (**c**) breaking of eastward subducting slab and development of a dextral shear zone, (**c'**) breaking of westward subducting block and upthrusting of crustal rocks on the Eastern block,	

	(**d**) crushing of the volcanic arc, upthrusting of the oceanic crust as ophiolite and accretion of Aravalli-Delhi fold belts. VA = Volcanic Arc	164
4.7	Model for the origin of Aravalli basins along RRF (after Roy, 1990)	165
4.8	Evolutionary stages for the Proterozoic Aravalli mountain (Aravalli and Delhi fold belts) in Rajasthan, NW India. (**a**) Plume-generated ductile stretching (rifting) of the Archaean crust of Rajasthan in Palaeoproterozoic, (**b**) deposition of Aravalli supracrustals, including volcanics of Delwara, in the rifted basin, (**c**) westward subduction and delamination of mantle lithosphere whereby mantle magma underplated the crust, forming granulites at the base and rifting of the softened crust above the delaminated region for deposition of the Delhi supracrustals, (**d**) extension of the continental crust causing a series of rift basins in the northern part of the evolving Delhi basin; sedimentary fill of the extensional basin by Delhi Supergroup; and decompression melting of the ascending asthenosphere and of the overlying crust whereby basic magmas emplaced in the rifted basins, including the initiated south Delhi basin, while granites intruded the crust below the rifted basins, (**e**) Enlargement of the extensional basins, including the basin initiated to the south of the rifted basins, (**f**) Convergence of the diverging blocks (due to depositional load, decay of plume responsible for lithosphere extension and change of stress direction with time), resulting into deformation of both Delhi and already deformed Aravalli fold belts, (**g**) Continued convergence of the continental blocks resulted into welding of the mantle lithosphere, accretion of the fold belts and tectonic excavation of the granulites that now occur as fragments in the BGC and Delhi metamorphic terrains. For details see text	167
5.1	Location map (**a**) and index map (**b**) showing the position of the different Proterozoic fold belts in central Indian region. 1 = Mahakoshal folds belt, 2 = Satpura fold belt, 3 = Sakoli fold belt, 4 = Dongargarh fold belt. Abbreviations: CITZ = Central Indian Tectonic Zone, CIS = Central Indian shear zone, SONA ZONE = Son-Narmada Lineament zone, SNNF = Son-Narmada North Fault, SNSF = Son-Narmada South Fault. (**c**) Geological map of central India (after Roy et al., 2002) showing the different Proterozoic fold belts, granulite occurrences and shear zones. M = Malanjkhand, Ts = Tan shear zone, TF = Tapti Fault. Granulites of Ramakona-Katangi (RKG), Bhandara-Balaghat (BBG), Bilaspur-Raipur (BRG) and Makrohar (MG) are all located	

	in the basement rocks (Tirodi gneiss) of the Sausar Group making up the Satpura fold belt	178
5.2	Geological map of the Central Indian Tectonic Zone (CITZ) (after Roy et al., 2002) sandwiched between Bundelkhand craton in the north and Bastar craton in the south (*top figure*). The location of the four fold belts and shear zones of central India as well as the four granulite localities (Fig. 5.1) are also shown in *bottom figure*. All abbreviations as in Fig. 5.1	188
5.3	Geological map of Dongargarh fold belt, central India (after Sarkar, 1957; Deshpande et al., 1990). 1 = Amgaon gneiss complex, 2 = Bijli rhyolite, 3 = Pitepani volcanics, 4 = Dongargarh granitoid, 5 = Bortalao formation, 6 = Sitagota volcanics, 7 = Karutola formation, 8 = Manjikhuta formation, 9 = Ghogra formation, 10 = Katima volcanics (5–10 lithounits belong to Khairagarh Group)	193
5.4	A plausible model for the evolution of the fold belts of the central Indian region. (**a**) Existence of a large Indian protocontinent (united cratons of Rajasthan-Bundelkhand-Bastar, RCBKCBC). (**b**) Plume-generated early Proterozoic igneous activity in the Indian protocontinent, initiating rifting of the protocontinent in the western region (Bundelkhand-BGC cratonic region) and also resulting into igneous activity of the Dongargarh bimodal volcanics (2.5 Ga old) with minor ultramafics and granite intrusions in eastern part (Bastar cratonic region) of the protocontinent. (**c**) Development of the rift basin that became the SONA Zone bounded by north and south faults (called Son-Narmada North Fault, SNNF, and Son-Narmada South Fault, SNSF, respectively). The SONA Zone became the site for deposition of the Mahakoshal supracrustal rocks, while sinking of the softened crust of the Bastar craton developed a basin for the Khairagarh supracrustals. (**d**) Deformation of Dongargarh-Khairagarh Group rocks due to basinal subsidence and also by compression due to tensional forces associated with the development of SONA Zone. (**e**) Formation of shear, the Central Indian Shear Zone (CIS), due to ductile stretching of the region of Bundelkhand-Bastar cratons, caused by sub-lithospheric mantle activity and also by deformation (F1) of the Mahakoshal supracrustals in the SONA Zone. (**f**) Development of an ensialic basin as a result of thinning of continental crust due to shearing (CIS) and deposition of the Sausar Group supracrustals therein. Magma underplating, due to crustal delamination, is believed to have developed granulites in the lower crust of the Bundelkhand-Bastar cratonic region. (**g**) Convergence of	

	the once rifted Bundelkhand and Bastar cratons, resulting in deformation (F2) of the Mahakoshal rocks and the Sausar Group rocks along with their basement gneisses	200
6.1	Location of Singhbhum fold belt between Singhbhum craton and Chhotanagpur Granite Gneiss Complex (Chhotanagpur craton) (after A.K. Saha, 1994). (A) = Dhanjori volcanics, (B) = Dalma volcanics, (C) = Jaganathpur volcanics, CB = Chaibasa formation, CH = Chakradharpur granite, DH = Dalma formation, GM = Gorumahisani formation, K = Kuilapal Granite, N = Nilgiri Granite, NM = Noamandi Iron-Ore formation, M = Mayurbhanj Granite, OB = Ongerbira volcanics, SM = Simlipal basin, SZ = Singhbhum shear zone	210
6.2	Diagram showing tectonic setting of Singhbhum marginal basin and associated morphostructural unit (redrawn from Bose, 1994)	222
6.3	Slab break off evolutionary model for the Singhbhum fold belt, based on available geological-geochronological data. (**a**) Ductile stretching and rifting of Archaean craton (SC+CGGC) in Palaeoproterozoic (ca. 2.1 Ga) and sedimentation of Singhbhum Group. (**b**) Development of ensialic crust and deposition of more sediments with minor tuffs (cf. Dunn and Dey, 1942). (**c**) Start of collision at about 1.6 Ga ago and folding of supracrustal and basement rocks. (**d**) Convergence (perhaps oblique) and subduction of southern block (SC) followed by narrow rifting and slab weakening. (**e**) Slab rupture, underplating and extrusion of mafic magma (Dalma volcanics) amidst sediments of Singhbhum Group. Additional volcano-sediment deposition and high level intrusion of mafic (gabbro-pyroxenite dykes at 1.6 Ga). (**f**) With underplating following slab rupture, mafic magma extruded and intruded. Heat from the underplated magma also induced partial melting within the crust to produce 1.6 Ga old granites (Kuilapal etc.) and granophyre (Arkasani). (**g**) Further slab break off and slab downgoing/sinking generated melt, giving rise to calc-alkaline magma that appeared as volcanics (Chandil) and acid tuff (Ankro area of Chandil) at 1487 ± 34 Ma (Sengupta and Mukhopadhyay, 2000). (**h**) Slab sinks away and melting at depth produced the last phases of granitoid liquid while dykes (Newer Dolerite; Saha, 1994) were injected from the underplate into the overlying crust. Abbreviations: CGGC = Chhotanagpur Granite Gneiss Complex, SC = Singhbhum Craton, SFB = Singhbhum Fold Belt, SSZ = Singhbhum Shear Zone, NSZ = Northern Shear Zone	225

7.1	Simplified geological map of the Eastern Ghats Mobile Belt (EGMB) (after Ramakrishnan et al., 1998) with megalineaments after Chetty (1995). MSZ = Mahanadi Shear Zone; NSZ = Nagavalli Shear Zone; SSZ = Sileru Shear Zone; VSZ = Vamsadhara shear Zone. *Inset* shows location of the mobile belt	232
7.2	Location map (**a**); the attempted correlation of EGMB with Napier Complex (NPCx) and Rayner Complex (Rayner Cx) of the Enderbyland, Antarctica (**b**); (**c**) shows a synoptic history of P-T paths (see text) in different domains of the Eastern Ghats Mobile Belt. Note that the UHT metamorphism with high pressure IBC and IBC heating-cooling paths are attributed to an earlier event of ∼1.6 Ga, whereas ITD path to a later, ∼1.0 Ga event (Dasgupta et al., 1994; Mezger and Cosca, 1999; Sengupta et al., 1999; Rickers et al., 2001; and others). The difference in metamorphic evolution in North and South EGMB may not be real (see text)	238
7.3	Age domains in the Eastern Ghats Mobile Belt (after Rickers et al., 2001) (see text for details)	251
7.4	Newer division of Eastern Ghats Mobile Belt into crustal provinces (after Dobmeier and Raith, 2003). See text for details	252
8.1	Simplified Geological map of the southern India (after GSI and ISRO, 1994), showing the major geological domains, the Western Dharwar Craton (WDC), Eastern Dharwar Craton (EDC), and Southern Granulite Terrain (SGT) along with the Cauveri Shear Zone System (CSZ). Abbreviations: AKSZ = Achankovil Shear Zone; AH = Anamalai Hills; AT = Attur; BS = Bhavani Shear zone; BL = Bangalore; BR = Biligirirangan; CHS = Chitradurga Shear Zone; CG = Coorg; CM = Coimbatore; EDC = East Dharwar Craton; K = Kabbaldurga; KL = Kolar; KKB = Kerala Khondalite Belt; MS = Moyar Shear zone; N = Nilgiri; OT= Ooty; PCSZ = Palgahat Shear Zone; PL = Pollachi; PMB = Pandyan Mobile Belt; SGT = Southern Granulite Terrain; SH = Shevaroy Hills; WDC = West Dharwar Craton; GR-Am = Isograd between Greenschist and Amphibolite Facies; Am-Gt = Isograd between Amphibolite and Granulite Facies; TZ = Transition Zone of amphibolite and granulite facies. *Inset* shows various identified crustal blocks	264
8.2	Simplified Geological map of part of the southern Indian shield (based on GSI and ISRO, 1994), showing the location of the Southern Granulite Terrain (SGT) to the south of the Am/Gt (Amphibolite-granulite facies) isograd (drawn	

| | after Pichamuthu, 1965). To the north of the isograd are the Western and Eastern Dharwar Cratons (WDC and EDC). The Highland massifs are: Bilgirirangan (BR), Coorg (C), Cardamom (CH), Madras (MA), Kodaikanal (KK), Malai Mahadeswara (MM), Nilgiri (NG), Shevroy (SH), Varshunadu (VH). The prominent shear zones: The Cauvery Shear Zone System (CSZ), comprising Moyar (MS), Bhavani (BS), Palghat-Cauvery (PCS), Salem-Attur (SA-At). Others are: Gangavalli (GA), Karur-Kambam-Painvu-Trichur (KKPT) and Achankovil Shear Zones (AKSZ), Korur-Odanchitram Shear Zone (KOSZ). The different crustal blocks recognized in the SGT *sensu lato* are: the Northern Block (NBSGT) and Southern Block (SBSGT), Madurai Block (MB). The block between the Palghat Shear Zone (PCS) and the Achankovil Shear zone (AKS) is designated Pandyan Mobile Belt (PMB). To the south of the AKS is the Kerala Khondalite Belt (KKB). Other abbreviations are: CM = Coimbatore; NC = Nagercoil; OT = Ooty; RM = Rajapalayam; TL = Tirunelveli. *Inset* shows the SGT, the KKB and the different shear zones traversing the SGT. Other abbreviations as in Fig. 8.1 . . . . | 266 |
| 8.3 | Litho-tectonic map of the Kerala Khondalite Belt (KKB) and a part of the Southern Granulite Terrain (i.e. Madurai Block), redrawn from T. Radhakrishna, 2004. Note that the gneiss/migmatite and granulite are cofolded by superposed deformation F1 and F2 with axial traces (delineated by the writer) which are cut across by the Achankovil shear zone, see text for details . . . . | 278 |

Chapter 1
Precambrian Terrains and Mountain Building Processes

1.1 Introduction

Orogenic (Gr. *Oros* means mountain and *genic* means birth) belts or orogens are some of the most prominent tectonic features of continents. These terms are, however, not synonymous to Mountain belt which is a geographic term referring to areas of high and rugged topography. Surely, mountain belts are also orogenic belts but not all orogenic belts are mountains. In this book, the orogenic belts, also called mobile belts, are termed fold belts because they are made up of rocks that show large-scale folds, and faults/thrusts and metamorphism with evidence of melting or high mobility in the core region during orogenesis. These belts are characteristically formed of (a) thick sequences of shallow water sandstones, limestones and shales deposited on continental crust and (b) deep-water trubidites and pelagic sediments, commonly with volcanoclastic sediments and volcanic rocks. Typical mobile belts, rather fold belts as titled here, have rocks that were deformed and metamorphosed to varying degrees and intruded by plutonic bodies of granitic compositions. Some fold belts are also characterized by extensive thrust faulting and by movements along large transcurrent fault zones. Even extensional deformation may be found in such belts. Most belts show a linear central region of thick multiply deformed and metamorphosed rocks bordered by continental margins, but some belts are also having oceanic margin on one side.

The application of the plate tectonic model to the study of the fold belts (or mobile belts) has revolutionized our ideas about how these belts form. We now believe that orogenic belts form at convergent margins because of long periods of subduction beneath the plate margin or because of the collision of two continents, of a continent with an island arc, or of a continent with an oceanic crust. Different types of fold belts form depending on the character of the colliding blocks and on which side one block overrides the other (Press and Siever, 1986).

The subduction of oceanic lithosphere gives rise to two different fold belts, depending upon the nature of the overriding plate. Subduction beneath oceanic lithosphere gives rise to the formation of island arc. Subduction under continental lithosphere gives rise to a linear fold belt on the overthrusting plate margin that runs along the subduction zone. Such fold belts are generally termed *Andean type* orogenic belts. The term Cordilleran type is no longer in use since the recognition that the Western Cordillera of North America formed by rather more complex processes than implied by this simple model (cf. Kearey and Vine, 1996). Another type of fold belts are *Collision fold belts* which develop when subduction adjacent to a continental margin brings another continent that make up an integral part of the subducting oceanic lithosphere. The resulting collision of two continents causes the creation of a collision fold belt by stacking of thrust slices of crust. Collision fold belts are complex since the geological record of collision always preceded by Andean type orogenesis. By studying the young orogenic belts, we can discover the relationship between orogenic structures and related plate tectonic activity. Similar structures in older orogenic belts can be used to infer the existence of similar plate tectonic activity in the geological past. It is commonly found that fold belts subdivide the stable Archaean cratons and show wholly ensialic deformation, with no rock associations that could be equated with ancient basins. In some Proterozoic fold belts, however, ophiolitic complexes have also been recognized, which can be attributed to processes associated with subduction zones, for example Wopmay orogen at the northwestern margin of the Laurantian shield (Hoffman, 1980). There are many examples in literature (Kroener, 1981; Condie, 1982) in which Precambrian orogenic belts have been explained in terms of modern plate tectonic processes. Gibb et al. (1983) have proposed that the structural province boundaries of the Canadian shield can be explained in terms of plate margin interaction during Proterozoic times. These structures making suture zones were formed by continent-continent collision of the juxtaposed continental masses, following the consumption of oceanic lithosphere. In this scenario we have to look for temporal and spatial occurrences of various major rocks as a function of plate motion or collision.

Orogenic belts are commonly at the margin of the stable shield or craton. For example, the Caledonian Mountain belt (Lower Palaeozoic) is seen on the western margin of the Baltic shield (mainly in Norway and Great Britain); the Hercynian or Variscan (Upper Palaeozoic) belt extends roughly from Western Europe to the Pacific with a notable N-S appendix in the Ural Mountains. The Tertiary fold belt of the Himalayan ranges is also trending E-W bordering the northern part of the Indian shield. The Alpine ranges occur mainly to the south of the Hercynian belt that it partly overlaps (Fig. 1.1). A similar overlap of the mountain belts is seen around the Siberian shield, with orogenic belts from Caledonian to Hercynian and to Tertiary, arranged more less concentrically around the shield; the ages of the fold belts decreasing radially outward. For example, the Hercynian belt of Western Europe includes Lower Palaeozoic rocks that were previously deformed in the Caledonian orogeny. In addition, the core of the Central Alps includes rocks metamorphosed in the Hercynian orogeny and re-metamorphosed in the Alpine orogeny. Since the older basement is rarely exposed, it is not possible to affirm that the Cenozoic cordilleras are not underlain by rocks as old as the Caledonian shield. Moreover, the older events may have been erased by later orogenies. These problems need to be recognized while discussing the evolution of the fold belts in the Indian shield.

1.2 Archaean Terrains

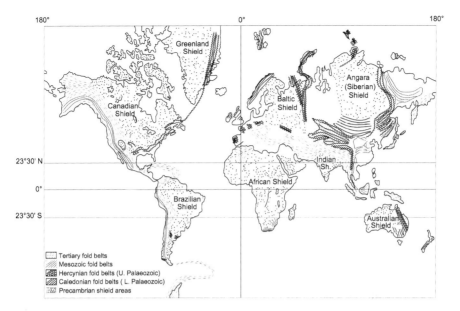

Fig. 1.1 Tectonic units of the continents: the Shields and Orogenic belts (after Umbgrove, 1947)

1.2 Archaean Terrains

The continents (including the shelf region) and oceans are the major features of the Earth surface. Continents are made of low-density sialic rocks and contain ancient (Precambrian) crystalline rocks-association of gneiss-granite-granulite/greenstone that forms the continental nucleus of Archaean age (2.5–3.8 Ga). These Archaean terrains are distinguishable because of metamorphic grade into two regions. One, high-grade granulite-gneiss regions that show amphibolite or granulite facies metamorphism, and second, greenstone belts that have low-grade (greenschist facies) metamorphosed volcanic rocks and associated sediments. The granulite-gneiss constitutes the bulk of the Archaean terrains. They are mostly quartzofeldspathic gneisses derived by deformation of felsic igneous rocks (e.g. granite becomes a gneiss on deformation) of tonalite-trondhjemite composition. There are minor metasedimentary rocks such as quartzites, metacarbonates, and metamorphosed iron formations. Deformed mafic-ultramafic complexes or layered gabbro-anorthosite complexes and orthoamphibolites make up the remaining gneissic regions. The high-grade gneisses show ample evidence of partial melting to produce migmatite and charnockite. The latter rock is characterized by the mineral hypersthene. These granulitic rocks range in composition from acidic, hypersthene granite (charnockite *sensu stricto*) through intermediate hypersthene-granodiorite (enderbite) to hypersthene gabbro (norite) or basic charnockite. The high-grade granulite-gneiss terrains are complexly mixed on a scale of tens to hundreds of kilometers with low-grade greenstone belts that contain mafic-ultramafic to silicic volcanic rocks and shallow intrusive bodies, volcanogenic and shallow water sediments.

The contact between greenstone belts and high-grade granulite-gneiss areas is complex. At places, the contacts are shear zones that modify or mask the original relationship. Elsewhere, greenstone rocks are deposited on older gneissic basement. In still other regions, gneissic/granitic rocks intrude the greenstone belt. The sedimentary rocks of the Archaean terrains have one of two broad associations: either they are immature volcanogenic sediments that are characteristic of the greenstone belts, or they are quartz-carbonate-iron formations assemblage occurring with deformed mafic-ultramafic layered igneous complexes in gneissic terrains. Another feature of the Archaean terrains is that their constituent rocks are intensely deformed and show more than one generations of folds (see Nisbet, 1987).

The oldest rocks on the Earth are as old as 3800 Ma and occur in widely separated Archaean terrains that are mainly tonalitic gneisses, metamorphosed basalts, and minor ultramafic and calc-alkaline volcanics. These metaigneous rocks occur as supracrustal sequences or greenstone belts. The two types of rocks are considered by some geologists as ancient vestiges of sialic continents and mafic oceanic crust (cf. Sharma and Pandit, 2003).

Greenstone belts of mafic to ultramafic volcanics with intercalated sediments generally occur in a synclinorium type structure, 40–250 km wide, and 120–800 km long. Most often, the greenstones are divisible into three stratigraphic groups. A lower group consists of komatiitic to mafic volcanics rocks with pillow structures, having a bulk composition similar to ocean ridge basalts. A central group of andesites and calc-alkaline volcanic rocks are similar in trace elements and REE to the rocks found on island arcs. An upper group comprises clastic sediments such as graywackes, sandstones, conglomerates, and banded iron formations with chert and limestones.

The origin of the greenstone belts is thought (non-uniformitarian model) to develop as a downwarp in the primitive crust in response to the load of injected high-Mg lavas and subsequent filling by sediments (Fig. 1.2). It is believed that the crust was thickened by compression of the trough and crustal melting at depths gave rise to diapiric granites (Windley, 1984). Most workers suggest a plate tectonic model for the formation of the greenstone belts and propose that they originated in subduction related setting, particularly in ensialic marginal basin. For Rocas Verdes complex in Southern Chile, Tarney et al. (1976) proposed that greenstone belts developed in ensialic marginal basins. Their model is shown in Fig. 1.3. According to this model, the greenstone belts developed on thin continental crust in a back-arc environment. They were covered by sediments that came from the flanking continents and volcanic arc. As the basin was subjected to compression, volcanics and overlying sediments were deformed and intruded by granites generated at depth. The back arc environment also explains how Archaean oceanic crust was preserved; if it originated in a normal ocean basin, it would probably have been destroyed by subduction.

Commenting on the granulite-gneiss belts, Windley (1984) pleaded that the geochemical characteristics of these belts could be replenished only when some form of plate tectonics was operative during Archaean times. It is by this mechanism that the mantle-derived continental crust was created during the Archaean (see also Sharma

1.2 Archaean Terrains

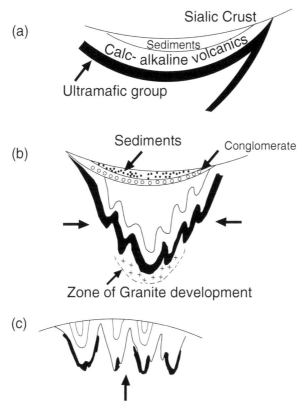

Fig. 1.2 Non-uniformitarian evolutionary model of Greenstone belt (after Windley, 1984). (**a**) Synformal structure; (**b**) crustal thickening; (**c**) uplift of deformed greenstones

and Pandit, 2003). Windley argues that this mechanism was probably not identical to present day plate tectonics as the Archaean time is documented to have high heat flow and thinner lithosphere. However, by combining the marginal-basin and subduction related tonalite batholith models, the Archaean plate tectonics could provide a mechanism for the complementary origin of both greenstone and granulite-gneiss belts. The proposed mechanism of Windley (1984) considers the Archaean crust being made up of a large number of small, relatively thin continental crustal blocks. Subduction of oceanic crust between these gave rise to back-arc spreading and formation of ocean-type crust (Fig. 1.4). It is by subduction only that calc-alkaline volcanics and tonalite magma were generated. It is suggested that by the end of Archaean, these small continental fragments welded into large sialic blocks characterized by alternate belts of greenstones and granulite-gneiss formed, respectively, from the oceanic crust and eroded tonalite batholiths. In this event, bigger and thicker plates dominated the Proterozoic lithosphere.

There are workers (see Moores and Twiss, 1995) who are reluctant to accept these simplistic applications of modern plate tectonic models to the Archaeans, due

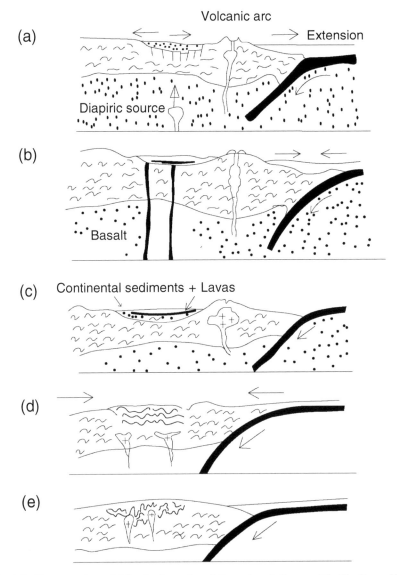

Fig. 1.3 Plate tectonic model for the formation of Greenstone belt exemplified by Rocas Verdes Complex, South Chile (after Tarney et al., 1976). (**a**) Extensional phase and basin formation; (**b**) volcanic phase and emplacement of basalt; (**c**) continental sediments and arc lavas as supracrustals; (**d**) deformation due to basin closure and intrusion of synorogenic tonalite; (**e**) emplacement of late potassic Granite

to several reasons. There are extensive differences between modern and Proterozoic tectonics. Although greenstone belts bear some resemblance to modern island arc or marginal sequences, significant differences cast doubt on a simple correlation. Ultramafic lavas (komatiites), abundant in the Archaean, are almost entirely

1.3 Proterozoic Terrains

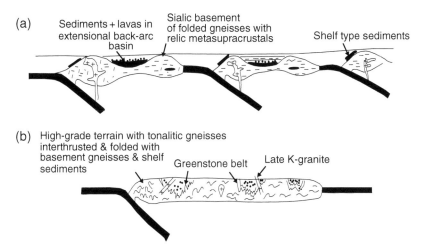

Fig. 1.4 Plate tectonic model for the evolution of greenstone-gneiss-granulite in the Archaean (redrawn from Windley, 1984). (**a**) Several small continental plates with ensialic back-arc basins. The Andean-type margins are invaded by mantle-derived tonalitic batholiths and plutons; (**b**) aggregations of the continents to produce a large continental plate consisting of greenstone and granulite belts

lacking in modern marginal basins or island arcs. If the greenstone belts represent marginal basins, they lack the accompanying mature arc that is ubiquitously associated with modern marginal basins. If they are island arcs, they differ in composition from modern island arcs and seem to lack the compositional zonation possessed by modern arcs that reflect the polarity of the subducting slab. The quartzite-carbonate-mafic-ultramafic association present in Archaean gneissic terrains has no clear modern equivalent. Many workers (e.g. Moores and Twiss, 1995) suggested that the sedimentary rocks represent shelf sequences and the associated mafic-ultramafic complexes may represent the Archaean equivalents of oceanic crust. Despite these differences, however, mantle convection almost certainly did occur and some forms of plate tectonics in early Proterozoic and Archaean times cannot be ruled out (cf. Sharma and Pandit, 2003).

1.3 Proterozoic Terrains

The Archaean terrains became stable for first time in Proterozoic (< 2500–600 Ma) and served as basement for the deposition of Proterozoic sediments. Precambrian regions of the Earth's crust that have attained tectonic stability are called the *cratons*. The evidence of this relative stable tectonic environment is documented in the occurrence of mature sediments such as quartzites, quartz pebble conglomerates, deposited on eroded, rather peneplaned Archaean basement rocks. These mature sediments define regionally extensive stratigraphic units. Quartzites are often intercalated with abundant iron formations (interstratified magnetite/hematite rocks, iron carbonates and iron silicates). These are mostly undeformed cratonic sediments, like the platform sequences, lying upon a pre-existing older basement, indicating the existence of large stable continental region in Proterozoic time. They host many vast Precambrian placer gold and uranium deposits as well as most iron deposits of the world. These sedimentary sequences extend over large areas in the Indian shield but are not the theme of this book. Interested reader may refer to the relevant source either in journals or in published books/memoirs. Another type of Proterozoic rocks making up the fold belts are either peripheral to the cratonic areas or are sandwiched between two cratonic regions within the Indian shield. In other words, the cratons were sutured or accreted by the Proterozoic fold belts. These fold belts with their characteristic geological setting and tectono-thermal evolution constitute the main theme of this book. They can be called fold belts to indicate major belts of pervasive deformation and tectonic mobility. Fold belts are most prominent tectonic features of the Indian shield. These belts form a linear or arcuate belt of deformed and metamorphosed rocks. They were the loci for the development of former intracratonic basins and rifts as they are formed of (i) thick series of shallow-water sandstones, limestones and shales, all deposited on continental crust and (ii) deepwater turbidites and pelagic sediments, commonly with volcaniclastic sediments and volcanic rocks deposited on ocean floor. Typical fold belts have been deformed and metamorphosed to varying degrees and intruded by plutonic rocks of granitic composition. Some fold belts may display structural symmetry on either side of the core made up of high degree of metamorphic rocks and plutonic bodies (Moores and Twiss, 1995).

The Proterozoic fold belts or orogenic belts are of two types. Some are multiply deformed regions of both supracrustals and Archaean basement rocks. Others exhibit thick sedimentary sequences deposited in linear troughs, presumably along ancient continental margins, subsequently deformed and recrystallized into linear fold-and-thrust belts, like those of Phanerozoic belts. In addition, large anorthosites occur as distinctive igneous-metamorphic rock-suite that appeared in Proterozoic time (1000–2000 Ma). These anorthosites are intimately associated with granulites and the Eastern Ghats mobile belt would represent a good example of this type of fold belt. Other Proterozoic fold belts that are covered in this book are the Aravalli fold belt of Rajasthan in western India, the Satpura, Mahakoshal, Sakoli, and Dongargarh fold belts in central India, the Singhbhum mobile belt in eastern India and finally the Pandyan mobile belt in southern Indian shield. For each of these fold belts of the Indian shield, we will consider the plate tectonics theory for their development, because these Proterozoic fold belts show deformation and metamorphism which can be attributed to collision tectonics. Being intensely eroded, these fold belts display multiply deformed and metamorphosed thick sequences of miogeosynclinal rocks and the basement in the core or root zone. High angle fault zones are also numerous in almost all these fold belts. Therefore, the study of these Proterozoic fold belts would give some information about plate tectonic interactions for the first 95% of Earth's history. By studying these features, we have an opportunity to decipher part of the tectono-thermal history of the Earth that is preserved nowhere else. Before the acceptance of plate tectonics, the formation of the fold belts was conceived essentially by vertical movements and it is interesting to know the early views on the origin of fold belts, known then as mountain ranges.

1.4 Mountain Building Processes

Earlier, the mountain belts were explained by *contraction hypothesis* according to which the wrinkles formed during cooling of the Earth and decrease of its radius became the mountains of the Earth's crust. According to this hypothesis, the continents were essentially stationary on the Earth's surface. However, vertical movement was accepted to explain the presence of marine rocks in high mountains like the Himalaya. The driving force for such vertical movements was lately thought to be isostasy—the theory initially proposed in 1865 by George Airy to explain gravity measurement in India.

1.4.1 Geosynclinal Theory

In the middle of the nineteenth century, the American geologist James Hall proposed the geosynclinal theory. According to this theory the vertical movements of tectonic origin developed a subsiding linear trough or a geosyncline in which thick, shallow water sediments were deposited along both sides of the trough (miogeosyncline) and deep water sedimentary and volcanic rocks in the centre (eugeosyncline). Following the deposition of these thick sequences, deformation occurred, with "eugeosynclinal" sediments becoming the most deformed and metamorphosed. Deformation was conceived to be symmetrical, with thrusting of all rocks in both directions away from the centre over the flanking undeformed continental platforms. The mountain building process (orogenesis) was conceived to have three main phases (Fig. 1.5).

1.4.1.1 Geosynclinal Phase

Here a linear depression appears in the crust into which sediments derived from surrounding regions of the crust were deposited, commonly accompanied by extrusion and intrusion of basic volcanic rocks (see Dickinson, 1971). Sedimentation continues in a more or less steady state, but not uniform over the whole basin in either composition or thickness due to variations in depth of water, proximity to source area, etc. In the shallower part of the basin, rocks such as limestones and sandstones tend to predominate, and in the deeper parts, a much thicker sequence of "flysch"-type sediments is generally found. It is in this deeper part that basic volcanic activity is generally concentrated (Fig. 1.5a). These differences from place to place in sedimentary type (or facies) are most useful to the geologist in helping him to reconstruct the form of the ancient basin, or, conversely, to establish directions and amounts of relative movements that have taken place during folding and thrusting.

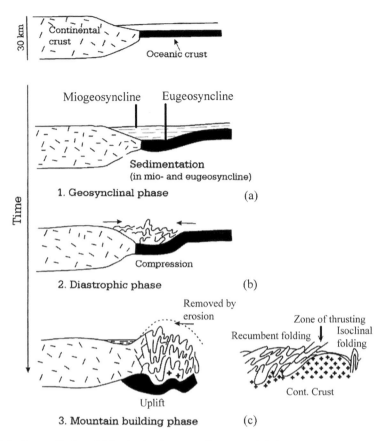

Fig. 1.5 Geosynclinal model for the development of an orogenic belt

1.4.1.2 Diastrophic Phase

The diastrophic or tectogenic phase involves strong deformation of the sediment-filled basin (Fig. 1.5b). The structures formed in this phase are in detail unique to each geosyncline; commonly occurring features are: first, wide spread formation of folds of various sizes from a few cm to many km, and second, thrust faulting in many scales. Flat-lying rock masses moved for very long distances along large faults. The folding and low-angle thrusting are intimately related and seem to proceed together. The deeper levels, including the basins in some of the pre-existing crust, become the sites of intense metamorphism, commonly invaded by granite bodies. At the upper and marginal levels, metamorphism is less intense or absent, but the pre-orogenic basement can become involved in the deformation and appear as slices of crystalline rocks, generally in the cores of the large folds in the covering sediments. The metamorphism of the basin sediments along with the basement have been classed into three *depth zones*: Epizone (characterized by minerals like chlorite, muscovite, epidote), Mesozone (characterized by garnet, staurolite, hornblende, etc), and Catazone

1.4 Mountain Building Processes

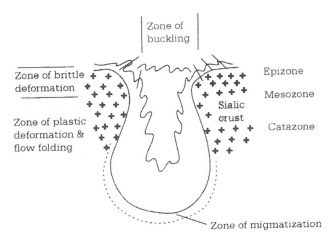

Fig. 1.6 Depth zones and distribution of deformed and metamorphosed rocks in a crustal section

or Katazone (characterized by sillimanite, cordierite, hypersthene). In the lower part, the geosyncline becomes mechanically weak or mobile during metamorphism. The anatectically-formed granite can rise in the upper or middle zones in form of a diapir or dome (Fig. 1.6).

1.4.1.3 Orogenic Phase

This is the mountain building phase characterized by uprising of the deformed and metamorphosed geosynclinal sediments and basement crust. Sometimes the upper folded layers seem to become detached from its position and slide under the influence of gravity as extensive thrust sheets to lie over the less deformed parts of the geosynclinal margin and even onto the undeformed "foreland", as in the Garhwal Himalaya (Gansser, 1964). During the later part of the orogenic phases, marginal and intermediate depressions form which rapidly fill-up with coarse sediments (molasse), produced by rapid erosion of the rising mountain range (Fig. 1.5c).

The greatest failing of geosyncline theory was that the tectonic features were classified without having an understanding of their origin. In this theory, both continents and oceans were considered permanent. A major weakness of this model was that there was no explanation for what caused geosynclines to develop and subsequently to deform. Another problem with the geosyncline model was that it could not explain the very large amount of shortening exhibited by the nappes in the Alps. This led the Austrian geologists Ampferer and Hammer in 1911 to formulate the hypothesis of "Verschluchung" (the downbuckling or subduction) of the European platform beneath the Alpine geosyncline. In 1905, the German geologist, Gustav Steinman, recognized the intimate association of serpentinites, pillow lavas and radiolarian cherts in the Alpine orogen and proposed that the sediments were laid down in deep water over a rapidly subsiding continental platform. This recognition led to the belief that geosynclines also had deep-water environment of deposition.

Lately, some authors like Kearey and Vine (1996, p. 7) do not recommend the use of the geosynclinal terminology, e.g. eugeosyncline and miogeosyncline for sediments with and without volcanic members respectively. They state that "the term geosyncline must be recognized as no longer relevant to plate tectonic processes." Nevertheless, it must be noted that the relation of sedimentation to the mobilistic mechanism of plate tectonics allowed the recognition of two specific environments in which geosynclines formed, namely passive continental margins and subduction zones. However, the terms mio- and eu-geosyncline are useful terms to indicate specific depositional environment and would be retained here.

1.4.2 Plate Tectonics Theory

A direct challenge to the permanency of continent and ocean basins came in 1915 when Alfred Wegener proposed his hypothesis of *continental drift*. This hypothesis witnessed long controversies. Nevertheless, two important findings are quite interesting. In 1930, a study of gravity at sea by the Dutch geophysicist F.A. Vening Meinesz disclosed that deep-sea trenches in the Caribbean and in Indonesia were associated with negative gravity anomaly to suggest downbuckling of the crust into the mantle—a concept that was called tectogene. The formation of tectogene, according to Arthur Holmes, is in response to a downgoing convection current in the mantle. At the same time in 1928, the Japanese seismologist, K. Wadati, recognized that earthquake sources beneath Japan were located along an inclined planar zone that extends from the trench east of Japan and dips westward under the islands. The revolution began in 1960 with the circulation by Hess of a manuscript entitled, "Evolution of ocean basin and History of ocean basins." In this manuscript, Hess (1962) argued that the oceanic crust was young and was created over rising limbs of mantle convection cells. Taking support of the palaeomagnetic data of Blackett, Irving and Runcorn and the drift theory of Wegener, R.S. Dietz coined the term *Sea floor spreading*. In short, the sea floor spreading can be likened to a conveyer belt in which lithospheric plates were transported laterally carrying the continents with them.

The development over the last four decades of the Plate tectonics theory gave us a way of understanding mountain building processes. Therefore, we must have a relevant knowledge of this theory that would form the required background about the evolution of fold belts of younger and older ages. In plate tectonics, unlike earlier orogenic hypothesis, the sediments deposited in widely separated depositional sites are telescoped and sutured onto the convergent plate margins. In plate tectonics it is not essential that the strata must be deformed and metamorphosed, as it depends on the location of the deposition, e.g. sediments deposited away from a trench will escape deformation during plate convergence. Furthermore, the regional metamorphism of the basinal sediments is not depth controlled but depends on the particular gradient that can prevail in the given crustal segment, irrespective of the depth.

1.4 Mountain Building Processes

The basic concept of the Plate Tectonics is that the rigid outer shell of the Earth, *lithosphere,* lying above the Low Velocity Zone (LVZ) or Asthenosphere, is divided by a network of boundaries. These boundaries were originally distinguished on their seismicity and later by the presence of mountain belts/ocean ridges, subduction zones/trenches, and by chain of volcanoes and earthquake belts, separating the rigid outer shell into separate blocks which are termed *Lithospheric Plates.* The plates (80–150 km thick) could be exclusively continental or oceanic or both. Today there are six major plates and a number of smaller plates. The hypothesis of sea-floor spreading, put forward by Hess (1962) led to establish that the lithospheric plates were transported laterally (carrying the continents over them) by convection currents in the mantle. The plates move away from *ocean ridges,* which are the sites of seismic and volcanic activity and abnormal high rates of heat flow. Because of tensional forces, the ridge develops a central rift valley, which is devoid of any sediment cover, but further away from the ridge, sediments become more abundant and also thicker and older.

1.4.2.1 Plate Margins

Three types of plate margins are recognized, depending on their movements. In following description, we examine the nature of the plate boundaries and their geological characteristics (Fig. 1.7).

Constructive Plate Margin

The constructive plate margins develop when continents rift and move away to form new ocean basins by upwelling and solidification of magma generated by decompression melting of the upper mantle underneath. These margins initiate at

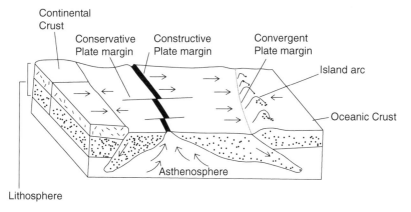

Fig. 1.7 Block diagram showing the different plate margins. Mantle material rises below the Constructive plate margins (ocean ridges) and plate material descends into the mantle at Convergent (destructive) plate margins (ocean trenches). At Conservative plate margins (transform faults) plates slide past each other, without getting created or destroyed

a divergent plate boundary or ocean ridges and are also referred to as divergent plate margins. The ocean ridges can be conceived as the sites of creation of new oceanic material and can be called Atlantic type margins, considering the geographic region where it is characteristically developed now. These are, in fact, passive margins and include a coastal plane and a submarine topographic shelf of variable width, generally underlain by a thick sequence (10–15 km) of shallow-water mature clastic or biogenic sediments. Shelf region passes into a steep slope toward ocean basin. Normal faults are characteristically found in sediments along these margins. The continental rift zones, e.g. African rift or Red Sea rift, represent incipient constructive boundaries.

Convergent Plate Margin

The convergent plate margin is located at ocean trench where two plates converge and when denser of the two plates sinks, along an inclined plane or *Benioff Zone,* below the other to be eventually resorbed into the mantle. These are also called the destructive plate margins, typified by the Andean type margin. These margins exhibit an abrupt topographic change from deep-sea trench to a high belt of mountain. Shelf region is very narrow or absent. Mountain chains along these margins are characterized by a chain of volcanoes of andesitic composition. Intense deformation near the trench results into thrust complex. Most subduction zones now are situated at island arc within the oceans but certain subduction zones border the continents. There are thus two types of convergent boundary now. The first follows the deep ocean trenches and the second follows the belt of young mountain ranges of Alpine-Himalayan chain.

Conservative Plate Margin

The conservative plate margins are faults where two adjoining plates move laterally past each other so that lithosphere is neither created nor destroyed. The direction of relative motion of the two plates is parallel to the faults. Conservative plate margins occur within both oceanic and continental lithosphere, but the commonest conservative plate margins are oceanic *transform faults* (Wilson, 1965). The best-documented example of a large continental transform fault is the San Andreas Fault of California, and therefore these margins are called the Transform or California type. These margins are characterized by sharp topographic differences between ocean and continents. They are marked by active strike-slip faulting, sharp local topographic relief, poorly developed shelf and irregular ridge and basin topography and many deep sedimentary basins.

The oceanic faults make a very prominent feature of the oceans. The ocean ridge crests are repeatedly offset by parallel sets of such faults. The displacement along these faults could be dextral (right lateral) or sinistral (left lateral). Since the spreading direction of the ocean ridges remains parallel to these faults, the divergent motion away from the ridge axis would be "transformed" to a transcurrent motion along such a fault, and then perhaps transformed again to convergent motion at a

1.4 Mountain Building Processes

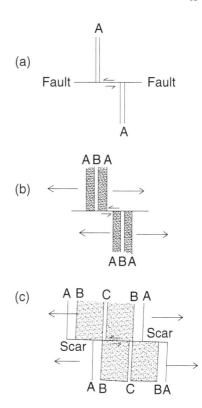

Fig. 1.8 Representation of development of a ridge-to-ridge transform (adapted from Verhoogen et al., 1970, Fig. 3.75). (**a**) Oceanic ridge (*double line*) along which new crustal material added is displaced by a sinistral strike-slip fault (*single line*); (**b**) addition of magma between line A and A and slip on the fault; (**c**) addition of material between lines B and B; scar of previously active sections of the fault extend laterally into the surrounding crust. Note that the sense of slip on the transform fault is opposite of that inferred from the displacement of the ridge by strike-slip faulting

trench (Fig. 1.8). Therefore, Tuzo Wilson (1965) called these faults *transform faults*, fundamentally different from the strike-slip faults on land. A transform fault must be parallel to the direction of relative motion of the lithospheric plates on either side and is therefore controlled by the relative velocity of the two plates. Strike-slip fault, on the other hand, is controlled by stress. By using transform faults and the spreading rates, the relative angular velocities of several plate pairs have been established (varying from 1 to 5 or more cm per year).

The plate margins also have distinct geologic characteristics. Constructive plate margins are characterized by basaltic volcanic activity along an axial rift. By contrast, destructive plate margins occurring at the junctions between oceanic versus oceanic or oceanic against continental plates are characterized by the presence of island arcs or continental volcanic chains, associated with intrusive magmas of intermediate composition and metamorphism at depth. Conservative plate margins, shown as transform faults, are generally devoid of volcanic activity.

Plate tectonics is an expression of thermal convection within the Earth. The convection occurs by the mass movement at surface and at depth. According to one hypothesis, the convection may be whole mantle whereas other hypothesis considers that the convection is shallow mantle in nature. Rising convection currents at ocean ridges (where the ocean plates move away from the ridge) brings heat to the surface.

The cold moving plates on encountering the opposing moving plate descends along a trench (due to descending convection currents) and heats up to be absorbed in the deep mantle. The trench dips below an adjacent continental margin or an island arc. This inclined zone is an activity of earthquake and is called a *Benioff Zone*. It is a critical piece of evidence in favour of the process of subduction.

In short, the ocean ridges can be conceived as the sites of creation of new oceanic material and the trenches as the loci of destruction.

1.4.2.2 Physiography of Plate Tectonic Units

This section gives a brief account of geological features and environment of different physiographic units of the plate tectonics, including the plate margins.

All convergent margins, regardless of their length, age, stage of development, display an overall similarity in topography. The orogens developed at plate convergence possess a number of features in common, for example a trough, a foreland, or undeformed plate on either side, fold-and-thrust belt, an internal crystalline core zone of deformed and metamorphosed sedimentary and volcanic rocks, granitic plutons, and a suture zone that may not be recognizable in all older fold belts. The *sutures* are the boundaries between crustal blocks that were carried on two different plates and that may have been juxtaposed by plate movements (Dewey, 1977). The recognition of sutures in the continental geology is essential for the older fold belts as it is for the Phanerozoic ones. Most sutures are recognized by the presence of ophiolites but these rocks may not always be present in the Proterozoic fold belts. In the ancient fold belts, sutures can be identified either by characteristics of the boundary itself or by a major discontinuity across the boundary, such as marked changes in lithology, geologic history, structural style, palaeomagnetic vectors, or faunal assemblage. Orogenic belts record the subduction of one plate beneath another, as well as collisions between crustal blocks, such as two continents, a continent and an island arc, or a continent and an oceanic plate. Because all oceanic crust older than 200 Ma (Early Jurassic) has been subducted, Proterozoic fold belts are the main repositories of information about early plate tectonic interactions. By studying these features, we can possibly decipher the tectonic history of the Earth preserved in these fold belts and nowhere else. The following description of physiography of the different elements of the plate tectonic environment will be helpful for a greater appreciation of the plate tectonic theory. Those interested in more details should consult Moores and Twiss (1995). The description given below starts across the margin of the downgoing plate (Fig. 1.9) and covers all elements successively.

Outer Swell

The Outer Swell is a topographic bulge that develops in the down-going plate before the plate bends into the mantle (see Fig. 1.9). This bulge or swell is a few hundred meters above the abyssal plain. It is suggested that this outer swell is the result of elastic bending of the plate as it starts descending into the mantle. It should obviously have normal faults parallel to the trench.

1.4 Mountain Building Processes

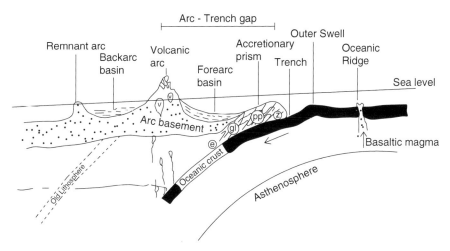

Fig. 1.9 Physiographic-cum-geologic units in a convergence of the Ocean-Ocean plates (see text for details). e = eclogite facies; gl = glaucophane schist facies; pp = prehnite-pumpellyite facies; z = zeolite facies

Trench

It is a deep topographic valley that forms downwards towards the boundary between the subducting and overriding plates (see Fig. 1.9). It is typically 10–15 km deep and is continuous for thousands of kilometers. Most trenches contain flat-lying turbidite sediments deposited by currents flowing either down into the trench away from the overriding plate, or along the axis of the trench.

Arc-Trench Gap

It includes the entire region between the trench and the Volcanic Arc. It consists of the steep inner trench wall that flattens into an area of gentle slope called the fore-arc basin (or Upper Trench Slope). The two areas, namely the trench and the fore-arc basin, are separated by a small topographic ridge—the Outer ridge, not to be confused with the outer swell, described earlier. The fore-arc region may be underlain either by a thick wedge of mostly deformed sedimentary rocks, known as the Accretionary Prism, or by deformed arc basement rocks covered in places by thin layer of sediments.

Accretionary Prism

It forms on the inner wall of an ocean trench. It develops when trench-fill turbidites (flysch), and perhaps also the pelagic sediments and underlying oceanic crust, are scrapped from the descending oceanic plate by the leading edge of the overriding plate to which they are welded. The accretionary prism is the main site of crustal deformation in a subduction zone. The deformation begins at the foot of the inner trench wall. Rocks in the accretionary prism are cut by numerous imbricate thrusts that are sympathetic to the subduction zone, i.e. they dip in the same direction. Because of these imbricate thrust faults, mostly listric, the accretionary sediments define a series of wedge-shaped pockets within which are developed complex folds verging towards the trench (see Fig. 1.9). The deformed rocks in the accretionary prisms are mostly sediments derived from either overriding or down-going plates. In some cases, seamounts from the down-going plate are being incorporated into the accretionary prisms. Sediments from the arc region are added to the top of the accretionary prisms when they are deposited in basins in the fore-arc area. They are also carried by the turbidity currents into the accretionary prism as subduction carries them back toward the arc and the basal thrust faults propagate out into the undeformed sediments (cf. Moores and Twiss, 1995). This process is called *offscrapping*. It is by scrapping that the deep-sea sediments that accumulated on the downgoing plate are incorporated into the accretionary prism. Offscrapping results in progressive widening of the accretionary prism, i.e. outward growth of the prism. With continued subduction, older thrust wedges are gradually moved upward and rotated arcward by addition of new wedges to the base of the accretionary prism. As a result, the older thrusts become more steeply dipping with time. Offscrapping, also called underplating or subcretion, is also responsible for decrease in the dip of the subduction zone as the Volcanic Arc becomes more mature.

Deformation structures in the accretionary prisms reveal more than one generation, more clearly visible in soft sediments. In some cases, the deformation in the accretionary prism is so intense that any pre-existing stratigraphic continuity is destroyed. Such chaotic deposits are called *mélanges*. During the deformation, a fore-arc basin may develop between trench and the island arc in a trough. At some subduction zones, continental basement or old rocks of the island arc can be traced out to the lower trench slope, nearly devoid of accretionary prism. Here, only thin sediments accumulate in basins on the basement rocks and very little, if any, sediment occurs in the trench. The arc basement is commonly cut by normal faults down thrown on the trench side. Many trenches show relatively undeformed sediments on down-going plate that extend for several kilometers beneath deformed rocks of the trench inner wall.

Fore-Arc Basin

It is a trough between trench and island arc (see Fig. 1.9). It covers the oldest rocks of the accretionary prism. The sediments deposited in this basin are derived from the volcanic arc and hence clastic and carbonates with some fine-grained turbidite deposits.

Volcanic Arc

Beyond the fore-arc region in the active volcanic arc itself, older rocks occurring below a topographically higher region constitute the Frontal Arc, also called the *Arc Basement*. This should not be confused with the fore-arc region. In the Island Arc, also known as Volcanic Arc or Magmatic Arc, the arc basement is a shallow marine platform or an emergent region of older rocks. In continental arcs, the continental platform of older rocks stands 1–5 km above sea level. Most island arcs consist of volcanoes that have an elevation of 1–2 km above sea level, irrespective of the elevation of the basement. In contrast, the elevation of the continental arcs depends on the elevation of the basement. The arc basement, i.e. Frontal arc consists of older, deformed and metamorphosed rocks on which modern Arcs are built. However, the character of the basement is quite variable. Continental arcs such as the Andes are built on a continental basement complex.

In Island arc, we have two types of sediments, namely the clastic and carbonate. The clastic sediments consist of debris from active volcanoes. They range from fine sand to coarse conglomerates and breccias deposited near the source regions. Pumice is the common constituent of these sediments. In continental arcs, sediments are predominantly subaerial and subordinate marine. Most island arc sediments are marine, reflecting deposition in deep sea fan complexes that are shed from active volcanic islands. In tropical regions, active island arc contains fringing carbonate reefs. These carbonate deposits are flooded intermittently with volcanogenic debris. Thus, carbonate and volcanic deposits interfinger together with each other. The volcanic arc in the region of active magmatic activity is marked on the surface by a chain of volcanoes, generally andesitic in composition. Large plutonic bodies of batholithic dimensions are unexposed in recent Arc but abundant in ancient continental arcs. In plan view, the volcanoes are spaced along the Arc at a fairly regular interval of approximately 70 km. In both continental and oceanic arcs, igneous complexes are correlatable with depth or to the subducting plate. Increase in Al, Na and K for a given rock type has been correlated with the depth of Benioff Zone in a number of Arc complexes. Such compositional variations are of great value for inferring the direction of dip of ancient subduction zone. However, for ancient arc, these compositional variations must be used with caution for a number of reasons, as stated by Moores and Twiss (1995). These are: (1) the same abundance in elements in a single magmatic system as it crystallizes and differentiates; (2) the same differences in the same place through time as the subduction changes its dip or its depth or as it migrates with respect to the magmatic center; (3) the high susceptibility in alteration of many major and minor elements in rocks during metamorphism and weathering. These processes severely affect the elements, namely Na, K, Si, and to lesser extent Al, that are used to characterize the volcanic rocks of the Arc; (4) some Arcs do not show any variation in composition with depth; (5) the cause of variation of chemistry with depth is not clearly understood because of unclear origin of the magmas and degree of contamination of magmas. Metamorphism in arc and arc basement (frontal arc) reflects the high heat flow that existed in the region. In continental arcs, the arc basement is a continental crust and is generally above the minimum melting temperature of the granite. This fact alone could explain the abundance of granitic batholiths in continental arc regions. Under the volcanic arc a high-temperature, low-pressure metamorphism occurs and when these rocks occur adjacent to the fore-arc region near trench with low T, high P environment, we have *paired metamorphic belts*. The paired metamorphic belts are often parallel to the plate boundary over extended distances and the two belts are of the same age and are typically separated by major faults. In Miyashiro's classic model for Japan, the high-pressure belt is located in the accretionary wedge between a subduction zone and an island (magmatic) arc, whereas the low-pressure belt is up to several hundred kilometers farther inland from the plate margin.

Back-Arc Basin

The back-arc basins are small ocean basins behind the volcanic arc (Fig. 1.9). They are also called marginal basins. In island arcs, these basins have oceanic crust (structure and abyssal depth) similar in composition to those of the ocean basins. Here the basins are bordered by remnant arc or Third arc that are linear topographic ridges composed of thicker crust, comparable to island arcs. These back-arc basins receive the sediments derived from the island arcs. In continental arcs, the sediments are deposited in the basins that form on top of the continental platform toward the continental interior. These epicontinental basins are called Retro-arc basins or foreland basins. Back-arc basins of oceanic island arcs have two characteristics. Some are composed of entrapped old oceanic crust; others are composed of younger lithosphere. The latter typically show seismic activity, high heat flow, and sea-floor spreading. The composition of volcanic rocks in modern back-arc basins is variable. Some rocks occurring away from the arc have similar chemistry as the mid-oceanic ridges. The volcanic rocks found near the arc are similar to the rocks of the main arc. The remnant arcs also show the same igneous and sedimentary rocks that characterize active volcanic arc, but they are older than the associated active island arcs and the marginal basins. The remnant arc occurring as ridges seems to be the remnants of a pre-existing arc that split to form the sides of a developing marginal basin. Therefore, the origin of remnant arcs is closely related to the origin of the back arc basins themselves. In the back arc basins, clastic sediments derived from the arc show interfingering relationship with the lavas and pelagic sediments. The back-arc sediments show folds and thrusts toward the interior of the continent, although the back arc region of the oceanic and continental arcs is marked by extensional tectonics and subsidence relative to the arc itself.

The back-arc basins are considered to have developed in response to tensional tectonics. Several origins of the back-arc basins have been proposed (see Moores and Twiss, 1995). Some authors believed that an existing island arc rifted along its length in response to tensional forces and that the two halves, corresponding to the island arc and the remnant arc, separated to give rise to the marginal basin (Fig. 1.9). Others suggested that the island arc, being most ductile, undergone initial rifting; subsequent widening gives rise to a back-arc basin.

Other back-arc basins owe their origin to the formation of new oceanic crust behind an island arc. There are three ways by which marginal basins could originate (Fig. 1.10). First, when forceful injection of a diapir from the subducting slab causes extension of the overlying oceanic crust (Fig. 1.10a). Second, when convection currents driven by the drag of the down-going slab cause extension of the oceanic crust behind the island arc (Fig. 1.10b). Third, when an overriding plate retreats by change of global plate motion, change of compression to extension of the oceanic

1.4 Mountain Building Processes

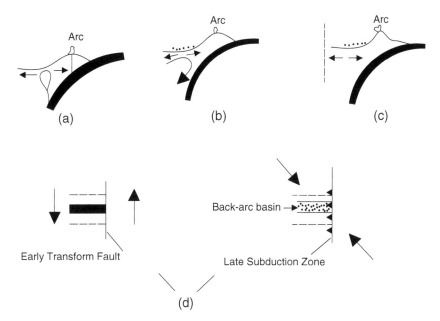

Fig. 1.10 Different mechanisms for the formation of Back-arc basin (after Moores and Twiss, 1995). (**a**) Forceful injection of a diapir from the subducting slab and extension of the overlying crust; (**b**) Convection currents driven by the drag of the downgoing slab resulting into extension of oceanic crust behind the island arc; (**c**) Retreat of overriding plate by global plate motion changes compression to extension of the oceanic crust behind the arc; (**d**) Back-arc basin formed by change of a Transform fault to a subduction zone

crust behind the arc gives rise to the formation of back-arc basin (Fig. 1.10c). Some basins may result from the entrapment of oceanic crust if a pre-existing fault zone becomes an oceanic subduction zone during a change of plate motion (Fig. 1.10d). The Aleutian basin may have formed in this way.

Some back-arc basins, familiarly known as the *marginal basins*, are formed due to rifting or secondary spreading of continental crust, resulting into Japan style marginal basin (Fig. 1.11). The resulting arc is the continental arc, which may be later added by magmatic material appearing as a melt material of the subducting

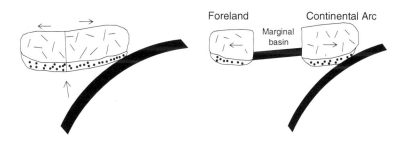

Fig. 1.11 Marginal Basin (Back-arc basin) formed due to secondary spreading (see text)

oceanic plate. The Japan Sea is a narrow ocean between the passive coast of Asia and the active volcanic arc of Japan. The continental arc formed by the secondary spreading is made up of the continental crust, which separated into an island arc and a continent near it. The supracrustal rocks in the marginal basin so formed are similar to those deposited in true geosyncline and are subsequently folded and metamorphosed by convergence of the two blocks that were separated earlier. Like the volcanic arc, the marginal basin is also an area of high heat flow.

1.5 Continental Rifting and Sedimentary Basins

James Hall of New York recognized long ago (1859) that mountains, as the most elevated parts of the Earth's crust, had risen by gigantic inversion of relief from the more depressed regions. Thus, it becomes obvious that a sedimentary basin or a trough (geosyncline) is the first requisite for the orogenesis. Some authors use the term rift for such basin and the process is called *rifting*. The term rift is applied to an elongated depression in the continental lithosphere because of applied extensional forces. Intracratonic basins are generally the consequence of extensional tectonics. In some cases, these extensional basins are asymmetric with an overall increase in thickness towards one of the bounding faults (Dunbar and Sawyer, 1988), indicating that the basin-bounded faults are active during sedimentation inducing subsidence and creating space for preservation of sediments. The basin-fill strata are also affected by intrabasinal gravity faults reflecting synsedimentary downward displacement that generate accommodation space for sediment deposition through out the history of basin evolution (Bally, 1980).

Even prior to the advent of plate tectonics it has been recognized that subsequent to the stabilization of the continental areas or cratons 2500 Ma ago, the Archaean crust was subjected to ductile stretching due to movement of mantle at depth and drag effects of sublithospheric convection currents (Bott, 1981, 1982). The crustal stretching or extension resulted in the formation of depressed regions or geosynclinal basins whose floor could be made of sialic crust or, if stretching was sufficiently large, of oceanic crust (Courtillot, 1982; Courtillot and Vink, 1983). In this depression, sediments start accumulating and eruption of magma results due to decompression melting of the ascending asthenosphere (Fig. 1.12).

Since the strength of the lithosphere under tension is least, the continental crust shows rifting associated with normal faults. The rift zone must be approximately normal to the regional tensile stress system. The rifts often contain volcanic rocks that are alkaline in the initial phase of rifting. The location of the rift is often controlled by pre-existing zones of crustal weaknesses (Virk et al., 1984). The rifting phenomenon occurs in a sequence of events in which first a normal fault develops in the upper crust under tensile stress (Fig. 1.13a). The fault movement warps the crust whose bending has the maximum tension at the location of greatest curvature. In this location, a second fault develops and a rift valley is thereby established (Fig. 1.13b). Continued tension gives rise to continuous subsidence of the keystone,

1.5 Continental Rifting and Sedimentary Basins

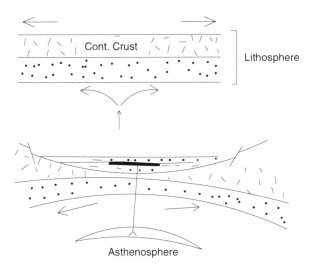

Fig. 1.12 Ductile stretching as a possible mechanism for rift basin formation (see text)

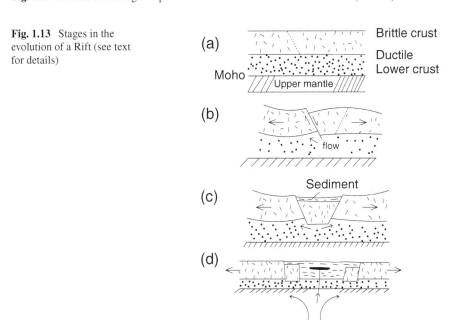

Fig. 1.13 Stages in the evolution of a Rift (see text for details)

aided by the sediment load. These movements are accompanied by compensatory flow in depth (Cronin, 1992; McKenzie and Morgan, 1969). Further tension separates the crustal blocks or lithospheric plates farther away and this horizontal separation of the crustal blocks indicates formation of true oceanic rift or trough. The floor of the rift drops below sea level. As the rift widens, sedimentation begins and

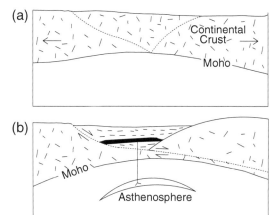

Fig. 1.14 Development of an Extensional basin in a continental lithosphere. (**a**) Rising convection currents generate tensional stresses, causing stretching of the crust; (**b**) deposition of sediments and basaltic lava in the extensional basin

marine conditions prevail. During the sedimentation emplacement of magma/lava also occurs due to melting of the asthenosphere.

The sedimentary basins, particularly in the intraplate environments, also develop by oceanward creep of the continental lower crust with a concomitant rift formation (Fig. 1.14).

1.5.1 Rifting and Triple Junctions

Rifting is commonly associated with domal uplift of the overlying crust because of impinging of mantle plume on the base of the lithosphere (Burke and Dewey, 1973). The plume is the consequence of some sort of thermal anomaly in the upper mantle, at the lithosphere-asthenosphere boundary. Any increase in temperature is likely to raise the boundary and consequently thinning of the lithosphere. Because of this activity and consequent doming of lithosphere, we have three-armed rifts originating from a triple junction, and also have basaltic magmas with the initiation of continental rifting, similar to the East African Rift, Gulf of Aden and the Red Sea that form a triple junction (Oxburgh and Turcotte, 1974; Patriat and Courtillot, 1984). If all three arms of the Rift continue to spread, the result is three separate continents and a mid-ocean Ridge-Ridge-Ridge (R-R-R) junction. On the Earth's surface that comprises of a mosaic of interlocking plates, there are several places where three plates are in contact and these are called triple junctions. The type of triple junction depends upon the various combinations of the basic types of plate boundary: divergent, convergent, and strike-slip. A convergent boundary can have either of two polarities, depending upon which is the overriding plate. A strike-slip boundary can have either of two shear sense; and a divergent boundary will have only one subduction direction. Thus, there are in all five different geometries for the plate boundary. Since three plate boundaries come together at a triple junction, so there

1.5 Continental Rifting and Sedimentary Basins

are 125 combinations of 5 boundary geometries, taken three at a time (5×5×5). Of these, only 16 are kinematically possible, and 14 can actually exist for any geological significant length of time. These 14 types of junctions are called *stable triple junctions*. Stability of a triple junction does not mean that the location of the junction is fixed on the earth's surface or on a plate boundary (Cronin, 1992). Triple junctions between three ocean ridges, such as that in the South Atlantic between the African, South American and Antarctica plates, are known as ridge-ridge-ridge or RRR triple junction (Fig. 1.15a). A similar notation can be used to recognize triple junctions involving trenches (T), or Transform Faults (F). Therefore, a trench-trench-trench junction is denoted as TTT triple junction (Fig. 1.15c, d) while a ridge-ridge-transform fault junction would be termed RRF triple junction (Fig. 1.15e).

Figure 1.15 also shows the evolution of these three triple junctions from one time to a later time (see right-side figures). The geometric configuration of RRR (Fig. 1.15a) is always stable and the magnetic anomalies within the surface area created have Y-shaped patterns around the spreading ridges (Fig. 1.15b). The triple

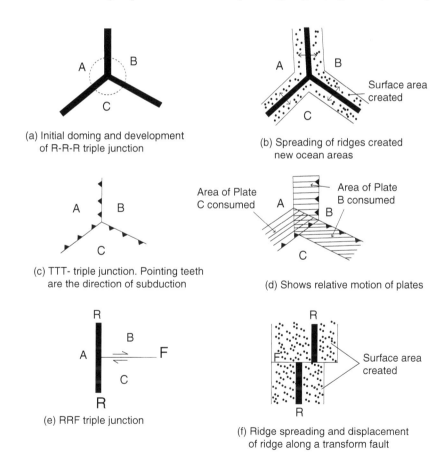

Fig. 1.15 Evolution of triple junction from one time to later time (after Kearey and Vine, 1996); see also text

junction between three trenches (TTT triple junction) is stable only when the relative motion of plates A and C is parallel to the plate boundary between B and C (Fig. 1.15c). That is, if the rates are the same and if the direction of subduction of plate C below plate A is exactly parallel to the boundary between plates B and C (Fig. 1.15d). When the relative motion of plate C is not parallel to the boundary between plates B and C, the triple junction is unstable. The triple junction RRF (Fig. 1.15e) is unstable because there is relative motion between plate B and plate C. Consequently, the RRF triple junction evolves at once to form an RFF junction (see Fig. 1.15f and compare with Fig. 1.8).. The RRR junctions are the only junctions that are always stable. Other triple junctions may be stable in certain restricted situations but are otherwise always unstable. Because RRR triple junctions are more stable than other forms, such triple junctions might still exist in old ocean floor, and can be recognized by bent magnetic anomalies, as in the Y shaped area of Fig. 1.15b.

The stability of the boundaries between plates is dependent upon their relative velocity vectors, as explained below (Fig. 1.16).

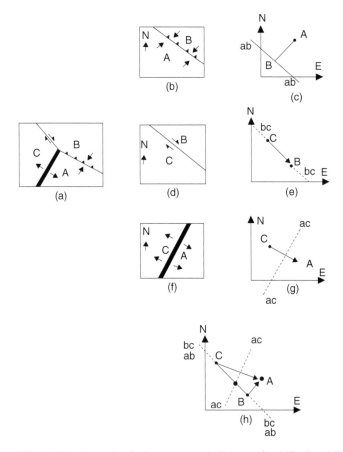

Fig. 1.16 RTF triple junction and velocity vectors as indicators of stability/instability of Plate boundaries (after Kearey and Vine, 1996); see text

1.5 Continental Rifting and Sedimentary Basins

When three plates come into contact at a triple junction the stability depends upon the relative directions of the velocity vectors of the plates in contact. Figure 1.16a shows a triple junction at ridge-trench-transform fault (R-T-F). In order to be stable, the triple junction must migrate up or down the three boundaries between the pairs of plates. To appreciate we ascertain this migration at each boundary in turn. In Fig. 1.16b, plate A is underthrusting plate B at a trench in NE direction. Now relative movement between A and B is shown in velocity space in Fig. 1.16c in which the velocity of any single point is represented by its N and E components. The line joining two points represents velocity vectors. Thus, the direction of line AB is the direction of relative movement between A and B, and its length is proportional to the magnitude of their relative velocity. Therefore, line *ab* must represent the locus of a stable triple junction, and B must lie on *ab* since there is no motion of the overriding plate B with respect the trench.

Next, we consider a transform boundary between plates B and C (Fig. 1.16d). Again, the relative movement between B and C is represented in velocity space (Fig. 1.16e). Between these two plates, the line BC is the velocity vector. The locus of a point travelling up and down the fault BC is in the same direction as vector BC, because the relative movement direction of B and C is along their boundary.

Finally, consider a ridge separating the plates A and C (Fig. 1.16f) and its representation in velocity space (Fig. 1.16 g). The relative velocity vector AC is now normal to the plate margin, and hence the line ac represents the locus of a point travelling the ridge. The ridge crest must pass through the mid-point of the velocity vector CA (provided the accretion is symmetrical).

By combining the velocity space representations, the stability of the triple junction can be determined from the relative position of the velocity lines representing the boundaries (Fig. 1.16 h). If they intersect at one point, it means that the triple junction is stable, because that point is able to travel up and down all three plate margins. If the velocity lines do not all intersect at a single point, the triple junction is unstable.

As stated already, only the RRR triple junction is stable for any orientation of the ridges. However, if one branch ceases to spread, it would form a failed arm, called *Aulacogen* (Fig. 1.17). Of the remaining two, one rift spreads more or less normal to the continental margin and the other spreads obliquely. Aulacogens are long-lived, deeply subsiding sedimentary troughs that extend at high angles to the fold belt (Burke, 1980; Hoffman et al., 1974). Since they occur from Late Proterozoic times

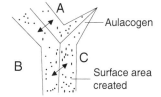

Fig. 1.17 Development of an Aulacogen from a RRR triple junction; see text for details

onward, some geologists have invoked their presence at places where they do not seem to be. Aulacogens are located at re-enterants on continental plate margins, and their initial formation is cotemporaneous with continental rupture. They are characterized by vertical tectonics. The sedimentary fill is about three times thicker than on the adjacent craton. The sediments are undeformed or weakly deformed, in contrast to the extreme tectonism experienced by adjacent orogenic fold belts. The aulacogens have a tendency to reactivation and often contain intrusion of alkaline igneous rocks. Aulacogens provide a complete igneous, sedimentary, and structural history of events associated with an orogenic belt. One should be therefore cautious to declare a rift within a continental plate as aulacogen (cf. Klein and Hsui, 1987; Beaumont and Tankard, 1987). The aulacogens are a very favourable location for the development of a river system, which carries detritus from the craton.

In India, the triple junction is located at 15 degree south latitude and 75 east longitude in the central Indian Ocean. From this junction a branch of the rift system, the Carlsberg ridge extends NW into Gulf of Aden to the Afar RRR triple junction, from which extend the Red Sea and the East African rifts. The second is the SE Indian Ridge that goes to south of Australia. The third is SW Indian ridge which extends in a series of short ridge and long transform fault segments around the southern tip of Africa into the South Atlantic Ocean to the Bouvet ridge-transform-transform triple junction (cf. Moores and Twiss, 1995).

For identification of the triple junction on or near the Indian shield, one has to consider criteria such as age, orientation and igneous activities in the three-armed rifts that are evolved by domal uplift of the crust, as described earlier.

It must be noted that rifting in the brittle zone of the crust, giving rise to *graben* (and associated horst), like the Rhine Graben or Godavari graben, is limited only to a small width (Bott, 1964, 1971). The amount of extension achieved by displacement on a steep normal fault is limited by depth to which crustal block can sink as it is compensated by lateral flow in the ductile lower crust. If the subsiding faultwedge is bounded by vertical faults, the subsidence is lower than if the fault wedge is bounded by inclined faults. This is easily understandable if one considers that a wedge-shaped block of wood would float lower in water than a block of the same thickness with parallel edges.

1.5.2 Pull-Apart Basin

Typical Pull-apart basins occur at different scales and they form in the sedimentary cover above strike-slip fault in the basement. The most popular mechanism is local extension between two en echelon basement strike-slip fault segments. These can be right-stepping with dextral shear or left-stepping with sinistral shear. The local extension is accommodated in the sedimentary cover by normal faults at the releasing oversteps and bends. In other words, when a strike-slip fault has notable curvature, the curved area, separating the ends of the faults, is thrown into tension

1.5 Continental Rifting and Sedimentary Basins

Fig. 1.18 Pull-apart basins due to strike-slip motion and extension. (**a**) Basin at a fault curvature during transtensional regime, (**b**) sharp pull-apart basin by strike-slip motion, (**c**) basins formed as a result of fault termination

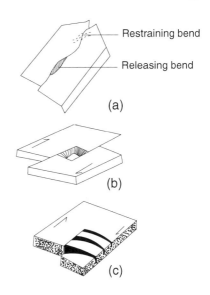

or compression. Compression gives rise to an elevated region by crustal shortening, typified by folds and thrust faulting, while tension gives rise to an extensional trough, known as a pull-apart basin (Fig. 1.18). The strike-slip margins are initially straight and parallel, but may sag with time (Fig. 1.18a, b). A similar basin would also develop when one fault terminates and sidesteps to an adjacent parallel fault (Fig. 1.18c). In a pull-apart basin, the faults should be steep near the surface and should show evidence of strike-slip movement. Pull-apart basins progressively grow in the same direction as the fault movement so that the oldest sediments occupy the margins only (Fig. 1.19). As the pull-apart basin grows, its floor is stretched and attenuated whereby igneous material is emplaced in the centre of the basin (revealed by high gravity). However, this may not happen in all circumstance, and some basins may not have been attenuated to an extent to cause rupture of the lithosphere and igneous rocks are completely absent. The pull-apart basin can have any shape. Pull-apart basins are excellent target for hydrocarbons.

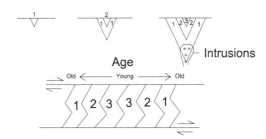

Fig. 1.19 Stages in the formation of Pull-apart Basins (see text for details)

1.5.3 Foreland Basin

It is a flexure depression peripheral to the mountain range (Fig. 1.20). An example of this is the development of the Indo-Gangetic plain basin adjacent to the Himalaya. This foreland basin extends much further into the surrounding craton and forms in advance of the thrust front of the orogenic belt. Close to the mountain range, the sediments are coarse-grained and deposited in a shallow water or continental environment. Farther away the sediments are fine-grained and often turbidite. The sediments thus form a wedge-shaped unit whose stratigraphy reflects the subsidence history of the basin as it grows and migrates outwards as convergence continues. The stratigraphy shows progressive overlap of sediments onto the foreland. This means that the stratigraphy is characterized by units, which thin laterally, overlap older members, or are even truncated by erosion.

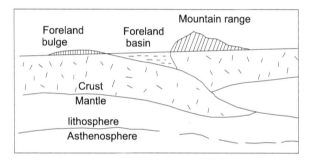

Fig. 1.20 Foreland basin formation as a flexural depression of the continental crust

1.6 Orogeny and Plate Tectonics

Careful study of fold belts has revealed that the period of orogenesis is generally quite long, often exceeding 100 million years. Orogenic belts or fold belts usually consist of roughly parallel ridges of folded and thrusted sedimentary and volcanic rocks, portions of which have been strongly metamorphosed and intruded by somewhat younger igneous bodies. Initiation of the orogenesis occurs with the development of a depositional trough or geosyncline because of rifting or ductile stretching of a pre-existing continental crust or lithosphere. The rifted basin becomes a wide depositional site by further stretching of the continental crust due to divergent subcrustal convection currents. The floor of this rifted basin sinks below sea level and is flooded. The separating margins of the rift thus become the coasts of a growing intercontinental sea. The floor of such oceans may be made of stretched continental crust or, if crustal extension was extreme, the floor may be of oceanic crust. This basin receives sediments, which are a vast accumulation of deep-water marine deposits that may exceed a few thousand meters in thickness as well as thinner shallow water deposits. The deep-water sediments deposited in eugeosyncline (seaward

1.6 Orogeny and Plate Tectonics

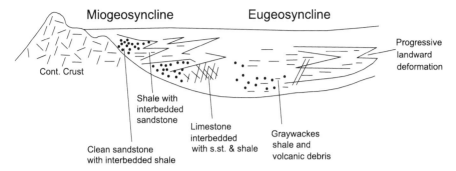

Fig. 1.21 Different sedimentary deposits in shelf, slope and deep-sea conditions within a geosyncline

of the miogeosyncline) consist primarily of graywackes, volcanic debris, and shale. The shallow water deposits of the miogeosyncline and those on the shelf region are toward the continental side. They consist of relatively clean sandstones, limestones, and shale (Fig. 1.21).

The plate tectonic cycle begins at the divergent boundaries along ocean ridges, causing the two lithospheric plates (overridden by continental plates) to move away and converge elsewhere with another lithospheric plate. With the seafloor spreading, the lithosphere is transported laterally, away from the ridge toward a convergent plate boundary, i.e. toward a trench. The migrating plates carry the sediments deposited on their margin and the deformation generally progresses in a landward direction so that the deep-water sediments are the first to be deformed. At the convergent plate boundary the oceanic lithosphere is bent and it moves down into the Earth's mantle, producing an ocean trench at the surface. The descending plate beneath the ocean trench is heated by friction and other heat sources at depth and causes partial melting of the down-going plate with some overlaid sediments. Water and other volatiles help in the melt generation and the melt rises to the surface, forming volcanic arc, as volcanoes parallel with the ocean trench. The oceanic trench becomes a trap for the ocean sediments and for additional sediments derived from the adjacent island arc.

During the plate movement, the leading edge of the overriding plate bulldozes material from the ocean floor into an accretionary wedge. Some sediments are scrapped into the wedge with time. The accretionary wedge grows with time and may include other materials such as sea mounts or islands scrapped from the sea floor.

Where two plates converge, and the leading edge of one plate is made of oceanic and the other of continental block, the intervening oceanic crust (being denser) of the oceanic plate is subducted or thrusted below the continental plate; a very small quantum of sediments is carried in the subduction. The subducted oceanic crust (with minor trapped sediments) undergoes melting at depth. The melt rises near or on the surface, resulting in the formation of volcanic arc or *magmatic arc* (commonly andesitic in composition). The rocks of the arc at depth crystallize into granitoids

(I-type). The magmatic arc is often located a few hundred kilometers seaward of the ancient coastline. The igneous rocks of the arc are eroded and deposited in the basins on either side of the arc and are called *fore-arc basin* (on the trench side) and *back-arc basin* on the other side of it.

With continued subduction the intervening oceanic crust disappears, and finally the colliding plates become continent versus continent. Because of the plate convergence, the sediments and the interbedded lava on the ocean floor are folded and even thrusted to increase in thickness and eventually affected by higher geothermal gradients. The complexly folded rocks at some stage of their deformation are recrystallized by heat brought upwards by hot fluids. These fluids are in fact liberated from subducted material or from the magmas generated below. The influx of heat is also notably due to heat production by radioactive elements in the sedimentary pile. Additionally, heat contribution is also due to transformation of mechanical work (used in rock deformation) into thermal energy. This temperature increment is due to simple compressional squeezing of rocks at depth due to weight of overlying material. A further increase in temperature can also be caused by friction, generated during lithospheric sliding into the mantle. The thickened crustal rocks undergo thermal relaxation and consequently a series of progressive metamorphic reactions take place in the rocks with increased depth or pressure and temperature. Amongst these, some critical reactions give rise to *metamorphic isograds,* which are distinguishable in the exposures by the appearance or disappearance of an index mineral or a specific mineral paragenesis. The isograds separate the metamorphic rocks into zones or facies or grades (Fig. 1.22). The series of metamorphic assemblages developed in different metamorphic zones are a reflection of T-depth profile at the time of metamorphism. The T-depth profile implies that the types of metamorphism, Barrovian or Buchan/Abukuma, are the result of different tectonic conditions. High T,

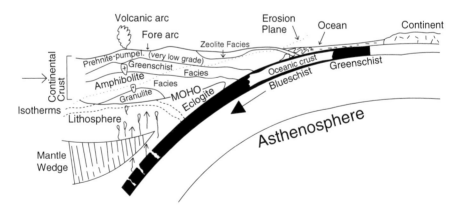

Fig. 1.22 Schematic cross section of converging ocean-continent plates, showing island arc and the distribution of metamorphic facies. High P/T facies series develops in the subduction zone while low P/T facies series is formed in the arc. *Dotted line* is trace of an erosion plane along which a progressive sequence of regional metamorphism is encountered in eroded fold belts

1.6 Orogeny and Plate Tectonics

low P metamorphism (Buchan type) implies that temperatures are elevated above a normal geothermal gradient of 25°C/km. The condition develops in contact aureoles around shallow igneous intrusions in a volcanic arc environment. On the other hand, high P, low T metamorphism generates blueschists, which implies that the temperatures of the rocks are significantly lower than would be in a normal geotherm. This occurs in subduction zones where cold rocks are carried to large depths faster than temperatures can be re-established. Such metamorphism is rare in most orogenic belts. Where present, it is overprinted by Barrovian metamorphism. Adjacent high T-low P metamorphism (Buchan type) and blueschists (high P–low T) metamorphism constitute a paired metamorphic belt. Barrovian metamorphism indicates a normal geothermal gradient and is widespread in all orogenic belts. For this reason it is also known as classical regional metamorphism. The Alps provide a good example of the Barrovian metamorphism. In the Himalaya, however, the metamorphism is inverted such that higher grade rocks occur in higher tectonic levels. Such an occurrence could indicate a primary inversion of isotherms, as would develop at a subduction zone above a down-going slab. Alternatively, it could indicate a tectonic inversion of the isotherms following metamorphism, like that on the inverted limb of a recumbent fold. This problem will be discussed in detail at the appropriate place when the evolution of the Himalaya is considered in Chap. 3. In areas of good exposure, we can obtain an idea of the shape of isograd surfaces in 3-D. Where they are undeformed or only weakly deformed, they appear to be gently curved surfaces that intersect the Earth's surface at small angles. The interpretation of metamorphic zones in terms of plate tectonics is complex. However, the rocks in the central portion of all orogenic belts are metamorphic.

The distribution of metamorphic zones in many deformed belts is roughly symmetrical, such that the highest grade rocks constitute the central portion of the fold belt, which is a part of the mountain root at depth, and less metamorphosed or unmetamorphosed rocks occur on the flanks. In the mountain building processes by plate tectonics, the different domains have a relationship which is typical for a collision fold belt, whether the collision occurred between an oceanic plate and a continental plate or between a continent and another continent. The metamorphic isograds or facies boundaries show trends similar to that held by the subduction zone (i.e. trench). A similar orientation is also exhibited by granitic rocks of the magmatic arc (Fig. 1.23). High-grade metamorphism may be symmetrical with respect to the arc, but uneven erosion shows regionally metamorphic rocks in a progressive manner in one direction only. The sediments lying on the continental margins may be intricately folded and are often thrust upon by metamorphic rocks of the fold belt.

In the orogenic belts, roots are regions of greater crustal thickness that underlie almost all continental orogenic belts in the world. Two hypotheses have been proposed to explain the formation of mountain root (cf. Wilson and Burke, 1972). One is related to subduction processes and the other to collision processes. According to the subduction model, the mountain roots form beneath an active continental margin by the rise of mafic magmas away from the subducting slab and intrusion of the magmas into the lower crust. These magma bodies heat the crustal rocks in the vicinity and thus form the mobile core. The thickened continental crust is raised due

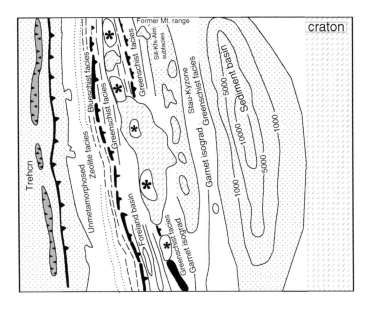

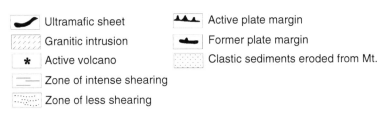

Fig. 1.23 Schematic map showing the development and location of plate tectonic elements and the distribution of metamorphic facies at a convergent continent-ocean plate boundary

to isostasy. The high topography of the orogenic belts exists because of the isostatic support given by the thickened crust "floating" in high-density mantle. Gravitational collapse of the thickened crust results in thrusting of the recrystallized rocks on the foreland. The second model explaining the formation of the mountain roots is the collisional model. According to this model, the root is formed by underthrusting of one continent by another. This causes crustal thickening and hence elevated topography of the mountain belt. The collision model seems more plausible for the formation of roots and high topograpahy of the orogenic belts. In young mountains, like the Himalaya, thickening of mountain root approximately reflects the amount of shortening that has taken place, but thickening should occur rapidly otherwise erosion and ductile collapse will hinder the formation of the root and high topography of the orogenic belt.

It is possible to map the isograds and successive structures, such as s-surfaces (S1, S2, S3 etc.), lineations (L1, L2, L3 etc.) and fold axes (F1, F2, F3 etc.), and relate them temporally with the metamorphic events of minerals (syn-, late-, or post-tectonic with respect to deformation phases D1, D2, D3 etc.). This may help in unraveling the geohistory of the fold belt. At certain locales, the rocks subjected to regional metamorphism also undergo anatexis (partial melting). This process gives rise to migmatites, granulites, and granitic plutons. The granitic plutons (S-type) are the products of accumulation of anatectic melt that intrudes the overlying rocks. The isostasy and the buoyancy of the anatectic magma and the heated metamorphic rocks are exhumed to form lofty mountain ranges or fold belts. During uplift of the orogen, some parts of the subducted oceanic crust may also be obducted as high-pressure rocks called *blueschists* and eclogites. The orogenic uplift is often accompanied by the formation of down-sliding nappes. With the elevation of orogenic belt, unconsolidated marine sediments begin to slide downwards along gentle inclines and fall off as *turbidite currents* into the marginal deeps where they often build thick chaotic *flysch* deposits. The weathered debris transported by streams is deposited as *molasses* into a marginal deep. It must also be noted that the sediments deposited on continental shelf are also complexly folded and may even be displaced inland along low-angle thrust faults and the sedimentary portion of the mountain chain.

1.7 The Wilson Cycle

As discussed in the previous section, the orogenic belts are characteristically formed of thick shallow-water sediments deposited on continental crust and of deep marine sediments that are subjected to deformation and metamorphism during plate tectonic operations. It is postulated that Proterozoic fold belts also originated by the opening and closing of the geosynclinal basin. The periodicity of ocean formation and closure is known as the *Wilson Cycle*, named after J. Tuzo Wilson, by Dewey and Burke (1973) in recognition to his contributions to the theory of plate tectonics. The Wilson cycle gave us a way of understanding mountain building processes. Therefore, we must have a good knowledge of this concept which is a necessary background for understanding the evolution of fold belts of younger and older ages. The Wilson cycle suggests a pattern for the developments of orogenic belts and the cycle is assumed to have operated through much of geological time. The Wilson cycle has three stages. (1) Separation of the continents and development of an ocean basin (geosyncline) with accumulation in separate areas of thick deposits of shallow marine sediments (miogeosynclinal) and deep-water marine sediments (eugeosynclinal); the latter with mafic igneous rocks of basaltic to andesitic compositions. (2) Movement of the continents towards each other, causing deformation of the eugeosynclinal and miogeosynclinal rocks together with emplacement of ophiolites from depth. (3) Continental collision attendant at some stage with intrusion of granite batholiths in the core region. The generation of most magma can be related to one of the three stages in the Wilson cycle. During separation of the continents, the principal igneous activity includes emplacements of tholeiitic lava flows, dykes and sills as well as alkaline intrusions—kimberlites and carbonatites—on the continents along a relatively wide zone parallel to the line of separation of the continents. Extrusion of tholeiitic basalts, and sometimes alkaline volcanics, occurs in the ocean basins. Again, extrusion of andesitic lavas occurs in island arcs along the leading edge of the continent adjacent to subduction zones. During the period when the continents move toward each other, the subduction occurs on either one or on both continental margins that are approaching. There is little magmatic activity along the continental margins devoid of subduction. Along the leading edge of the continents there are both extrusions and intrusions of granitic composition. When continents collide, the principal igneous activity is the intrusion of granitic rocks along the fold belts formed by collision. Sometimes extrusion of rhyolites also occurs along the belt. Following the collision of the continents, this process is repeated because the continents are not consumed due to buoyancy.

The Wilson cycle can include many events leading to orogeny, and one example of the events, adapted from Dewey and Bird (1970), is described below (Fig. 1.24).

The beginning of the Wilson cycle starts with the initial rupture of a continent and development of an accreting plate margin (Fig. 1.24a). As separation of continents proceeds, there develops a rift valley (Fig. 1.24b) and small ocean (Fig. 1.24c). These are the early stages of opening of the ocean. The African Rifts and Red Sea are the present day examples. With the separation of the continents, continent-derived sediments accumulate at the continental shelves at the ocean-continent interfaces. The Atlantic-type ocean is fully developed with a central ridge (Fig. 1.24d). It may be noted that the two plates are drifting away from the ocean ridge but they are not get separated because the upwelling magma is added to the trailing edge of the plate. In the following stage, the oceanic lithosphere uncouples and descends along a trench into the mesosphere. Figure 1.24e represents a consumption stage of oceanic plate margin and is the initial stage of ocean closing in the Wilson cycle. If the uncoupling takes place at a considerable distance oceanward from the continental margin, an island arc, like Japan, forms. If the uncoupling and subduction of the oceanic lithosphere or oceanic plate occurs adjacent to the continental margin, it develops Andean-type orogen because of tectonic and volcanic processes (Fig. 1.24f). As the ocean continues to contract, the opposing continental margin ultimately arrives into the consumptive regime. This is the early stage of continent-continent collision (Fig. 1.24 g). Because of buoyancy constraints of the continental lithosphere, plate consumption ceases in the later stages of continent-continent collision (Fig. 1.24 h). The lithospheric plate detaches beneath the Himalayan-type orogen, and the final-stage of the Wilson cycle is marked now by the fully evolved mountain belt as the place of suture between the newly joined continental masses.

Strictly speaking, the Wilson cycle is not a cycle because there is no evidence of exactly the same sequence of events repeating itself in the same region. In addition, whenever two continents are moving apart, they are moving toward each other on the opposite side of the Earth or toward another continent. The movements of the continents in the Wilson cycle are taken relative to a Pangaea-like landmass that appears to have formed repeatedly in geologic time because of the joining of the continents. Moreover, there are instances where the orogenesis involves only

1.7 The Wilson Cycle

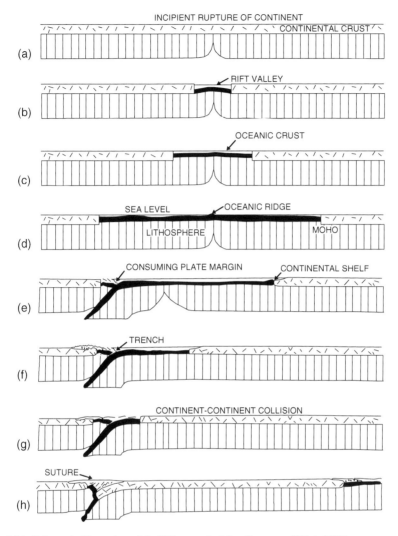

Fig. 1.24 Schematic illustration of the Wilson cycle (after Dewey and Bird, 1970)

intracratonic rifting due to ductile stretching and thinning, and the depositional basin having sialic/continental basement, and not oceanic crust. The basin becomes wider due to subcrustal convection currents and the ongoing sedimentation is intervened by volcanic lava flows, resulting from decompression melting at depth. Accumulation of the supracrustal material results in sinking of the basin. With deepening of the basin, decay of the underlying plume with time, and a change in the vertical component of stress, the diverging continental blocks are likely to reverse their movement direction to cause shortening of the crustal segment (Sharma, 2003). Consequently, the deformed and metamorphosed basinal rocks would give rise to fold belt or orogenic belt along which the crustal blocks are welded or sutured. Structures in these accreted crustal blocks are virtually unaffected (albeit overprinted) by the younger mobile belt. In contrast, the continental blocks welded together by plate tectonic processes show markedly different stratigraphy or tectonic history and have a discontinuity in the orientation or style of structures.

1.8 Suspect Terrains

During the evolution of orogenic belts, the plate tectonics motions not only result into collision of the continent(s) or arc(s) but at times some orogenic belts include terrain(s) that have distinct geology and structures, without showing any relationship to the subduction zone. That is, the relationship to the main crustal blocks is suspect. These terrains are called *Suspect* or Exotic *terrains*. They are allochthonous in the term of tectonics, and they have accreted to the continent over a considerable period, after travelling great distances. The process of amalgamation or welding has been referred to as accretionary tectonics or mosaic tectonics.

A terrain is an area surrounded by sutures and characterized by rocks having a stratigraphy, petrology, or palaeo-latitude that is different from that of the neighbouring terrains or continents. In a collisional fold belt, a terrain is a remnant of crust that has had a history different from that of the subducted oceanic crust or that of the main crustal block. Those who doubt a suspect terrain must work out detailed geological histories of the terrains, especially on the time of docking of two terrains. A suspect terrain is a mappable unit and Jones et al. (1983) have given several criteria to distinguish the identity of separate terrains. These are:

1. The stratigraphy and sedimentary history
2. Petrogenetic affinity and magmatic history
3. The nature, history and style of deformation
4. Palaeontology and palaeo-environments
5. Palaeopole position and palaeodiclination.

Based on these criteria, Jones et al. (loc. cit.) recognized four major types of suspect terrains in western North America. These are:

I. Stratigraphic terrains which are characterized by distinct stratigraphies.
II. Disrupted terrains that contain a heterogeneous assembly of flysch, serpentinites, shallow water limestones, and graywackes with occasional exotic blocks of blueschists.
III. Metamorphic terrains, in which a metamorphic overprint has destroyed the original stratigraphy.
IV. Composite terrains, which contain two or more such terrains that had amalgamated before accretion to the continent.

The chronological sequence of accretion of terrains to the continent can be known from geological events that indicate accretion and link adjacent terrains. These include the deposition of sediments across terrain boundaries; the appearance of sediments derived from an adjacent terrain; and "stitching" together of terrains by plutonic activity. Again, the Apparent Polar wandering paths (APWP) of two separate terrains should be different until their accretion, after which they should exhibit a single APW path.

Considering the diversity of terrains, especially the terrain of fold belt itself and the flanking continental blocks, it seems that we may become over suspicious on the problem of suspect terrains. However, the recognition of suspect terrains in fold belts requires a major modification of the Wilson cycle and of our ideas about how the orogenic belts develop. Moores and Twiss (1995) state that "the motion of major crustal blocks in a collision may represent only a part of the activity that constructs an orogenic belt, and many other relative plate motions may be represented by the numerous sutures among different exotic terranes".

References

Bally, A.W. (1980). Basins and Subsidence: A Summary. In: Bally, A.W. et al. (eds.), Dynamics of Plate Interiors, Geodynamic Series 1, vol. **1** AGU-GSA, Washington D.C., pp. 5–20.

Beaumont, C. and Tankard, A.J. (eds.) (1987). Sedimentary basins and basin-forming mechanisms. Can. Soc. Petrol. Geol. Mem., 12 and Atlantic Geoscience Soc. Spl. Publ. 5, 527p.

Beaumont, C. and Tankard, A.J. (eds.) (1987). Sedimentary basins and basin-forming mechanisms. C.S.P.G. Mem. 12/A.G.S. Spl. Publ. 5, Can. Soc. Petrol. Geol.

Bott, M.H.P. (1964). Formation of sedimentary basins by ductile flow of isostatic origin in the upper mantle. Nature, vol. **201**, pp. 1082–1084.

Bott, M.H.P. (1971). Evolution of young continental margins and formation of shelf basins. Tectonophysics, vol. **11**, pp. 319–327.

Bott, M.H.P. (1981). Crustal doming and the mechanism of continental rifting. Tectonophysics, vol. **73**, pp. 1–8.

Bott, M.H.P. (1982). The mechanism of continental splitting. Tectonophysics, vol. **81**, pp. 301–309.

Burke, K. (1980). Intracontinental rifts and aulacogens. In: Burchfiel, B.C. et al. (eds.), *Continental Tectonics*. National Academy of Sciences, Washington, DC., pp. 42–49.

Burke, K. and Dewey, J.F. (1973). Plume generated triple junctions: key indicators in applying plate tectonics to old rocks. J. Geol., vol. **81**, pp. 406–433.

Condie, K.C. (1982). Early and middle Proterozoic supracrustal successions and their tectonic settings. Am. J. Sci., vol. **282**, pp. 341–357.

Courtillot, V. (1982). Propagating rifts and continental breakup. Tectonics, vol. **1**, pp. 239–256.

Courtillot, V. and Vink, G.E. (1983). How continents break up. Sci. Am., vol. **249**(1), pp. 40–47.

Cronin, V.S. (1992). Types of kinematic stability of triple junctions. Tectonophysics, vol. **207**, pp. 287–301.

Dewey, J.F. (1977). Suture zone complexities: a review. Tectonophysics, vol. **40**, pp. 53–67.

Dewey, J.F. and Bird, J.M. (1970). Mountain belts and the new global tectonics. J. Geophys. Res., vol. **75**, pp. 2625–2647.

Dewey, J.F. and Burke, K.C.A. (1973). Tibetan, Variscan, and Precambrian basement reactivation: products of continental collision. J. Geology, vol. **81**, pp. 683–692.

Dickinson, W.R. (1971). Plate tectonic models of geosynclines. Earth Planet. Sci. Lett., vol. **10**, pp. 165–174.

Dunbar, J.A. and Sawyer, D.S. (1988). Continental rifting at pre-existing lithospheric weaknesses, Nature, vol. **333**, pp. 450–452.

Gansser, A. (1964). *Geology of the Himalaya*. Wiley Interscience, New York, 289p.
Gibb, R.A., Thomas, M.D. and Mukhopadhyay, M. (1983). Geophysics of proposed Proterozoic sutures in Canada. Precambrian Res., vol. **19**, pp. 349–384.
Hall, J. (1859). Description and Figures of the Organic Remains of the Lower Hederberg Group and Oriskany Sandstone. Natural History of New York; Palaeontology. Geol. Surv., Albany, NY, 544p.
Hess, H.H. (1962). History of the ocean basins. In: Buddington, A.F. (ed.), *Peterologic Studies*. Geol. Soc. Am., New York, pp. 599–620.
Hoffman, P. (1980). Wopmay orogen: a Wilson cycle of early Proterozoic age in the northwest of the Canadian Shield. In: Stragway, D.W. (ed.), *The Continental Crust and its Mineral Deposits*. Geological Association of Canada Special Paper 20, pp. 523–549.
Hoffman, P., Burke, K.C.A. and Dewey, J.F. (1974). Aulacogens and their genetic relation to geosynclines, with a Proterozoic example from the Great Slave Lake, Canada. In: Doth, R.H. and Saver, R.H. (eds.), *Geosyncline Sedimentation*. Soc. Econ. Pal. Mineral Spl. Publ., vol. **19**, pp. 38–55.
Jones, D.L., Howell, D.G., Coney, P.J. and Monger, J.W.H. (1983). Recognition, character and analysis of tectonostratigraphic terranes in northwestern North America. In: Hashimoto, M. and Uyeda, S. (eds.), Accretion Tectonics in the Circum-Pacific Regions. Terra Scientifica Publ. Co., Tokyo, pp. 21–35.
Kearey, P. and Vine, F.J. (1996). *Global Tectonics*. 2nd ed. Blackwell Science, Cambridge, 333p.
Klein, G. deV. and Hsui, A.T. (1987). Origin of cratonic basins. Geology, vol. **15**, pp. 1094–1098.
Kroener, A. (1981). Precambrian plate tectonics. In: Kroener, A. (ed.), *Precambrian Plate Tectonics*. Elsevier, Amsterdam, pp. 57–90.
McKenzie, D.P. and Morgan, W.J. (1969). Evolution of triple junction. Nature, vol. **224**, pp. 125–133.
Moores, E.M. and Twiss, R.J. (1995). *Tectonics*. W.H. Freeman Company, New York, 415p.
Nisbet, E.G. (1987). *The Young Earth: An Introduction to Archaean Geology*. Allen & Unwin, Boston.
Oxburgh, E.R. and Turcotte, D.L. (1974). Membrane tectonics and the East-African rift. Earth Planet. Sci. Lett., vol. **22**, pp. 133–140.
Patriat, P. and Courtillot, V. (1984). On the stability of triple junction and its relationship to episodicity in spreading. Tectonics, vol. **3**, pp. 317–332.
Press, F. and Siever, R. (1986). Earth. 4th ed. W.H. Freeman Company, New York.
Sharma, R.S. (2003). Evolution of Proterozoic fold belts of India: a case of the Aravalli mountain belt of Rajasthan, NW India. Geol. Soc. India Mem., vol. **52**, pp. 145–162.
Sharma, R.S. and Pandit, M.K. (2003). Evolution of early continental crust. Current Sci., vol. **84**(8), pp. 995–1001.
Tarney, J., Dalziel, I. and de Witt, M. (1976). Marginal basin 'Rocas Verdes' complex from S. Chile: a model for Archaean greenstone belt formation. In: Windley, B.F. (ed.), *The Early History of the Earth*. Wiley, London, pp. 131–146.
Umbgrove, J.H.F. (1947). *The Pulse of the Earth*. Martinus Nijhoff, The Hague, 358p.
Verhoogen, J., Turner, F.J., Weiss, L.E. and Wahrhaftig, C. (1970). The Earth. Holt, Rinehart and Winston Incl., New York, 748p.
Virk, G.E.W., Morgan, W.J. and Zhao, W.L. (1984). Preferential rifting of continents: source of displaced terranes. J. Geophys. Res., vol. **89**, pp. 10072–10076.
Wegener, A. (1915). Die Entstehung der Kontinente und Ozeane. Viehweg, Braunschweig, 231p. 3rd ed. of 1922 translated into English as *The Origin of Continent and Oceans*. Methune & Co. Ltd., London, 1924.
Wilson, J.T. (1965). A new type of faults and their bearing on continental drift. Nature, vol. **207**, pp. 343–347.
Wilson, J.T. and Burke, K. (1972). Two types of mountain building. Nature, vol. **239**, pp. 448–449.
Windley, B.F. (1984, 1995). *The Evolving Continents*. 2nd & 3rd ed. Wiley, New York.

Chapter 2
Cratons of the Indian Shield

2.1 Introduction

The Indian shield is made up of a mosaic of Precambrian metamorphic terrains that exhibit low to high-grade crystalline rocks in the age range of 3.6–2.6 Ga. These terrains, constituting the continental crust, attained tectonic stability for prolonged period (since Precambrian time) and are designated *cratons*. The cratons are flanked by a fold belt, with or without a discernible suture or shear zone, suggesting that the cratons, as crustal blocks or microplates, moved against each other and collided to generate these fold belts (Naqvi, 2005). Alternatively, these cratons could be the result of fragmentation of a large craton that constituted the Indian shield. In either case, rifting or splitting of cratons is documented by the presence of fold belts that are sandwiched between two neighbouring cratons. The cratons or microplates collided and developed the fold belts that occur peripheral to the cratonic areas of the Indian shield. The rocks making up the fold belts were the sediments derived from crustal rocks and volcanic material derived from the mantle, all deformed and metamorphosed during subsequent orogeny(s) brought about by collision of crustal plates (cratonic blocks) that are now flanking the fold belts. There are six cratons in the Indian shield with Mid- to Late- Archaean cores or nucleus (Fig. 2.1). These cratons are: the Dharwar or Karnataka craton, Bastar (also called Bhandara) craton, Singhbhum (-Orissa) craton, Chhotanagpur Gneiss Complex (which is arguably a mobile belts of some workers), Rajasthan craton (Bundelkhand massif included), and Meghalaya craton. The last named craton is located farther east and is shown separately (see Fig. 2.10). The name Rajasthan craton is more appropriate than the term Aravalli craton used by some authors, because the term Aravalli is used for the Proterozoic Aravalli mountain belt made up of the supracrustals rocks of the Aravalli Supergroup and Delhi Supergroup, both of which were laid upon the Archaean basement—the Banded Gneissic Complex (BGC), including the Berach granite. The Bundelkhand granite located a few hundred kilometers in the east in the adjoining State of Madhya Pradesh is similar in age and petrology to the Berach granite, despite their separation by Vindhyan basin. These three Archaean domains of the BGC, Berach and Bundelkhand granites are, therefore, considered a single Protocontinent (named here Rajasthan craton) to the north of the Son-Narmada lineament (SONA Zone). To the south of this E-W trending lineament, there are

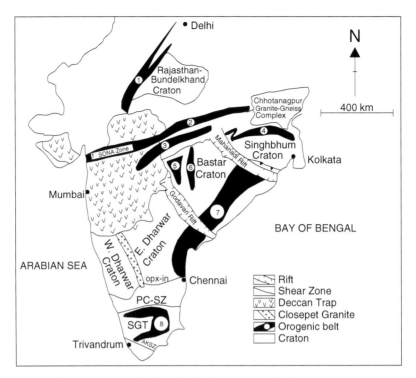

Fig. 2.1 Outline map of the Indian shield showing the distribution of cratons, including Chhotanagpur Granite Gneiss Complex, Rifts and Proterozoic fold belts. Megahalaya craton (formerly Shillong Plateau) is not outlined here but is shown in Fig. 2.10. The fold belts are: 1 = Aravalli Mt. belt, 2 = Mahakoshal fold belt, 3 = Satpura fold belt, 4 = Singhbhum fold belt, 5 = Sakoli fold belt, 6 = Dongargarh fold belt, 7 = Eastern Ghats mobile belt, 8 = Pandyan mobile belt. Abbreviations: Opx-in = orthopyroxene-in isograd; AKS = Achankovil shear zone; PC-SZ = Palghat Cauvery Shear Zone; SGT = Southern Granulite Terrain; SONA Zone = Son-Narmada lineament

three other cratonic regions, namely the Bastar craton, the Dharwar craton, and the Singhbhum craton (SC) which collectively constitute the southern Protoconinent of the Indian shield (Radhakrishna and Naqvi, 1986). The Chhotanagpur Gneiss Complex (CGC) is located to the north of the Singhbhum craton (Fig. 2.1) and the North Singhbhum fold belt or simply called Singhbhum fold belt (SFB) sandwiched between the Singhbhum craton and CGC in all probability is the outcome of the collision of SC and CGC (discussed later).

Each of these six cratons shows different geological characteristics. In this chapter, we enquire into the age, composition, and structural architecture of these cratonic masses to which the fold belts had accreted. In general, the cratons are dominated by granite and metamorphic rocks, mainly gneisses, which imply a series of intense mountain making episodes (deformation and metamorphism) in the Precambrian time before the stable conditions set in. A common feature of these cratonic regions is the occurrence of greenstone-gneiss association, as found in other

Archaean cratons of the world. Geochronological data have disclosed that rocks, especially the grey tonalitic gneisses, range in age from 3.4 to 2.6 Ga old, which may be taken to indicate that all these regions contain continental nucleus (cf. Mukhopadhyay, 2001). Another feature of these cratons is that they are often bordered by a shear zone or a major fault system and the intervening fold belt is composed of metamorphosed, deformed Proterozoic rocks. This implies that the stable Archaean cratons subdivided by mobile belts or fold belts had split or rifted during the Proterozoic and the resulting basin was wholly ensialic, with no rock associations that could be equated with ancient ocean basins. In most fold belts, as shown in subsequent chapters, one observes that gneiss-amphibolite-migmatites are exposed as the dominant cratonic rocks, suggesting that the supracrustals sequences rested upon the Archaean gneissic rocks of the cratons and that both basement and cover rocks were deformed and recrystallized in the subsequent orogeny.

In the following pages, these cratonic blocks are described with respect to their lithology, geology, geochronology, and structural characteristics in the given order:

(1) Dharwar craton (also called Karnataka craton) in the south
(2) Bastar craton (also called Bastar-Bhandara craton) in the central part
(3) Singhbhum craton (also called Singhbhum-Orissa craton) in the northeast
(4) Chhotanagpur Gneiss Complex in eastern India
(5) Rajasthan (-Bundelkhand) craton in the north
(6) Meghalaya craton in far east Indian shield

The following account is highly variable for each craton because all the cratons of the Indian shield have not been studied with equal intensity and the available geochronological and structural data are meager for some but sufficient for other cratonic regions, depending upon various reasons.

2.2 Dharwar Craton

2.2.1 Introduction

The Archaean Dharwar craton (also called Karnataka craton) is an extensively studied terrain of the Indian shield. It is made up of granite-gneiss-greenstone (GGG trinity) belts. The craton occupies a little less than half a million sq. km area. It is limited in the south by the Neoproterozoic Southern Granulite Belt (SGT) or Pandyan Mobile belt of Ramakrishnan (1993); in the north by the Deccan Trap (late Cretaceous); in the northeast by the Karimnagar Granulite belt (2.6 Ga old) which occupies the southern flank of the Godavari graben; and in the east by the Eastern Ghats Mobile Belt (EGMB) of Proterozoic age (Fig. 2.2 inset). The boundary between the Craton and the SGT is arbitrarily taken as Moyar-Bhavani Shear (M-Bh) Zone (Fig. 2.2) while the boundary between the Craton and the EGMB is demarcated by the Cuddapah Boundary Shear Zone. Besides these shear zones at the contact between the craton and the stated terrains/belts, there are many sub-parallel

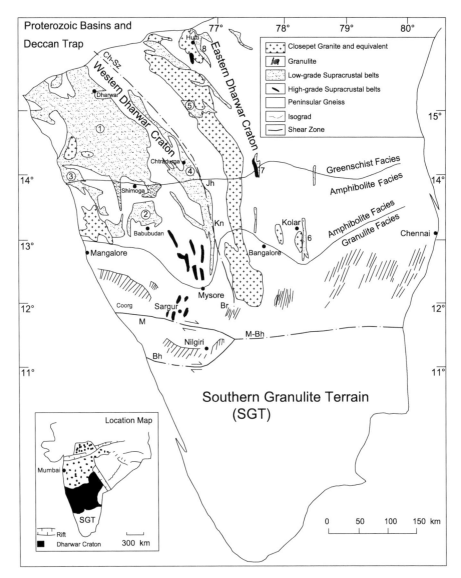

Fig. 2.2 Simplified geological map of the southern Indian shield (after GSI and ISRO, 1994) showing the Dharwar (-Karnataka) craton and its Schist belts, Southern Granulite Terrain (SGT) and shear zones. Metamorphic isograds between greenschist facies and amphibolite facies and between amphibolite and granulite facies (i.e. Opx-in isograd) are after Pichamuthu (1965). Greenstone (Schist) belts in the Dharwar craton (Eastern and Western blocks) are: 1 = Shimoga, 2 = Bababudan, 3 = Western Ghats, 4 = Chitradurga, 5 = Sandur, 6 = Kolar, 7 = Ramagiri, 8 = Hutti. Locality: BR = Biligiri Rangan Hills, Ch = Chitradurga, Cg = Coorg, Dh = Dharwar, Jh = Javanahalli, Kn = Kunigal, S = Shimoga. Shear zones: Bh = Bhavani; Ch-Sz = Chitradurga Shear Zone, M = Moyar; M-Bh = Moyar-Bhavani

NNW to N-S trending shear zones within the main Dharwar Craton, mostly at the eastern boundaries of major schist belts. Although the relationship between these nearly N-S shear zones and the E-W shear zones is uncertain, the shear zones should throw significant light on the crustal evolution of the southern Indian shield (Vemban et al., 1977).

Early studies on the Dharwar craton were ambiguous and controversial in regard to the status of gneisses and schistose (greenstone belts) rocks of "Dharwar System" (see review in M.S. Krishnan, 1982). In the early nineteenth century, Bruce Foote stated that crystalline granitoid gneiss was Fundamental gneiss and the schistose rocks of Dharwar System unconformably overlie the gneiss. This view was opposed in 1915–1916 by W.F. Smeeth who suggested that the gneisses were intrusive into the schists and hence not the Fundamental Gneiss. Smeeth designated the Fundamental Gneiss of Bruce Foote as Peninsular gneiss in view of its vast development in Peninsular India. The controversy continued for over three decades until geochronological data were generated along with relationships of rocks of the Dharwar craton. Detailed field work by the team of Geological Survey of India backed with large scale maps (1: 50,000) finally established that the Peninsular gneiss in western craton is the basement (infracrystalline) for the schist belts and that the intrusive relationship was due to re-melting of the Pre-Dharwar granite gneiss (for details see Ramakrishnan and Vaidyanadhan, 2008). The discovery of a regional unconformity defined by the presence of quartz-pebble conglomerate by the GSI confirmed the given conclusion. The Dharwar schists in western part of the craton was also divided into two Groups, the older Sargur Group (3.1–3.3 Ga) and the younger Dharwar Supergroup (2.6–2.8 Ga) because the Sargur and Dharwar successions are separated by angular unconformities at several localities in western Dharwar craton.

Dharwar craton, as the northern block of southern Indian shield (Fig. 2.2, inset), is a dominant suite of tonalite-trondhjemite-granodiorite (TTG) gneisses which are collectively described under the familiar term Peninsular gneisses. The TTG suite is believed to be the product of hydrous melting of mafic crust and a last stage differentiates of mantle, accounting for crustal growth, horizontally and vertically. The available geochronological data indicate that the magmatic protolith of the TTG accreted at about 3.4 Ga, 3.3–3.2 Ga and 3.0–2.9 Ga (see Meen et al., 1992). The Pb isotope data of feldspar suggests near Haedian (>3.8 Ga) juvenile magmatism (cf. Meen et al., 1992). The second category of rocks in the Dharwar craton is greenstones or schist belts with sedimentary associations. The greenstones are mainly voluminous basalts with subordinate fine clastics and chemical sediments and in certain areas with basal conglomerate and shallow water clastics (e.g. ripple-bedded quartzites) and shelf sediments (limestone and dolomite). In the Dharwar craton, both volcanics and sediments as supracrustal rocks were laid upon the basement of Peninsular gneiss (>3.0 Ga). During Dharwar Orogeny (2500 Ma) the volcanics have been metamorphosed into greenschist (chlorite schist) and amphibolite and even higher grade basic granulites while the associated sediments have recrystallized into quartzite, crystalline marbles, metapelites with index minerals denoting a particular metamorphic grade (Ramiengar et al., 1978).

The greenstone belts together with the intercalated metasediments designated as the *Dharwar Schist Belts* have N-S trend and show a gradual increase of metamorphic grade from N to S. The N-S trending compositional layering in the belts are transected by east-west running isograds (Fig. 2.2), suggesting that conductive/thermal relaxation of the superposedly deformed Dharwar supracrustals (along with their basement gneisses) occurred when their regional foliation (axial plane S2) had a steep disposition with respect to the rising isotherms during regional metamorphism. This is in contrast if the structurally duplicated (or tectonically thickened) crust at depth had gentle inclination of the compositional layering or dominant foliation and subject to uprising isotherms during thermal relaxation. In the latter situation the metamorphic isograds would be nearly parallel to the compositional layering and the small obliquity between them is considered to suggest that regional metamorphism outlasted deformation as in the Swiss Alps and other fold belts.

The metamorphic isograd between greenschist facies (low grade) to amphibolite facies (medium grade) is defined by a line (outcome of intersection of isothermal surface with the ground surface) on a regional geological map, shown in Fig. 2.2. The Opx-in isograd between the amphibolite and granulite facies is not a line but a zone of up to 30 km wide in which mineral assemblages of both amphibolite facies and granulite facies (characterized by the presence of hypersthene (a mineral of orthopyroxene group) are found together. This zone is called the Transition Zone (TZ). It is in this zone that excellent outcrops of "charnockite in-making" (Pichamuthu, 1960; Ravindra Kumar and Raghavan, 1992, 2003) or arrested charnockite (also called incipient charnockite) are seen to have formed from gneisses due to reduced water activity at same prevailing P and T conditions (Peucat et al., 1989). From the disposition and wavy nature of the isograds, it is inferred that the isothermal surfaces had gentle inclination towards north. Further south of the TZ occurs the main granulite terrane, called the Southern Granulite Terrane (SGT) (Fig. 2.2), where massive charnockite-enderbite rocks dominate amidst high-grade supracrustals.

In the Dharwar craton, the Peninsular Gneisses are found to contain tonalitic gneisses of 3.56–3.4 Ga age and enclaves of older metavolcanic-metasedimentary rocks categorized under name Sargur Group (3.2–3.0 Ga). Thus, it is likely that the Peninsular gneiss with its enclaves of older tonalitic gneiss and Sargur Group of rocks served as the basement upon which the Dharwar supracrustals were deposited.

Swami Nath and Ramakrishnan (1981) divided the Dharwar Craton into two blocks—the *Western Block* and the *Eastern Block*, separated by the Chitradurga Shear Zone (ChSz) (Fig. 2.2). Later, Naqvi and Rogers (1987) designated these blocks as Western and Eastern Dharwar Cratons. According to Ramakrishnan (2003), the Chitradurga shear zone, separating the Eastern and Western blocks of the Dharwar craton, extends via Javanahalli (Jh) through the layered mylonite of Markonahalli near Kunigal (Kn) and the western margin of the BR Hills into the Moyar Bhavani (M-Bh) shear zones (Fig. 2.2). The stated shear zone occurs all along the western margin of the 2.5 Ga old Closepet Granite. It is interesting to note that the schist belts in both cratonic blocks show the same N-S trend with almost constancy of strike and dip of the foliation (regional foliation).

However, the greenstone belts of the Western Block are characterized by mature, sediment-dominated supracrustals with subordinate volcanism and are recrystallized in intermediate pressure (kyanite-sillimanite type) Barrovian metamorphism. On the other hand, the greenstone belts of the Eastern block are often gold bearing and show low-pressure (andalusite-sillimanite type) metamorphism, perhaps due to profuse granite intrusions (2600–2500 Ma old) that invade the schists and metasediments of the Eastern block. Geophysical work (see, for example, Singh et al. 2003) revealed that the Western Block has a thicker crust of 40–45 km while the Eastern Block has a thinner crust of 35–37 km (Srinagesh and Rai, 1996). These differences between the two blocks, as well as the occurrence of lower crustal rocks (charnockite-enderbite) along the Western Ghats, suggest that the southern Indian shield had tilted towards the east in addition to generally accepted northward tilt which gave rise to huge exposure of granulite rocks (see Mahadevan, 2004).

The Western Block is characterized by the presence of 3000 Ma old TTG (tonalite-trondhjemite-granodiorite) gneisses—the Peninsular Gneiss Complex. As stated earlier, the Peninsular Gneisses contain 3400–3580 Ma old basement tonalitic gneiss enclaves (Nutman et al., 1992). From a careful fieldwork, Swami Nath and Ramakrishnan (1981) showed that the Peninsular gneisses also enclose relics of older supracrustals in form of narrow belts and enclaves or synclinal keels that show amphibolite facies mineral assemblages. They were called Ancient Supracrustals by Ramakrishnan (1990) and later designated by the term *Sargur Group* (3000–3200 Ma old) by Swami Nath and Ramakrishnan (1990) who also considered them equivalent to Waynad Group in Kerala. The metamorphic fabric of these Ancient Supracrustals (the Sargur Group rocks) and the surrounding gneiss complex is truncated by the low-grade Dharwar Schists (Bababudan Group) and both the former rocks are unconformably overlain by the main schist belts of the Dharwar Supergroup (2600–2800 Ma). Because of these features Swami Nath and Ramakrishnan (1990) proposed an orogeny (Sargur orogeny) before the deposition of the Dharwar supracrustals. The type Sargur Group, developed around Hole Narsipura, is represented by high-grade pelitic schists (kyanite-staurolite-garnet-mica schists), fuchsite quartzite, marble and calc-silicate rocks, besides mafic and ultramafic bodies. The mafic rocks (amphibolites) show typical tholeiitic trend and are essentially low K-tholeiites with normative olivine and hypersthene. The ultramafic rocks appear to be komatiite flow. Detrital zircons from metapelites yielded $\sim$3.3 Ga evaporation age) and 3.1–3.3 Ga (SHRIMP U-Pb age). Ultramafic-mafic rocks yielded Sm–Nd model age of $\sim$3.1 Ga.

The main Dharwar Schist Belts of the Western Block are: (1) Shimoga-Western Ghat-Bababudan, and (2) Chitradurga (Fig. 2.2). Their western margin preserves the depositional contact marked by basal quartz-pebble conglomerate and abundance of platform lithologies, whereas the eastern contact is tectonized and marked by mylonitic shear zones. Besides Chitradurga shear zone, other prominent shear zones are Bababudan and Balehonnur. The metabasalts of Bababudan Group are low-K tholeiites, showing flat HREE and moderate LREE enrichment to indicate mantle source. The negative Eu anomalies in these rocks suggest plagioclase fractionation (see in Naqvi, 2005). The regional stratigraphy of the WDC (after Swami Nath and Ramakrishnan, 1981) is given below:

Proterozoic mafic dykes
Incipient charnockites (2.5–2.6 Ga)
Younger Granites (Closepet etc.) (2.5–2.6 Ga)
Dharwar- {Chitradurga Group: alluvial and shallow marine sedim. and volcanics
Supergroup {Bababudan Group: shallow marine sedim. and basaltic volcanics
-------Deformed angular unconformity-----
Peninsular gneiss, TTG (>3.0 Ga)
-------------Intrusive/Tectonic contact-----
Sargur Group (3.1–3.3 Ga)
-----Intrusive/Tectonic contact---------
Basement Gorur Gneiss (3.3–3.4 Ga)

The *Eastern Block* consists mainly of volcanic (greenstone) belts of Dharwar group (2600–2800 Ma old) that are contemporaneous with the volcano-sedimentary schist belts of the Western Block. The prominent schist belts of the Eastern Block are Sandur, Kolar, Ramagiri, and Hutti (see Fig. 2.2). These volcanic belts have a narrow platformal margin marked by orthoquartzite-carbonate suite (e.g. Sakarsanhalli adjacent to Kolar belt, which is intruded by 2500–2600 Ma old granites. This caused dismembering of the supracrustals as screens and enclaves at the contact. These screens render it difficult to recognize pre-Dharwar (Sargur) enclaves in this block. The eastern margins of these schist belts are extremely sheared but could not be shown in geological maps for want of systematic geological mapping. There is no recognizable basement to these schist belts because these belts are engulfed on all sides by 2500–2600 Ma old granites and gneisses. The prominent meridional *Closepet Granite* occurring close to the boundary between the Eastern and Western blocks is one such younger granite, running parallel to the trend of the schist belts. According to Narayanaswami (1970), there are numerous elongated granites running parallel to the schist belts of the Eastern Block but they have not been fully delineated on geological maps. These granites occur in a series of parallel plutonic belts which have been collectively designated as Dharwar Batholith (Chadwick et al., 2000) in order to distinguish them from the widely used term Peninsular gneiss, scarcely exposed in the EDC. The TTG suite gneisses in the EDC show LREE enrichment and negative Eu anomaly. The gneisses span the age from 2.7 to 2.8 Ga (SHRIMP U–Pb and Rb–Sr methods). The protolith ages are >2.9 Ga, corresponding to the main thermal event of WDC. Amongst the younger granites, the Closepet Granite is an elongated (600 × 15 km) body intruding the Bababudan schist belts as well as the basement gneiss. It is a biotite granite or granodiorite to quartz monzonite composition. Recent studies by Chadwick et al. (2007) have shown that the granite splits into two segments near Holekal and has possibly emplaced as a sheet. The granite yielded well-defined Pb–Pb and Rb–Sr isochron of ∼2.6 Ga which is also the U–Pb zircon age of the granite (see Chadwick et al., 2007). The involvement of gneissic basement in the generation of the Closepet Granite and other late to post-tectonic granites (Arsikere, Banovara, Hosdurga) of Dharwar craton is evident from Sm–Nd model ages (T_{DM}) of the granite at ∼3.0 Ga.

Deformation of Dharwar Schist Belts show superpose folding wherein the N-S trending folds superposed on an older E-trending recumbent folds, giving rise to the prevalent NNW to NNE Dharwar Trend with convexity toward the east. But this view became untenable because the older E-W folds superposed by N-S folds could not be found on a regional scale. Recent structural studies show that the N-S trending tight to isoclinal upright folds have been coaxially refolded during D2 deformation. D3 phase produced more open folds. Crustal-scale shear zones at the eastern margin of most schist belts are now considered as the outcome of sinistral transpression in the late stage of tectonic evolution. Structural studies of Dharwar craton by Naha et al. (1986) showed that the Sargur, Dharwar and Peninsular gneiss (and Dharwar Batholith) have similar style, sequence, and orientation of folding (named Structural Unity). This view, however, overlooked the fact that "Peninsular Gneiss, where exposed in areas of less intense regional strain, preserves the original angular unconformable relations with the Dharwar schists. In areas of intense deformation, the Dharwar folds and fabric are superposed on Sargur folds and fabric, resulting in apparent parallelism as a result of rotation of earlier Sargur fabric into parallelism during the younger Dharwar deformation" (Ramakrishnan and Vaidyanadhan, 2008, p. 172).

Progressive regional *metamorphism* is convincingly documented in the Dharwar craton, especially in WDC (Fig. 2.2) where greenschist facies, amphibolite facies and finally granulite facies are noticed from N to S direction. The different isograds run nearly parallel and transect the compositional layering (S2 foliation) at high angle, suggesting that metamorphism outlasted deformation D2. Geothermobarometry along with fluid inclusion studies demonstrated that the P-T conditions in the low grade rocks in the north are 500°C/4 kb which increases to 600°C/5–7 kb in middle grade and 700–750°C/7–8 kb in the granulite grade rocks of the Dharwar craton. The regional metamorphism of EDC also occurred in a similar way, but is of low-pressure facies series. P-T conditions of EDC range from 670°C/3 kb in the Sandur belt in the north through 710°C/4–5 kb in the middle to 750°C/6–7 kb in the Krishnagiri-Dharmapuram area in the south (Jayananda et al., 2000). The low-pressure metamorphism of EDC is caused by the abundance of younger granites that seem to have supplied advective heat. The younger granites are almost absent in the WDC.

As stated already, the *Dharwar Schist Belts* are Late Archaean (age range of 2.8–2.6 Ga). They occupy a vast terrain of the Dharwar craton and are geologically significant for their linear occurrences like in other Precambrian fold belts of India (Gopalkrishnan, 1996). The Dharwar Schist belts appear to have evolved in ensialic intracratonic basins, subsequently subjected to regional compressions resulting in the deformation and recrystallization of these rocks, finally exposed by erosion in the present form. These Schist belts can thus be compared to the Proterozoic fold belts found in ancient shield areas of Canada, Africa, and Australia. One may also think that the Dharwar schist belts initially existed as mafic lava, like the Deccan Trap, covering a vast area of peneplaned Archaean gneisses. This basement-cover association, along with their eroded sediments, was involved in superposed folding, resulting into different schist belts that now characterize the Dharwar craton. Structural data suggest that the Dharwar craton deformed into tight to upright folds of two generations (with or without kinematic interval). Before we discuss evolutionary models for the origin of these greenstone belts of the Dharwar Supergroup, we need to know the agreed observations and facts about these Schist Belts. They are stated as follows:

1. The Dharwar Schist Belts (familiarly known as greenstone belts) rest on the Archaean gneisses with an unconformity, denoted by basal quartz-pebble conglomerate.
2. The Schists belts extend over a length of 600 km and width of 250 km or less.
3. The entire Dharwar succession is ~10 km in thickness.
4. The Eastern greenstone belts have larger volumes of mafic to ultramafic flows and intrusives at the base; fine clastic and chemical sediments are subordinate.
5. The Dharwar greenstone belts of Western Block have oligomictic conglomerate overlain by mafic, orthoquartzite-carbonate (platform)-cum-iron ore deposits which became geosynclinal (graywacke-argillite-mafic volcanics) in the upper succession.
6. The Dharwar Schist Belts are in the age range of 2.8–2.6 Ga.
7. The Eastern block shows profuse granite intrusions.
8. The Eastern schist belts are gold-bearing and show low-pressure regional metamorphism (andalusite-sillimanite type), whereas the Western schist belts show intermediate regional metamorphism (kyanite-sillimanite type).
9. The Schist belts record two major deformation events out of which the first event (Dh1) is most widespread and responsible for the regional NWN-SES trend of the greenstones (Roy, 1983).
10. The belts show an arcuate N-S or NWN-SES trend with convexity towards the east, which is due to later tectonic event (Dh2) (Roy, 1983).
11. The regional foliation in the Peninsular gneisses and schist enclaves (of Sargur rocks) is largely parallel to the trend of the Dharwar schist belts, attributed to rotation of earlier structures in these pre-Dharwar rocks (Mukhopadhyay, 1986).
12. The schist belts are wide in north and taper down to the south, with younging direction towards north, perhaps due to northward tilt of the southern Indian shield
13. Banded iron formation (BIF) is a characteristic component of these schist belts.
14. Regional metamorphism related to the Dharwar orogeny affecting these schist belts is post-tectonic with respect to Dh1 (Mukhopadhyay, 1986) and is progressive from N to S and ranges from greenschist, amphibolite to granulite facies (Pichamuthu, 1953).
15. The second deformation, Dh2, is older than 2500–2600 Ma (Mukhopadhyay, 1986).

2.2.2 Evolution of Dharwar Schist Belts

Considering the foregoing account, we can infer that the Dharwars are the remains of a great supracrustal sequence that once had covered a very large area of the southern Indian shield. Because of different lithology of the schist belts of the two blocks we shall consider the Western and Eastern greenstone belts separately for their evolution.

The depositional environment for the *Western Dharwar Schist Belts* was marine and shallow. Moreover, since the Dharwar succession comprises basic flows and sediments, the original attitude was mainly horizontal. The deposition was evidently on a peneplaned basement of the Peninsular Gneiss (with its older enclaves) because of the supermature basal quartz-pebble conglomerate of the Dharwar schists. It is proposed that the Archaean crust of the Western Dharwar craton rifted to give rise to linear ensialic intra-continental basins. In these basins, the supracrustals were deposited, not necessarily continuously. Because of rifting and thinning of the crust/lithosphere, the underlying mantle underwent decompression melting, pouring out basaltic lavas which intervened the rift-sedimentation. Sagging of the basin allowed further accumulation of volcanic material and sediments, especially graywacke in northern parts of the rifts. These Dharwar supracrustals along with their infracrystallines (i.e. basement) were subsequently involved in the Dharwar orogeny that produced superposed folds in the Dharwar Schist Belts characterized by N-S regional foliation. Field evidence suggests that post-Dh1 recrystallization affected both cover and basement rocks that at depth were also remobilized to generate granites (2600 Ma old) which intruded the overlying rocks of the Dharwar Supergroup. Later, faulting and erosion caused the present isolation of the schist belts. The Dharwar greenstone belt of Western block cannot be considered to have formed in a marginal basin because the depositional margin represents an ancient shoreline and a more stable environment with a proximal sialic crust. Consequently, the Western Block schist belts seem to have evolved in a rifted continental margin containing abundant detritus from the stabilized continental crust of 3000 Ma age and progressed through an unstable shelf to an ocean basin.

The *Eastern Block schist belts,* on the other hand, are likely to have developed away from the continental influence, and their development conforms to the marginal basin model. The metabasics (amphibolites) of the Eastern Block, are enriched in LIL and LREE, compared with the marginal basin basalts of the Scotia arc (Anantha Iyer and Vasudevan, 1979). The possible cause of interlayered komatiitic amphibolites and tholeiitic amphibolites (2700 Ma old) is attributed to melting of asthenosphere (>70 km depth) and subcontinental lithosphere (~50 km depth) respectively (Hanson et al., 1988). The schist belts in the Eastern Dharwar craton are narrow and elongated and are dominated by volcanics with subordinate sediments. The Kolar schist belt is similar to other schist belts of the Eastern Block, and the gneisses on its E and W sides are dated at 2632, 2613, and 2553 Ma (Fig. 2.3). These are interpreted to have inherited zircons from an older sialic basement (Peninsular Gneiss and/or older basement gneiss) that was subjected to spreading and served as a basement in the marginal basin region. Chadwick et al. (2000) correlated the greenstone succession of the Eastern Dharwar craton with the Chitradurga Group. Their age data range from 2.7 to 2.55 Ma (Krogstad et al., 1991; Balakrishnan et al., 1999; Vasudevan et al., 2000). The Eastern Block has been profusely intruded by granites which could have been originated from partial melting of the continental crust during subduction (that caused secondary spreading) or during

Fig. 2.3 Geological sketch map of central part of the Kolar Schist Belt (after Hanson et al., 1988)

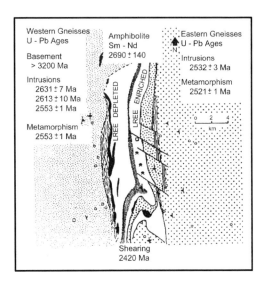

rise of mantle plume, resulting in the heating and rifting of the overlying continental crust (Jayananda et al., 2003). The profuse granite intrusion may also be the cause of andalusite-sillimanite type metamorphism found in the rocks of the Eastern Block. Rogers et al. (2007) have presented SHRIMP U–Pb zircon ages together with whole-rock geochemical data for granitoid adjacent to Hutti-Muski schist belt in the EDC. They show two phase of intrusion into the belt; the syntectonic porphyritic Kavital granitoid of 2543 ± 9 Ma, followed by post-tectonic fine grained Yelagat granite defining minimum age of 2221 ± 99 Ma.

2.2.3 Geodynamic Models for the Evolution of Dharwar Craton

As stated already, the basement for the schist belts was sialic, because basal conglomerate defines the angular unconformity (deformed) between the Archaean gneisses and the supracrustals of Dharwar (formerly basic lavas and sediments). An important implication of this observation is that the geodynamic models for the evolution of Dharwar greenstone belts must have started with continental rifting. Furthermore, the occurrence of Closepet Granite between WDC and EDC of the Dharwar craton probably indicates a geosuture (Ramakrishnan and Vaidyanadhan, 2008). The different models need to consider the following geological-geochemical characteristics.

1. The basement of the greenstone was continental and hence Peninsular Gneiss.
2. Geochemistry suggests that the greenstone volcanics are not the source for the surrounding granites.

3. The Closepet Granite is late to post-tectonic body at the boundary of WDC and EDC.
4. Western margin of most schist belts has thin marine sediments.
5. The NNW trend of the schist belts is the result of superposed deformation of the schist belts which originally had nearly E-W orientation.
6. The progressive metamorphism in both WDC and EDC increases from N to S and the E-W running isograds are unbroken in the entire Dharwar craton.

With these facts and observations, we now evaluate different geodynamic models.

Model 1 (Newton, 1990)

This model considers that the greenstone belts are back-arc marginal basins formed as a result of E-W rifting of Archaean continental gneisses (i.e. Peninsular gneiss). The marginal basin was developed by N-verging subduction of an oceanic plate (now vanished). The subduction zone was located at the contact of Dharwar craton and the Neoproterozoic Pandyan mobile Belt (non-existent then at present location). The contact is now marked by a crustal-scale shear zone, the Palghat-Cauvery shear zone. The oceanic subduction toward north caused rifting in the overlying continental crust, generating the E-W greenstone belts. A reference to Fig. 1.11 will help the reader to understand this model. The E-W trending greenstone belts were later deformed (by E-W compression or transpression) to produce the present N-S configuration of the schist belts.

The model fails to explain the origin of low pressure greenstone belts of the EDC and cannot also explain the paucity of the Archaean gneisses in the EDC.

Model 2 (Chadron et al. in Ramakrishnan and Vaidyanadhan, 2008)

Non-uniformitarian "sagduction" model which involves gravitational sinking of greenstone pile into the gneissic basement together with tectonic slides at the margin.

This model cannot account for the large-scale top-to-SW overthrusting reported in many schist belts. The model requires a huge amount of lava to be extruded and then superposedly folded to acquire the present orientation.

Model 3 (Chadwick et al., 2000)

It is a convergence model of schist belt evolution, like the Andes. It can be easily followed if Fig. 1.11 is referred to. The WDC is regarded the foreland which was subject to secondary spreading due to subduction of an oceanic crust beneath it. This resulted into marginal basins and a continental arc (Dharwar Batholith) that later accreted onto the craton around 2750–2510 Ma. The EDC greenstone belts are considered to have developed as intra-arc basins. Later, as a result of arc-normal compression there occurred NE-SW shortening which resulted in the sinistral transpressive shear system at the margin of most schist belts. The oblique convergence of an oceanic plate (subducting toward WNW) resulted in the Archaean plate tectonic evolution of the Dharwar craton. The Closepet Granite in this model is regarded as a part of the Dharwar Batholith

There are, however, some objections to this model. It fails to identify the fore-arc accretionary prism in the east. It also cannot account for the marginal marine stable continental sediment along the western margin of many schist belts. It is also a question as to how did the convergence occur for the amalgamation of the WDC and EDC when the foreland of WDC and the Arc of EDC were drifting apart during the formation of marginal basin.

Model 4 (Rajamani, 1988; Hanson et al., 1988; Krogstad et al., 1989)

Based on the observations that the Eastern Block is characterized by low-pressure high-temperature metamorphism (unlike the kyanite-sillimanite type of Western Block) and is devoid of older basement, Rajamani (1988) proposed a tectono-magmatic model from the geochemical and geochronological data on the gneisses and amphibolites of the Eastern Block. The gold-bearing schist belts or Kolar-type belts of the Eastern Block are predominantly metamorphosed basalts with no exposure of gneissic basement. However, they are surrounded by reworked gneisses and migmatites containing their (greenstones) enclaves or tectonic slices. The schist belts show diapiric intrusion of granites along their margins. The absence of clastic sediments and the presence of pillowed and variolitic structures of the mafic rocks/schist belts led Rajamani (ibid) to propose a tectonic setting (i.e. oceanic) for them, different from the model given above for the well-defined Western Dharwar-type schist belts. He named them "Older greenstone" or Kolar-type greenstone belts—representing pieces of oceanic crust welded together to form composite schist belts. Being host rocks for gold, as at Kolar and Hutti at the southern and northern ends of the belts, the schist belts are also named Gold -bearing Schist Belts of Eastern Karnataka (Radhakrishna and Vaidyanadhan, 1997).

A detailed petrological and geochemical study, particularly of the Kolar schist belt, shows that the metabasic rocks (amphibolites) are characterized by four textural types viz. schistose, granular, massive and fibrous. The study of Rajamani (1988), Hanson et al. (1988), and Krogstad et al. (1989) suggests that the Kolar Schist Belt, as a type belt, is divisible into an E and W part with respect to a central ridge made up of a fine-grained metavolcanic unit (Fig. 2.3). To the west of the belt, granodioritic gneisses are found to have ages of 2631 ± 7, 2613 ± 10, and 2553 ± 1 Ma. These gneisses are found to have inherited zircons of 3200 Ma age or older, suggesting that an older basement gneiss existed >3200 Ma ago. To the east of the schist belt, the gneisses were emplaced at 2532 Ma ago. According to Hanson et al. (1988), these gneisses have mantle-like signatures, as revealed by Pb, Sr and Nd isotope data.

The age difference of about 100 Ma between the western and eastern gneisses is taken to indicate existence of two gneissic terranes, presumably with a separate evolutionary history, with metamorphic ages of 2521 ± 1 Ma for the eastern terrane and 2553 ± 1 Ma for the western terrane. Within the gneissic terranes, there are komatiitic amphibolites that to the east are LREE enriched whereas those of the west-central part are LREE depleted. The komatiitic amphibolites of the west-central part give Sm/Nd age of 2690 ± 140 Ma (Fig. 2.3). The interlayered tholeiitic amphibolites are older having Pb–Pb isochron age 2733 ± 155 Ma, suggesting that

2.2 Dharwar Craton

some of the amphibolites of the Kolar Schist Belt are older than the 2530–2630 Ma old gneisses to the E and W of the belt. Furthermore, the other two types of amphibolites from west-central part also have different Pb isotope ratios, suggesting that the parental magmas of the tholeiitic amphibolites derived from sources with a different U–Pb history than the parental magmas of the komatiitic amphibolite. Their depth of melting was also considered different, about 150 km for the komatiitic and about 80 km for the tholeiitic amphibolites (cf. Rajamani, 1988). It is proposed that the tholeiitic and komatiitic amphibolites were tectonically interlayered (Rajamani et al., 1985). The eastern amphibolites have not been dated, but from their Nd and Pb isotope characteristics, Hanson et al. (1988) suggest an age of 2700 Ma. Because of the REE characteristics and the geological setting, these authors argue that the eastern amphibolites formed in different tectonic settings from that of the west-central amphibolites. The authors (Hanson et al., 1988) propose a tectono-magmatic evolutionary model for the Kolar Schist belt in light of plate tectonics. This model is enumerated below and shown in block diagram (Fig. 2.4), after Hanson et al. (1988).

1. Initially the western terrane consisted only of 3200 Ma or older basement; the eastern terrane did not exist (Fig. 2.4a).
2. At about 2700 Ma, the parental rocks of the eastern and west-central parts of the Kolar schist belt developed in widely separated environments (Fig. 2.4b). The parental rocks of the schist belt of west-central part are believed to have formed either as Mid-Oceanic Ridge basalts (komatiites due to high heat flow), or as back-arc basin basalts (if both tholeiitic and komatiitic basalts were formed at the same time and place).
   The parents for the eastern komatiitic amphibolites could be oceanic island basalts or island arc basalts.
3. Westward subduction of the ocean floor brought the komatiitic-tholeiitic basalts in juxtaposition (as they developed in separate environments) (Fig. 2.4c) and developed a magmatic arc on the eastern edge of a continent, supported from the characters of plutonic rocks of the western gneisses and their setting upon an older basement. The eastern gneisses with mantle signatures are similar to many Archaean granitoids, but their tectonic setting remains unexplained.
4. Around 2550 Ma the gneisses to the west and east of the Kolar schist belt were welded together. The accreted terranes were subsequently involved in superposed isoclinal folding due mainly to E-W subhorizontal shearing, followed by longitudinal shortening.
5. The last event is the N-S left-lateral shearing found in all rocks of the Kolar schist belt (Fig. 2.4d). The age of shearing is about 2420 Ma, as deduced by Ar-Ar plateau age for muscovite developed in the shear zones.

The model of Hanson, Rajamani and coworkers would be challenged if gneisses of 3.3 Ga or older age could be found in the eastern terrain of the Kolar area. In this situation, a more viable model proposed for the schist belts of the Western Block may also be applicable for evolution of the Eastern Block. Whether or not >3.2 Ga old gneisses existed in the eastern part, the low pressure- high temperature

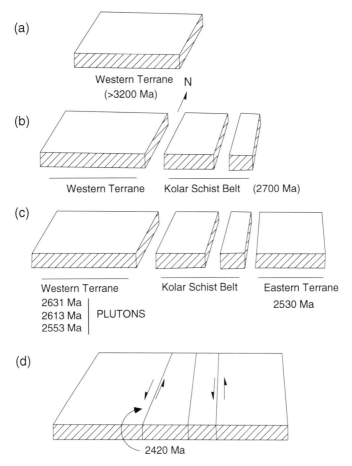

Fig. 2.4 Block diagram showing the crustal evolution in the Kolar area (after Hanson et al., 1988). See text for details

metamorphism of 2550 Ma for the Kolar area corresponds with the age of regional metamorphism (2.5 Ga) that affected both supracrustals and the infracrystallines (Archaean gneisses) of the Dharwar craton and developed the E-W trending isograds Pichamuthu, (1953, 1961). The E-W disposition of the regional metamorphic isograds is not compatible with the Westward subduction model of Hanson et al. (1988). The andalusite-sillimanite type metamorphism in the Eastern Block seems to be the result of profuse granite intrusions emplaced following peak metamorphism during which the Eastern Block was subjected to a greater upliftment relative to the Western Block. In support of this proposition, a more careful petrography could enable one to find kyanite relics in suitable compositions of the so-called andalusite-sillimanite type terrane of the Eastern Block. As stated earlier, the variation of facies series between the Western and Eastern Blocks is the outcome of eastward tilting of the southern Indian shield. This is manifested in the thicker crust in the Western Block compared to the Eastern Block and in the occurrence of vast exposures of charnockites (as lower crustal rocks) in Western Ghats.

We must now examine what light does the Late Archaean granite intrusions throw in regard to the geohistory of the two blocks, since granites have important bearing on crustal evolution. As we know, the granite intrusions represent a larger part of the EDC, since Peninsular Gneiss is absent or scantly reported from this Block. The granites are similar to those of the Western Block due to the similarity of outcrop pattern, relationship with the country rocks, geochemical characteristics and their isotopic ages (2.5 Ga, Friend and Nutman, 1991). Again, studies by Moyen et al. (2001) show that the granite intrusions of the E and W blocks were emplaced in the same transcurrent tectonic setting as the main phase of the Closepet Granite further south. The granites are clearly formed by anatexis of crustal rocks (Peninsular gneisses!) at depth, but the basement rheology was responsible for the upward movement of this low-viscosity magma to fill deformation-controlled pockets in the upper crust (Moyen et al., 2001). These considerations lead us to believe that the evolution of the E and W Blocks witnessed similar granite activity during the Late Archaean, presumably by their similar evolutionary history.

Finally, the model of Hanson et al. (1988), Rajamani (1988) and Krogstad et al. (1989) does not get support from a regional tectonic framework of southern India in terms of plate tectonics, given for the first time by Drury and Holt (1980) and Drury et al. (1984), based mainly on the Landsat imagery. They suggested that the Southern Granulite Terrain (SGT) underthrusted the Dharwar Craton along the E-W shear zone in Proterozoic. In light of the plate tectonic model, the 2663 or 2613 Ma old granites intruding the metavolcanics are considered to represent a magmatic arc (see Hanson et al., 1988). Nevertheless, it is a question whether the entire period starting from rifting to collision was an isolated event in the Eastern Block, not connected with the rifting of the crust in the Western Block of the Dharwar craton. It needs to be explained as to how both schist belts were affected by the same progressive regional metamorphism of Dharwar time (~2550 Ma) if they had developed separately in space and time. Ramakrishnan (2003) observes that preservation of the platformal margin and the sheared eastern margins of the schist belts in both blocks favour the view that all the Dharwar Schist Belts were evolved nearly contemporaneously on a continental crust but with different tectono-sedimentary environments. If so, the proposition by Chadwick et al. (2000) that the Western schist belts are foreland basins situated on a granitic crust and that the Eastern schist belts are intra-arc basins within the Dharwar batholith, is untenable. Also, the lack of fore-arc accretionary prism and the absence of suture zone to the east of Dharwar batholith do not support the proposed model of Chadwick et al. (ibid.) that is akin to Mesozoic-Cenozoic convergent setting (cf. Ramakrishnan, 2003, p. 12). It must be recalled that the greenstone belts from both E and W blocks are parallel and have similar structural styles, despite having different geotectonic settings. The structural similarity between the Peninsular gneiss, Sargur and Dharwar Schists is interpreted either as a single post-Dharwarian deformational episode or as due to rotation of older structures into parallelism with the younger Dharwar structures during the Dharwar orogeny.

The above account is thus helpful to answer critical questions whether the greenstone belts evolved diachronously or several cycles of greenstone belts occurred before Dharwar craton stabilized.

2.3 Bastar Craton

2.3.1 Introduction

The Bastar craton (BC) is also called Bastar-Bhandara craton. It lies to ENE of the Dharwar craton (DC), separated from the latter by the Godavari rift (see Fig. 2.1). Located to the south of the Central Indian Tectonic Zone (CITZ) the Bastar craton is limited by three prominent rifts, namely the Godavari rift in the SW, the Narmada rift in the NW and the Mahanadi rift in the NE. Its southeastern boundary is marked by the Eastern Ghats front. The western limit of the Eastern Ghats mobile belt overlying the Bastar craton is demarcated by a shear zone, which in fact is a terrain boundary shear zone (Bandyopadhyay et al., 1995). The Bastar craton is essentially formed of orthogneisses with enclaves of amphibolites, vestiges of banded TTG gneisses of 3.5–3.0 Ga, and low- to high-grade metasediments as supracrustals. Ancient supracrustals consisting of quartzite-carbonate-pelite (QCP) with BIF and minor mafic-ultramafic rocks, collectively called Sukma Group in the south and Amgaon Group in the north, occur as scattered enclaves and narrow belts within Archaean gneisses and granites. The succeeding metasedimentary belt is called Bengpal Group. The sequence is intruded by Archaean granites and the Bengpal Group is therefore considered Neoarchaean (2.5–2.6 Ga). Unconformably overain on the Bengpal Group Schists are BIF (Bailaddila Group). The next succession is felsic and mafic volcanics with pyroclastics (Nandgaon Group) intruded by 2.3–2.1 Ga old granites (Dongargarh, Malanjkhand etc.). The Nandgaon Group is overlain unconformably by basalts alternating with sediments, classed under Khairagarh Group.

H. Crookshank and P.K. Ghosh mapped the geology of southern Bastar during 1932–1938. Ghosh described the charnockites of Bastar. Recent geological summaries of Bastar craton is by individual workers, viz. Chatterjee, Ramakrishnan, Abhinaba Roy, Basu, Ramchandra, Bandyopadhya et al. (see in Ramakrishnan and Vaidyanadhan, 2008). S.N. Sarkar worked the geology of Kotri-Dongargarh belt.

The gneiss/migmatites and amphibolites, constituting the early crustal components of the Bastar craton, are grouped under the *Amgaon gneiss* that resembles the Peninsular Gneiss Complex of the Dharwar craton. It ranges in composition from tonalite to adamellite. Amgaon gneisses occur in the north of Bastar craton and south of Central Indian Shear zone (CIS). They were geochemically studied by Wanjari et al. (2005) who showed high Al_2O_3 trondhjemite along with cal-alkaline and peraluminous granites, similar to the 2.5 Ga old granites of Dharwar craton (Sarkar et al., 1981). The tonalite gneisses of Bastar craton yielded interesting age data. Geochronology of single grain U–Pb zircon ages gave 3580 ± 14 and 3562 ± 2 Ma for tonalite gneisses to the east and west of Kotri-Dongargarh linear belt (Ghosh, 2003, 2004). These ages represent the oldest Archaean crust not only in

Bastar craton but also in the Archaean cratonic regions of the Indian shield. The basement complex also contains granulite facies rocks and intrusion of granites of different ages.

The TTG gneisses enclose rafts of continental sediments (QCP facies) together with minor BIF and mafic-ultramafic rocks (Sukma Group of Bastar and Amgaon Group of Bhandara), as stated already (Wanjari and Ahmad, 2007). Large bodies of younger granites intruding the gneisses and supracrustals are notable components in the craton. Enclaves of granulites and high-grade metasediments also occur in the craton.

In *Bastar craton* the gneisses are classified into 5 types. These are: the Sukma granitic gneiss (Group 1), Barsur migmatitic gneiss (Group 2), leucocratic granite (Group 3) occurring as plutons with migmatitic gneiss, pegmatoidal or very coarse granite (Group 4), and fine-grained granite (Group 5) occurring amidst the Sukma gneisses. The gneisses of Groups 1 and 2 are chemically and mineralogically similar to the Archaean TTG, while the gneisses of Groups 3, 4 and 5 are of granitic nature.

Pb-Pb isotope dating of Group 1 gneisses yielded 3018 ± 61 Ma age. The intrusive granites, particularly the leucocratic granite of Group 3 yielded 2573 ± 139 Ma (Stein et al., 2004). However, Rb/Sr age of this granite is 2101 ± 32 Ma with initial Sr ratios $= 0.7050$. The fine-grained granite (Group 5) gave Rb/Sr age of 2610 ± 143 Ma with initial Sr ratios $= 0.7056$ (Sarkar et al., 1990). It must be noted that most of these ages are errochrons but they definitely point to the presence of >3000 Ma old gneissic rocks in the Bastar craton. Also, high alumina trondhjemite gneisses, occurring as enclaves within the granite of Bastar craton near Markampura, yielded U–Pb zircon age of 3509 Ma, which is considered as the age of crystallization of tonalitic magma (Sarkar et al., 1995).

In the Bastar craton, three Archaean supracrustal units are recognized (Ramakrishnan, 1990), as stated already. First is Sukma metamorphic suite consisting of quartzites, metapelites, calc-silicate rocks, and BIF with associated metabasalt and ultramafic rocks. Second is Bengpal Group which is also characterized by the similar rock association as that of the Sukma unit. Hence, no distinction can be made between the two groups except that the Sukma suite shows a higher grade of metamorphism characterized by cordierite-sillimanite in the metapelites. These two rock groups, with associated granite-gneiss, have a general strike of WNW-ESE to NW-SW and appear to form a synclinorium in the west where the Third Group, the Bailadila Group is seen to overlie them. This Group contains BIF, grunerite-quartzite, and white quartzites. Migmatitic leucosomes from the Bengpal Group rocks yielded an errochron of 2530 ± 89 Ma with initial Sr ratio $= 0.70305$ (MSWD $= 18.61$) (Bandyopadhyay et al., 1990; Sarkar et al., 1990). The granites and associated pegmatites from Kawadgaon, intruding the Archaean Sukma and Bengpal formations of Bastar craton yielded whole rock Rb–Sr isochron of 2497 ± 152 Ma (Singh and Chabria, 1999, 2002; Sarkar et al., 1983).

On the NW of the Bastar craton there also occurs a vast exposure of gneissic complex, known in literature as *Tirodi gneiss*, which is a two-feldspar gneiss with biotite and occasionally garnet. Accessory minerals are zircon, apatite magnetite and sphene. Straczek et al. (1956), while mapping the Central Indian manganese belt, recognized biotite gneisses of all varieties surrounding the Proterozoic Sausar belt, which are often intercalated with amphibolites and hornblende gneiss (Subba Rao et al., 1999). They named this stratigraphic unit as Tirodi biotite gneiss and provisionally placed it at the base of the Sausar Group. Narayanaswami et al. (1963) considered the Tirodi gneiss as the basement to the Sausar Group. Although the contact between the Tirodi gneiss complex and the Sausar is mostly tectonized at most places, recently a polymictic conglomerate has been reported at the contact of Sausar and Tirodi gneiss from the locality of Mansar (Mohanty, 1993), confirming that the Tirodi gneiss is a basement to the Proterozoic Sausar Group. This rejects the hypothesis of some workers (Fermor, 1936; Roy et al., 2001) who proposed the Tirodi gneiss as migmatized Sausar Group rocks. Since the metasediments of the Sausar Group attained upper amphibolite facies (Brown and Phadke, 1983), the basement Tirodi gneiss should have equal or higher grade than sillimanite-almandine-orthoclase subfacies of the amphibolite facies. This accounts for the intermingling of granulites and gneisses so often seen in the Tirodi gneiss complex of the Bastar craton (cf. Bhate and Krishna Rao, 1981). Because of the occurrence of two or more phases of migmatization the Tirodi gneiss complex is considered a reworked Amgaon basement complex. It is in this context that the Rb/Sr isochron age of 1525 ± 70 Ma for the Tirodi gneiss and its mineral ages around 900 Ma (Sarkar et al., 1988) need to be considered. The Tirodi gneiss, although occurs to the north of the Central Indian Shear zone, abbreviated CIS (Fig. 2.5), is considered equivalent to the Amgaon gneiss described above. Both the Tirodi gneiss complex and the Sausar Group rocks are intruded by granite pegmatite and quartz veins of different generations. A detailed mapping showed that the Tirodi gneiss terrain contains mafic granulite, porphyritic charnockite, cordierite granulite and amphibolites—all occurring as rafts and lenses within the migmatized and banded gneisses of the Tirodi biotite gneiss complex (Bhowmik et al., 1999; Bhowmik and Roy, 2003). Considering the nature of the occurrences of the granulites, it seems quite probable that the granulite pods or lenses are the restites formed as a result of partial melting of the Tirodi gneiss during the Proterozoic Sausar (Satpura) orogeny. Alumina-rich pelites yielded garnet-cordierite-sillimanite gneisses, while normal metapelites and gneisses yielded hypersthene from biotite-involving dehydration melting reactions. The possible reactions are:

Biotite + sillimanite + quartz = garnet + cordierite + K feldspar + H_2O;
Biotite + quartz = hypersthene + K feldspar + H_2O ± garnet;
Hornblende + quartz = hypersthene + clinopyroxene + plagioclase + H_2O.

The Tirodi gneiss complex shows different periods of crustal melting which produced different types of migmatites, mostly felsic migmatites (Bhowmik and Pal, 2000). Some authors consider the granulites within the Tirodi gneiss as mylonitized and deformed tectonized rock in view of their occurrence as lensoidal and sigmoidal shape two-pyroxene granulites and their marginal retrogression. Due to this occurrence, the granulite is mistakenly interpreted as the remnant oceanic crust or the obducted tectonic slices/mélanges while the Tirodi basement gneiss as the Crustal block had subducted under the Bastar craton along the CIS (cf. Yedekar et al., 1990, 2000). It is interesting to observe that the boundary between the granulites and

2.3 Bastar Craton

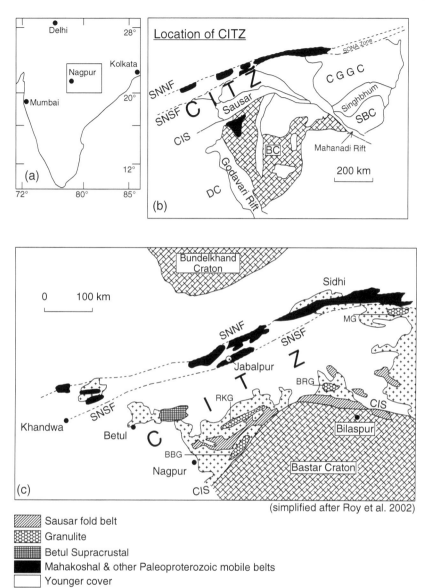

Fig. 2.5 Central Indian fold belts and cratons. (**a**) Location of central Indian fold belts. (**b**) Geological setting of Bastar Craton in relation to adjacent cratons and Central Indian Tectonic Zone (CITZ). Abbreviations: BC = Bastar craton, CGGC = Chhotangapur Granite Gneiss Complex, CIS = Central Indian shear zone, DC = Dharwar craton, SBC = Singhbhum craton, SONA ZONE = Son-Narmada Lineament zone bounded by Son-Narmada North Fault (SNNF) and Son-Narmada South Fault (SNSF). (**c**) Simplified geological map (*bottom sketch*) shows the CITZ sandwiched between Bastar craton in the south and Bundelkhand craton in the north. The four localities of granulites described in the Satpura fold belt are: BBG = Bhandara-Balaghat granulite, BRG = Bilaspur-Raipur granulite, MG = Makrohar granulite, and RKG = Ramakona-Katangi granulite

gneisses is lined up with sheet-like bodies of younger granitoids, suggesting that genesis of granulites is the outcome of partial melting phenomenon that produced the associated granitic bodies in the terrain.

There are four main occurrences of the granulites within the basement gneiss complex. (1) Ramakona-Katangi granulite domain (RKG) in the NW of the Bastar craton; (2) the Bilaspur-Raipur granulite (BRG) domain in the middle; (3) Makrohar granulite domain (MG) further east of the Satpura belt, near the SONA zone (Fig. 2.5c). These occurrences are in the form of boudins and pods within the Tirodi gneiss complex. The fourth occurrence of the granulite is the Bhandara-Balaghat Granulite belt (BBG). The granulite belt (4) is near the Central Indian shear Zone (CIS) and has a controversial location with respect to the shear zone. In geological maps, some workers (e.g. Bhowmik et al., 1999) show the BBG granulite within the Tirodi gneiss complex to the north of the CIS, whereas others (e.g. Abhijit Roy et al., 2006) place it within the Amgaon gneiss complex of the Bastar craton, south of the CIS.

The Ramakona-Katangi granulite domain (RKG) is dominated by basic type that includes mafic granulites and amphibolites. The host gneisses also experienced high-grade metamorphism and the garnet-amphibolites are not considered as retrogressed granulites. Bhowmik and Roy (2003) deduced a metamorphic history from garnet amphibolite and mafic granulites from this belt. The peak metamorphism (M1) for garnet-amphibolite has been estimated at 9–10 kbar/750–800°C while lower P-T value (8 kbar/675°C) was obtained for amphibolites. The M1 event in the rocks was followed by isothermal decompression phase (M2), estimated at 6.4 kbar/700°C, after which the rocks underwent isobaric cooling (M3) at 6 kbar/650°C (Bhowmik and Roy, 2003). The authors argue that the decompression in the mafic granulites is not continuous, but punctuated by a distinct heating event (prograde). The event is related to an extension phase marked by emplacement of mafic dykes. The combined P-T path is clock-wise. A near clock-wise P-T loop has also been established from studies of pelitic granulites of the RKG domain (Bhowmik and Spiering, 2004). On the consideration of growth zoning preserved in garnets and based on heterogeneous distribution of diverse inclusions of mineral assemblages in the porphyroblasts of pelitic granulites from RKG, Bhowmik and Spiering (2004) deduced a clockwise P-T path. The path is claimed to have started with an early prograde amphibolite facies metamorphism (identified by staurolite-biotite-quartz ± kyanite), followed by peak metamorphism (M1) at $\sim$9.5 kbar/$\sim$850°C, documented by dehydration melting of biotite and formation of pelitic migmatite. The path terminated by isothermal decompression (M2) at 6 kbar/$\sim$825°C documented by plagioclase corona and spinel-plagioclase-cordierite symplectite around garnet. The final P-T trajectory, according to the authors (Bhowmik and Spiering, 2004), was marked by isobaric cooling (M3) which terminated at 5 kbar/600°C. From this depth, the rocks were possibly emplaced in the shallower depth (equivalent to the depth of the Al-silicate triple point) by upthrusting along the CIS to be later acted upon by erosion to expose them in the present state. According to these authors, the deduced P-T loop is consistent with a model of crustal thickening due to continental collision, followed by rapid vertical thinning.

However, there is considerable debate on the extent of the collision event. Bhowmik et al. (1999) considered the collision orogeny to be related to an older pre-Sausar tectono-thermal event. The overprint of an amphibolite facies event was taken by these authors as evidence for the younger Sausar event (Mesoproterozoic). Bhowmik and Spiering (2004) derived 880 Ma age for the monazite included within a garnet of RKG domain. This age is close to the 860 Ma Rb/Sr mineral isochron age from the basement of Tirodi gneiss (Sarkar et al., 1986), but nearly 70 Ma lower than the $^{40}Ar/^{39}Ar$ age for cryptomelane derived from Ramakona granulite area (Lippolt and Hautmann, 1994). Bhowmik and Dasgupta (2004) consider this age to mark the timing of post-decompression cooling history in the RKG domain. These authors claim that the timing of the collision and peak metamorphism is likely to be older and the metamorphic history of the RKG domain is an outcome of a single Grenville-age tectonothermal event (Bhowmik and Dasgupta, 2004).

The Bilaspur-Raipur granulite domain (BRG) occurs in the eastern extremity of the CITZ bordering the Bastar craton (Jain et al., 1995; Bhattacharya and Bhattacharya, 2003). This domain represents an ensemble of supracrustals, granite, gneisses, and granulites, each separated by tectonic contacts. Jain et al. (1995) reported mafic granulites from Ratanpur area. The granulites are boudin type. The geological history of this belt is poorly understood.

The Makrohar granulite domain (MG) occurs to the south of the low-grade Mahakoshal supracrustal belt. Previous workers have identified three distinct lithologies in the MG (Pascoe, 1973; Solanki et al., 2003). These are: (a) felsic gneiss-migmatite, (b) supracrustal lithopackage of sillimanite, quartzite, meta-BIF, calc-silicate gneiss, calcite marble, pelitic schists, garnetiferous metabasics, including amphibolites and hornblende schists, (c) metaigneous rocks comprising gabbro-anorthosites and porphyritic granitoid. The latter intruded the supracrustal rocks. Solanki et al. (2003) have documented garnet-cordierite-biotite-sillimanite assemblage in pelitic granulites and garnet-hornblende-plagioclase-epidote-rutile-sphene assemblage in metabasic rocks. Pitchai Muthu (1990) previously reported corundum-bearing sillimanite schists from the same area. Based on textural, mineral-chemical and geothermobarometric studies, Solanki et al. (2003) established P-T conditions of three stages in the development of pelitic granulites and garnet-metabasic rocks. These stages are: (1) 9 kbar/800°C for peak metamorphic condition, (2) 6.5 kbar/740°C for early stage of retrogression, and (3) 685°C for final re-equilibration. The granulite events, however, remain isotopically undated.

The lithological ensemble of the Bhandara-Balaghat granulite (BBG) domain is subdivided into 4 distinct components: (i) a large migmatitic felsic gneiss terrain, locally with garnet, (ii) enclaves or isolated bands of garnet-cordierite gneiss, BIF, quartzite, corundum-bearing and felsic granulite within the Tirodi gneisses, (iii) a mafic-ultramafic magmatic suite of metagabbro-metanorite and gabbro metanorite-metaorthopyroxenite, occurring as concordant sheets in the felsic gneisses, and (iv) metabasic dykes and amphibolites. The gabbroic suite of rocks is particularly dominant in the southern part of the BBG domain where it is interlayered with felsic and aluminus granulites. By contrast, norites and meta-orthopyroxenites are quite common in the northern part where they are associated with garnet-cordierite gneiss.

Bhowmik et al. (2005) recognized 5 phases of deformation, D1 to D5. The D1 has caused banding demarcated by alternate light and dark migmatitic layers. D2 is a foliation parallel to banding. This was followed by a strong ductile shear zone that produced south-verging isoclinal folds, D3, and a strong mylonitic foliation S3. This phase deformed mafic dykes that were emplaced in the felsic granulites. The σ-type asymmetrical orthopyroxene porphyroclasts indicate that the southerly tectonic transport was a high T phase during D3. Subsequent deformation produced narrow steep ductile shear zone fabrics that affected the amphibolites. The D1 to D3, according to Bhowmik et al. (2005), pre-date the Sausar orogeny and the cross folds in the granulites due to D5 are also found in the Sausar Group rocks.

The granulite facies metamorphism is dated at 2040–2090 Ma by Bhowmik and Dasgupta (2004). Ramchandra and Roy (2001) have reported Sm/Nd and Rb/Sr ages from charnockitic gneisses and two-pyroxene granulites from BBG domain. They show three distinct age clusters, at 2672 ± 54 Ma; 1416 ± 59 to 1386 ± 28 Ma; and 973 ± 63 to 800 ± 16 Ma. These ages are correlated with two temporally separated phases of granulite facies metamorphism of Archaean and Mesoproterozoic ages, finally overprinted by the Sausar orogeny (Ramachandra and Roy, 2001). However, in the absence of detailed information on the type of rocks and mineralogy being dated and the methodology being used, it is difficult to use these dates to constrain tectono-thermal events in the Sausar mobile belt (see Chap. 5).

Petrological work by Bhowmik et al. (1999) showed that mafic granulite (metagabbro) is the dominant component of the BBG that also has enderbite gneiss, charnockite, cordierite gneiss and meta-ultrabasites (gabbro-norite-pyroxenite). P-T estimates by the authors revealed an anti-clockwise path characterized by heating and isobaric cooling. Charncokites from the BBG are found to show peak metamorphism at $\sim$10.5 kbar/775°C and a lower P-T ($\sim$5 kbar/700–650°C), which are considered to correspond to Pre-Sausar and Sausar orogeny (Bhowmik et al., 1991; Roy et al., 2006). Recent geochronological data given by Roy et al. (2006) reveal quite interesting ages on the charnockites and mafic granulites from BBG. Sm–Nd isochron of charnockite whole rock (WR) and its three mineral separates gave an age of 2672 ± 54 Ma but Rb–Sr isotope systematics yielded an age of 800 ± 171 Ma. The high error in age and higher MSWD of the samples are attributed to limited spread among Rb–Sr ratios of mineral phases and whole rock, besides high Rb/Sr ratios. Again, the garnet-bearing as well as garnet-free mafic granulites yielded WR and mineral separates Rb/Sr and Sm/Nd isochron ages in the range 1400–1420 ± 70 Ma. The coincidence of Sm/Nd and Rb/Sr ages indicate the crystallization ages of these mafic bodies. Rb/Sr isochron for two of the analyzed samples of mafic granulites from BBG, however, gave an age of about 970 Ma which Roy et al. (2006) consider as re-set ages by a thermal imprint.

Two more granulite belts are reported in the Bastar craton (Ramakrishnan and Vaidyanadhan, 2008). One is the Bhopalpatnam granulite belt at the northern shoulder of Godavari graben that conceals the actual contact between Dharwar craton and Bastar craton. Zircon U–Pb dating report 1.6–1.9 Ga for the granulites (Santosh et al., 2004). Another belt is the Kondagaon granulite belt which occurs in the middle of the craton. Here, charnockites and leptynites dominate the western part of the granulite belt. The Kondagaon granulite belt indicates ages around 2.6 Ga, similar to those of Karimnagar granulite belt on the southern flank of the Godavari graben.

Recently, Srivastava et al. (1996) have identified three sets of mafic dyke swarms in the southern Bastar craton. Two sets are sub-alkaline tholeiitic whereas the third set is boninite-like mafic rock (high SiO_2 and high MgO). The earliest dyke (D-1) is an amphibolite, mostly intruding Archaean granite gneisses but not the Proterozoic granites since it is cut by 2.3 Ga old granite. The second set of dyke cuts all rock-types, including the 2.3 Ga old granite. The age of this latter dyke (D-2) collected from Bastanar gave an age of 1776 ± 13 Ma (Srivastava and Singh, 2003). French et al. (2008) have carried out U–Pb (zircon and Baddeleyite) dating of two NW-SE trending mafic dykes from southern Bastar that gave 1891 ± 0.9 Ma and 1883 ± 1.4 Ma ages.

Geodynamic Evolution of Bastar crtaon cannot be modeled for want of thermo-tectonic and sufficient geochronological data on the rocks of the craton. However, the following account would form the foundation for future attempt in this direction.

The grey gneiss complex of Bastar craton contains relicts of early continental rocks of 3.5–3.6 Ga. The Archaean crust of Bastar craton seems to have witnessed rifting in the Neoarchaean time for the supractrustals of Sukma and Amgaon Groups (~3.0 Ga) which now occur as enclaves in the basement TTG gneisses of the craton. With deposition of the supracrustals the craton was presumably stabilized. On the stabilized crust, a WNW-ESE rift was formed and subsequently filled with amygdule basalt in association with clastics of Bengpal Group. The Bengpal Group witnessed widespread granite activity at the Arcchaean-Proterozoic transition (2.5–2.6 Ga). After Bengpal event, there occurred another continental rifting for the deposition of N-S trending Kotri-Dongargarh fold belt (Sect. 5.5). After a break in sedimentation, a new cycle of felsic and mafic volcanics (Nandgaon Group) occurred, with a basal conglomerate and quartzite. A major granite pluton (Dongargarh), perhaps comagmatic with the felsic volcanics, intruded Nandgaon Group at about 2.3 Ga ago. Thereafter there was another volcanic-clastic cycle (Khairagarh Group) which seems to have concluded the main felsic to mafic volcanism in this stage of continental rifting. Finally, the younger succession of dominantly fine clastic plus chemical sediments (Chilpi Group) occurred which were subsequently involved in the Proterozoic orogeny (Satpura fold belt). Bastar craton was completely stabilized by about 1800 Ma to have deposition of the Purana basins of central Indian region.

2.4 Singhbhum Craton

The Singhbhum craton (SBC) is also called Singhbhum-Orissa craton in eastern India. It is made of Archaean rocks that are exposed in an area of ~40,000 km^2 in Singhbhum district of Jharkhand (formerly Bihar) and northern part of the State of Orissa. The craton is bordered by Chhotanagpur Gneissic Complex to the north,

Eastern Ghats mobile belt to the southeast, Bastar craton to the southwest, and alluvium to the east. Much of the geological information about Singhbhum craton (SC) or Singhbhum Granite Complex (SGC) is due to Saha (1994). The following rock-suite constitute the Singhbhum craton (Fig. 2.6):

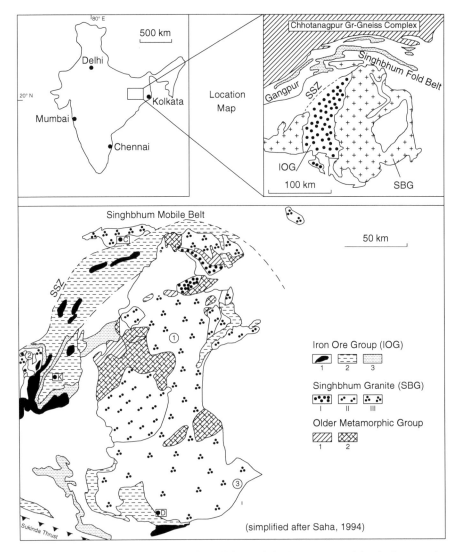

Fig. 2.6 Location and Geological map of Singhbhum (-Orissa) craton comprising Archaean rocks of Older Metamorphic Group (1) and Older Metamorphic Tonalite Gneiss (2), Singhbhum Granite Group (SBG) with three phases (I, II, & III) of emplacement, and Iron-Ore Group (IOG) made up of: 1 – lavas and ultramafics, 2 – shale-tuff and phyllite, 3 – BHJ, BHQ, sandstone and conglomerate. Abbreviations: C = Chakradharpur, D = Daiteri, K = Koira, SSZ = Singhbhum shear zone. (1) = Singhbhum Granite, (2) = Bonai Granite, 3 = Mayurbhanj Granite

2.4 Singhbhum Craton

1. Singhbhum Granite (I, II, III phases) with enclaves of (i) Older Metamorphic Group (OMG), and (ii) Older Metamorphic Tonalite Gneiss (OMTG).
2. Iron Ore Group (IOG,) dominantly Banded Iron formation (BIF) at the margin of the Singhbhum Granite
3. Volcanics or greenstone belts (Simlipal, Dhanjori, Dalma etc.)

The Older Metamorphic Group (OMG) (Sarkar and Saha, 1977, 1983) is a supracrustal suite of rocks composed of amphibolite facies pelitic schists, garnetiferous quartzite, calc-magnesian metasediments and metabasic rocks. Komatiitic lavas are missing from the OMG, unlike other Archaean terrains of the world. Radiometric dating of rocks of the OMG does reveal ages older than 3300 Ma. The OMG rock-suite is perhaps the oldest supracrustals of the eastern Indian shield, and perhaps equivalent to older Dharwar Schist Belt (Sargur). The OMG rocks are synkinematically intruded by tonalite gneiss grading into trondhjemite and designated as the Older Metamorphic Tonalite Gneiss (OMTG). The OMTG (whole rock) has been dated both by Sm–Nd and Rb–Sr systematics. The Sm–Nd age of 3800 Ma for the OMTG is considered as the crystallization age of the magma, although it might also represent the time of generation of mafic melt in the mantle with crystallization of the melt at the base of the crust (Basu et al., 1981). The Rb–Sr isochron age of ~3.2 Ga for the OMTG (Sarkar et al., 1979) is interpreted as the age of melting associated with metamorphism of the OMG. The OMTG is considered as an anatectic product of the OMG into which the OMTG intruded at depth during metamorphism. The OMG is estimated to have been metamorphosed at 5.5 kbar and 660–630°C, indicating substantial crustal thickening during Archaean. The OMTG is characterized by gently sloping REE patterns without Eu anomaly (Saha and Ray, 1984b), that are comparable to the pattern in the Archaean Gneissic Complex and the tonalitic diapirs of Barberton. The geochemistry and REE patterns suggest derivation of the OMTG by partial melting of OMG amphibolites (Sharma et al., 1994) and even by partial melting of low-K basalt or mantle peridotite. Bose (2000), on the other hand, considers that the OMTG and the amphibolites of the OMG may represent Archaean bimodal magmatism (Saha et al., 1984).

The Older Metamorphic Group (OMG) as a supracrustal must have been initially deposited as sediments and volcanics on an earlier basement, now unrecognizable in the region. This is perhaps due to their transformation and remobilization generating granitic rocks, now seen as OMTG, and possibly also as several later generation granites that are now grouped in the Singhbhum Granite Complex (SGC). This basement to the OMG appears to be granitic, in view of the presence of muscovite and quartz-rich bands as well as the occurrence of zircon in the OMG. At Rairangpur locality a group of dark-coloured tonalite rafts are reported "floating" in the Singhbhum Granite. These rafts are believed to have been basement for deposition of the volcano-sedimentary sequence of the OMG (Dey, 1991). Both OMG and OMTG document two phases of deformation and exhibit a structural accordance, suggesting that the Archaean deformation occurred after the emplacement of the OMTG. The general strike of the OMTG is NE-SW (Saha, 1994), but folding has changed this to NW-SE. According to Saha (1994) the OMG intruded by the OMTG are the oldest formations in the Singhbhum region (Sarkar and Saha, 1983). They have been folded first about steep NE plunging axes and later about the SE plunging axes, with variable dips, seen in the northern region of the Archaean craton. The Older Metamorphic Group in the type area has the axial planes of the first generation strike NW, with axes plunging NE. These have been affected by folds plunging towards SE (Sarkar and Saha, 1983).

The formation of OMG and OMTG was successively followed by the emplacement of the Singhbhum Granite (SBG) that intruded in two phases (SBG I and SBG II), deposition and folding of the Iron Ore Group (IOG) supracrustals and emplacement of Singhbhum Granite phase III (SBG III) (Saha, 1994). The OMG banded amphibolite and OMTG are commonly seen as rafts and inclusions in the vast expanse of the Singhbhum Granite Complex, indicating that the OMG supracrustals and OMTG originally covered a wide area now occupied by the Singhbhum Granite batholith. The structural elements in the OMG area oriented oblique to those occurring in the adjacent supracrustal envelope forming the Iron Ore Group (see below).

The oval-shaped Singhbhum Granite Complex (SBG), together with other smaller plutons of more than one generation, constitutes major part of the Archaean craton. This granite complex is part of the earliest continental segment to cratonise and together with the Archaean supracrustals have been designated the Archaean Cratonic Core Region (ACCR) by Mahadevan (2002). In the granite complex, the supposedly primary foliation shows swirling patterns which led Saha (1972) to identify 12 separate magmatic bodies, domical or sheet-like in shape, which appear to have been emplaced in three successive but closely related phases. Each phase of the magma is believed to have been derived independently from the same source region in the crust, because of distinctive trace element distribution patterns (Saha, 1979). The phase I is K-poor granodiorite to trondhjemite, occurring only as small patches at the northern and eastern parts of the craton. The phase II is dominantly granodiorite, mainly confined to the southern contact of the OMTG. The phase III is mainly granite. The phase III covers the largest part covering an area of ~ 1800 km^2. However, samples of phase I and II with gently sloping REE patterns and weak or no Eu anomaly are similar to the patterns shown by OMTG. The samples of phase III show similar patterns (LREE enrichment, flat HREE and negative Eu anomaly). Comparable to the Singhbhum Granite are other granites, namely the Chakradharpur granite gneiss in the NW and the Bonai Granite body in the west, occurring at the perpheral region of the granite nucleus (ACCR of Mahadevan, 2002). Granitic rocks of proven diverse ages within the Singhbhum Granite massif (e.g. Mayurbhanj Granite), however, suggest remobilization of the basement during later orogenesis (Naha and Mukhopadhyay, 1990).

Rb-Sr isochrons (Sarkar et al., 1979) for the Singhbhum Granite are less precisely dated than those for the OMTG, indicating a great heterogeneity in the samples. Poorly constrained ages preclude use of existing geochronology for correlation of magmatic/metamorphic events. However, careful geochronological data (Moorbath et al., 1986; Sharma et al., 1994; Sengupta et al., 1996) suggest two crust-forming events in the region during the Archaean, with their ages at 3.4 and 3.2 Ga. The isochron age of 3200 Ma for the tonalite gneiss is considered to represent melting associated with metamorphism of the OMG (Sarkar et al. 1979). The 3800 Ma Sm/Nd age of the OMTG, determined by Basu et al. (1981), is considered the crystallization age of the magma, although it might also represent the time of generation of a mafic melt in the mantle with crystallization of the melt at the base of the crust (Basu et al., 1981). The metasedimentary enclaves in the 3.2 Ga old Singhbhum Granite from southern Singhbum are dated at about 3.5 Ga (see Saha, 1994). The Granite phase III (ca. 3.1 Ga) is believed to be intrusive into the IOG. The emplacement dates of the three phases of Singhbhum Granite (Saha, 1994), underwent a major revision when Sharma et al. (1994) gave the time span of ~3.3–3.1 Ga for the Granite phases I, II and III (Saha and Ray, 1984a and b).

Recent geochronological data reveal (see Misra, 2006) that the equivalent granite bodies of the Singhbhum Granite are the Bonai Granite, Nilgiri Granite and Chakradharpur Granite Gneiss, occurring respectively at the western, southeastern and northern margins of the Singhbhum Granite (Fig. 2.6). The Bonai Granite is dominantly porphyritic granite, intimately associated with less abundant equigranular variety. The pophyritic granite ranges in composition from granite to granodiorite and rarely tonalite, whereas the equigranular variety is a two-mica trondhjemite. The porphyritic granite intrudes the IOG and contains large xenolithic blocks of these metalavas. Therefore, the Bonai Granite is younger than the IOG supracrustals and stratigraphically equivalent to SBG-III. The Bonai granite is separated from the main Singhbhum Granite batholith by a belt of IOG supracrustals (Jamda-Koira horse-shoe synclinorium) (Sengupta et al., 1991; Saha, 1994).

The Nilgiri Granite (Saha, 1994) forms the southeastern part of the Singhbhum (-Orissa) craton, and is separated from the main Singhbhum Granite by a narrow strip, 3–8 km wide of IOG phyllite and greenschist (see Fig. 2.6). The Nilgiri Granite varies in composition from TTG to granite. The Nilgiri pluton is seen intruded by the Mayurbhanj Granite (anatectic product of SBG-III) along its margin.

The Chakradharpur Granite gneiss (Bandyopadhyay and Sengupta, 1984; Sengupta et al., 1983, 1991) is an isolated body of the Singhbhum Granite and occurs amidst the supracrustal rocks of the Singhbhum Group (see Fig. 2.6). This granite gneiss forms the basement to the overlying Singhbhum Group, while its pegmatoid phase intrude both the older gneiss and the enveloping supracrustals. Geochemically, the older tonalite gneiss of the Chakradharpur Granite Gneiss is considered equivalent to the SBG-I or SBG-II, whereas the pegmatoid phase is considered equivalent to younger granite bodies (e.g. Arkasani Granite, Mayurbhanj) that intrude the Singhbhum Group (Saha, 1994).

The Iron Ore Group (IOG), consisting of banded iron formation (BIF) and metasedimentary and metavolcanic rocks, occurs in the western, eastern, and southern flanks of the Singhbum Granite massif. The BIF's are interbanded with lavas and pyroclastics, and even basic volcanics interbedded with rhyodacite and trachytic volcaniclastics (Banerjee, 1982). Along the eastern border zone of the Singhbhum Granite the BIF is intruded by a group of unmetamorphosed gabbro, norite, and anorthosite. The IOG rocks in the Singhbhum craton occur either as linear narrow intra-cratonic belts or as more extensive peripheral bodies. The IOG (formerly

Iron Ore Series; Krishnan, 1935) is believed to have deposited on the OMG. Three major basins of IOG are recognized along the fringes of the ACCR. These are: (1) Gorumhasani-Badampahar basin in the eastern part, (2) the Tomka-Daiteri (D) basin along the southern part, and (3) the West Singhbhum-Keonjhar basin or the Jamda-Koira (K) basin in the western flank of the ACCR. The basin (3) contains the most spectacular occurrence of the BIF in the Jamda-Koira valley to the west, where its outcrop defines a major horse-shoe shaped syncline. This syncline shows transverse folding and variations of plunge. Both features, according to Sarkar and Saha (1983), are considered synchronous with the main folding event. However, Chakraborty and Mazumdar (1986) have recognized three sets of folds within BIF near Malangtoli south of Jamda, and the fold interference is found to have resulted in type 1 and type 2 patterns of Ramsay (1967). The Gorumahisani basin (1) has volcanics and BIF whose strike of the axial planes of the dominant folds is nearly N-S, almost parallel to the elongation of the belt. The Gorumahisani belt represents either a pinched-in synclinal cusp within the basement or a faulted graben (Mukhopadhyay, 1976; Banerjee, 1982). Interestingly, this belt is parallel to one set of regional fractures within the Singhbhum Granite. In the Tomla-Daiteri basin (2) to the south, the regional folds show E-W striking axial planes. Thus, the axial traces of the regional folds within the different basins of the IOG curve round the Singhbhum Granite, suggesting that the fold in these supracrustals envelope were moulded around the rigid basement block.

Sarkar and Saha (1983) place these three Archaean basins in one stratigraphic group, termed the IOG, as interconnected basin, while Iyengar and Murthy (1982) believe that the Gorumahisani basin belongs to much older orogenic cycle during which the IOG (3) welded to the Archaean craton. A.K. Banerji (1974) also suggests that the Noamandi basin of IOG (3) is stratigraphically distinct. He recognized two cycles of iron formations, an older Gorumahisani Group and a younger Iron-ore Group, separated by the Jagannathpur volcanic flows. Dating of the intrusive granite phase-III (3.17 ± 0.1 Ga; Saha, 1994) of the Singhbhum Granite into the IOG suggests that the IOG are older than 3.1 Ga. The supracrustal of the stated IOG is surrounded on its east and north by a younger Dhanjori eugeosyncline in which about 10 km sediments and lava flows were deposited.

The *Dhanjori Group* of volcano-sedimentary formations rest unconformably on the Singhbhum craton and, in turn, are overlain unconformably by the Chaibasa Group representing the lower unit of the Singhbhum Group. The relationship between the Dhanjori volcanics and the argillaceous Chaibasa Formation is a debated topic. According to Dunn and Dey (1942), the Dhanjori is thrust over by the Chaibasa Formation and hence older than the latter. Sarkar and Saha (1962, 1977) regard the Dhanjori to be younger than the Chaibasa. However, on the basis of northward younging of the successive members of the Dhanjori formations and of the juxtaposed Chaibasa formation, Mukhopadhayay et al. (1975) concluded that the Dhanjori is older and is succeeded by the Chaibasa formation. In support of this view, Sinha et al. (1997) reported that there is a conglomerate at the base of the Chaibasa Formation exposed in the Jaduguda mines, separated from the underlying conglomerate of the Dhanjori sequence, rendering the Chaibasa Fm younger than the Dhanjori. The 2.2 Ga old (Pb—Pb age; Sarkar et al., 1985) Soda Granite that syntectonically intruded the Chaibasa Fm fixes the upper age limit of the Singhbhum Group. The age of the Dhanjori volcanics is poorly constrained, but Sm—Nd isotopic studies on Dhanjori basalt from Kula Mara and Kakdha area (S of Rakha Mines) gave an errochron of 2072 ± 106 Ma (Roy et al., 2002b). Recently, Misra and Johnson (2003 in Acharyya, 2003) report whole-rock Pb—Pb ages of 2787 ± 270 Ma while Sm—Nd whole-rock yielded 2819 ± 250 Ma for the Dhanjori basalts. From this we may conclude that IOG and the Dhanjori Group represent the Late Archaean (3000–2500 Ma) supracrustals, and hence should be included in the cratonic region of the Singhbhum region. It is for their older age that the Dhanjori volcanics differ from Dalma volcanic in their trace element and REE pattern (Bose, 1990). Dhanjori basalt has a fractionated REE whereas Dalma basalt (and their intrusive gabbro-pyroxenite) is depleted in LREE and show almost flat HREE (see Saha, 1994).

In the southwestern region of the ACCR the IOG is intruded by 3.16 Ga old Bonai Granite (see Fig. 2.6) and is unconformably overlain by a meta-psammitic cover sequence with conglomerate horizon at base and metavolcanics and tuffites in middle part. This overlying sequence is named *Darjin Group* (Mahalik, 1987). The Darjin Group is intruded by Tamperkola Granite (Mazumder, 1996; Naik, 2001) that has yielded in-situ Pb—Pb zircon age of 2809 ± 12 Ma (Bandyopadhyay et al., 2001). Interestingly, the Darjin rocks and its intrusive Granite are devoid of deformation and metamorphism, despite of their being late Archaean in age, as stated already. The deformed and recrystallized Singhbhum Group exposed further east is intuded by 2.2 Ga old Soda Granite (cf. Acharyya, 2003). The Darjin Group is overlain by a sequence of thick metavolcanics, tuffs, and metasediments (Bhaliadihi Formation of Naik, 2001) with a basal unit of BHJ and BHQ-bearing breccia and conglomerate. The metavolcanic dominated Bhaliadihi Formation resembles the Dalma and/or Chandil volcanics (Proterozoic).

The northern and northwestern boundary of the Archaean nucleus is not well defined, although the northern and eastern side of the cratonic nucleus is demarcated by a broad shear zone, called the Singhbhum Shear Zone (SSZ) (Mukhopadhyay et al., 1975). This zone of variable width consists of extremely deformed clastics with down dip lineation, BIF, wedges of granite gneiss from the Archaean nucleus, lithicwackes of the cover as well as mylonites and phyllonites. The stratigraphic sequence within this zone cannot be worked out (Sengupta and Mukhopadhyay, 2000). Discontinuous sheets of smaller and linear bodies of granites (syntectonic 2.2 Ga old Soda Granite, Arkasani granophyres and Mayurbhanj granite), some of which also occur as basement wedges, often in variable states of deformation and are exposed in the vicinity of the shear zone. According to Dunn (1929) and Mukhopadhyay (1984) the SSZ with its narrow belt of mylonites dic out westward. Extension of the SSZ is also reported along the NW margin of the Singhbhum Granite and the southern margin of the Chakrdharpur Granite to appear as the largest tectonic wedge of the basement granitoid in the north (Sarkar and Saha, 1962; Gupta and Basu, 2000). This occurrence led these authors to infer that the SSZ separates the domain of Singhbhum mobile belt and the Archaean craton hosting the Ongerbira

Volcanics. However, some workers recorded a lithological similarity of rocks across this supposed extension of the SSZ (cf. Mukhopadhyay et al., 1975). A continuity of structure and absence of mylonite belt was also noted across the supposed extension of the SSZ (cf. Mukhopadhyay et al., 1990). This controversial situation could be the result of discontinuity of the shear zone between Singhbhum mobile belt in the north and the cratonic region overlain by Ongerbira volcanics in the south.

2.4.1 Geodynamic Evolution

The earliest crust in the Singhbhum craton is found to be the Older Metamorphic-Group (OMG) which yielded the same age as the Old Metamorphic Tonalite Gneiss (OMTG). Zircon ages in politic rocks of OMG suggest that sialic crust had existed at about 3.6 Ga ago, prior to the formation of OMG. It has been suggested that this early continental crust was formed by recycling (through mantle convection) of mafic-ultramafic crust into the mantle whereby TTG gneisses generated, possibly in two-stage melting process. The early melting stage was at about 3.3–3.4 Ga which is the oldest component component, Singhbhum Granite-I in which xenoliths of OMTG are found. The Mesoarchaean continental crust of Singhbhum Granite I with its enclaves OMG and OMTG, rifted and formed greenstone belt of Badampahar Group which is characterized by sphinifex textured komatiites, pillowed tholeiites, BIF and fuchsite quartzite—a common association of Archaean greenstone belts. The closing of the greenstone basins was followed by granite intrusion (Singhbhum Granite phase II and III). The next cycle is marked by the volcanics and sediment association of 2.8 Ga (Roy et al., 2002a). Based on the new radiometric ages on the different rocks of the Singhbhum (-Orissa) craton, the evolutionary history of the craton (after Misra, 2006) is given in Table 2.1.

2.5 Chhotanagpur Granite-Gneiss Complex

The Complex is considered here as a cratonic region whose convergence with the Singhbhum craton to the south gave rise to the Singhbhum mobile belt. It covers an area of about 100,000 km^2 and extends in the E-W across the states of Chattisgarh, Jharkhand, Orissa and West Bengal. The Chhotanagpur Granite-Gneiss Complex (CGGC) is bordered on the north by Gangetic alluvium, on the south by Singhbhum mobile belt and on the northeast by the Rajmahal basalt (Fig. 2.7). Early regional studies on the CGGC are by Ghose (1983), Mazumder (1979, 1988), Banerji (1991) and Singh (1998), but no regional map of the entire cratonic region was available until Mahadevan compiled a geological map (Mahadevan, 2002).

The CGGC comprises mainly granitic gneisses, usually migmatized, and porphyritic granite, besides numerous metasedimentary enclaves. Gneisses from central and granites from the western CGGC gave nearly similar ages of 1.7 Ga (see in Chatterjee et al., 2008). The migmatites and granulites and granitic gneisses from northeastern CGGC (Dumka area) gave Rb-Sr ages of 1.5–1.6 Ga (see in Chatterjee et al., 2008). These dates correspond to the ages obtained by Pb-isotope on galena

2.5 Chhotanagpur Granite-Gneiss Complex

Table 2.1 New radiometric ages on the rocks of the Singhbhum (-Orissa) craton, after Misra (2006)

S. No.	Geological events	Rock types	Age (Ga)
8.	Stabilization of Singhbhum (-Orissa) craton Thermal metamorphism of OMG due to emplacement of SBG-III and Bonai Granite		$\sim3.09^5$ $3.16–3.10^1$
7.	Emplacement of SBG-III and possible upper age limit of formation of SSZ Emplacement of Bonai Granite pluton	SBG-III, Granodiorite to granite Bonai granite-granite to granodiorite and rarely two-mica trondhjemite	3.12^2 3.16^3
6.	Metamorphism of OMG, OMTG and IOG Formation of IOG and its folding-Meta morphism	Mafic lava, tuff, felsic volcanics Tuffaceous shale, BHJ-BHQ with iron ore, local dolomite, quartzitic sandstone and minor conglomerate	$3.24–3.24^{.5}$ Between 3.33 and 3.16^5

Unconformity

	Thermal metamorphism of OMG and OMTG due to emplacement of SBG-II and Nilgiri Granite		$3.30^{6,7}$
5.	Emplacement of Nilgiri Granite Pluton Emplacement of SBG-II	Nilgiri pluton, Tonalite to granite SBG-II, granodiorite Emplacement of tonalitic rocks occurring as enclaves within Bonai Granite Pluton	3.29^8 $3.33–3.3^4$ 3.38^3

Table 2.1 (continued)

S. No.	Geological events	Rock types	Age (Ga)
4.	Emplacement of SBG-I	SBG-I, tonalite to granite	3.44^2
3.	Folding of OMG supracrustals and syn-kinematic tonalite intrusion of OMTG	OMTG-a bio-hb tonalite gneiss grading to granodiorite	$3.44–3.42^{4,10}$
		Oldest xenocrystic zircon recovered from tonalite xenoliths within Bonai granite	3.44^8
2.	Deposition of OMG sediments upon unknown basement, with intermittent extensive volcanism and plutonism	Pelitic schists, Qz-Mt-Cumm-schist, banded calc-gneiss, para- and ortho-amphibolites	$3.55–3.44^5$
1.	Formation of unstable sialic crust not preserved in the present geological record	Represented by sialic nature of overlying sediments and presence of xenocrystic zircon in them	$\sim3.55–3.6^{10}$

[1] Sarkar et al. (1969).
[2] Ghosh et al. (1996).
[3] Sengupta et al. (1991).
[4] Misra et al. (1999).
[5] Misra (2006).
[6] Moorbath et al. (1986).
[7] Sharma et al. (1994).
[8] Saha (1994).
[9] Sengupta et al. (1996).
[10] Goswami et al. (1995).

2.5 Chhotanagpur Granite-Gneiss Complex

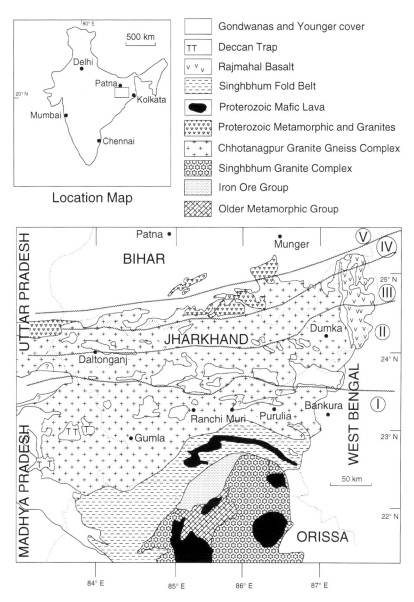

Fig. 2.7 Location and distribution of Chhotanagpur Granite Gneiss Complex (CGGC). The five geological belts (I–V) outlined by zigzag line in the CGGC terrain are drawn after Mahadevan (2002). *Dotted lines* are boundaries of the States

from metasediments in this part of the CGGC (cf. Singh et al., 2001). The metasedimentary rocks, occurring inside and outside the granitic gneisses, show varying degree of metamorphism, ranging from greenschist (mostly in SE part of the CGGC terrain) to granulite facies (central and eastern part of the terrain). The occurrence of metasedimentary and other rocks inside the gneisses of the CGGC suggests that the gneisses were the basement and that both these rock-units were subjected to metamorphism accompanied by profuse granite activities. It is for this reason that the CGGC acquired polymetamorphic character (Sarkar, 1968), like the Archaean gneissic complexes of most cratonic areas. The oldest age of ≥ 2.3 Ga of the gneisses of the CGGC from Dudhi area, west of Daltanganj (Mazumder, 1988) suggests that some of the metasedimentary enclaves within the gneisses were formed in Archaean. The enclaves of pelitic granulites, quartzites etc. within the basement gneisses of the CGGC indicate that the CGGC rocks have undergone partial melting in the Precambrian time, generating granitic rocks as intrusives and granulites as melanosome or restite.

The granulite assemblage in the CGGC is commonly represented by khondalite (garnet-sillimanite ± graphite), calc-silicate granulite (scapolite-wollastonite-calcite-garnet ± quartz), charnockite (hypersthene-granite), two-pyroxene granulite with or without garnet, hornblende granulite—all occurring as dismembered bands within migmatitic granitic gneisses (Roy, 1977; Sarangi and Mohanty, 1998). Apart from these, there are also enclaves of ultramafic bodies (hypersthene-spinel-hornblende ± olivine), lenticular or elliptical massif of anorthosite and syenite bodies (Mahadevan, 2002). The gneissic layering in granulite enclaves is similarly folded with the surrounding gneisses (Mahadevan, 2002; Mazumder, 1988). It is not clear whether the entire terrain shows a progressive regional metamorphism from greenschist through amphibolite to granulite facies or whether the gneisses are retrograded granulites. Anorthosites and alkaline rocks are also locally abundant in the granulites (reviewed in Mahadevan, 2002). The anorthosite in the eastern part of the CGGC was emplaced during or after the second phase of deformation, since these intrusives lack F1 deformation, and possibly F2 (Bhattacharya and Mukherjee, 1984). The granulites in general record at least three main deformation events of which F1 and F2 are intense and coaxial and their axial planes trend E-W to NE-SW. The F3 is a cross fold and is represented by broad warps along N-S trending axes. Although gneisses, migmatites and high-grade pelitic schists are present in the CGGC, attention has been given mainly on the granulites in the CGGC terrain.

The granulites and high-grade gneisses are reported from metamorphic belts located on the north and south of the median Gondwana outcrops that were deposited in intracratonic rifts of the CGGC during Upper Palaeozoic time. The southern belt is named South Palamau-Gumla-Ranchi-Purlia belt and the northern belt is designated Daltonganj-Hazaribagh-Dhanbad-Dumka belt by Mahadevan (2002) who divided the CGGC rocks into five major divisions (Fig. 2.7), based on a broad lithological ensemble (see last section). Three broad centres of the granulite pods are (i) the Palamau district in west (Ghose, 1965); (ii) the Purulia-Bankura region in the SE (Sen, 1967); and (iii) the Dumka-Mayurakhsi Valley in the NE (Bhattacharya, 1976; Ray Barman et al., 1994). Of these, only the granulites from Purulia-Bankura are located in the southern belt of Mahadevan. Other minor occurrences of granulitic rocks in the CGGC are to the east of Ranchi (Sarkar and Jha, 1985), between Parasnath and Madhupur.

2.5 Chhotanagpur Granite-Gneiss Complex

According to Acharyya (2003), the More Valley granulite terrain to the west of Dumka is of igneous parentage with very minor supracrustals showing metamorphic imprints. Euhedral zircon from the high-grade gneiss yielded U–Pb upper intercept age of 1624 ± 5 Ma while from massive charnockites in the vicinity the upper intercept age is 1515 ± 5 Ma. In both cases the lower intercept yielded ca. 1000 Ma age, suggesting a thermal overprinting on both gneiss and charnockite. The Upper intercept age signifies the time of crystallization of hypersthene gneiss at high temperature and medium pressure or midcrustal depth (Acharyya, 2003). Ray Barman et al. (1994, cited in Acharyya, 2003) believe that these rocks cooled isobarically in a P-T regime of around 750°C at or above the pressure of 4–7 kbar. Rounded zircons separated from basic granulites that occur as sills and dykes within the hypersthene gneiss yielded 1515 Ma age (Acharyya, 2003, p. 16), suggesting that the hypersthene gneiss and intrusive basic granulites were metamorphosed together at 1.5 Ga and the whole sequence then cooled isobarically. The granitic intrusions of ca. 1000 Ma age indicate Grenville thermal event documented by the lower intercept of the discordia. To the east, the granulites from Dumka also show an isobaric cooling. The age of 1457 Ma obtained by Rb–Sr systematics of syenite in the Dumka sector is taken to indicate the time of anatexis of paragneisses (Ray Barman et al., 1994). Well foliated granitoid from the Ranchi Shear zone containing granulite facies supracrustals yielded Rb–Sr whole-rock isochron ages of 1.7–1.5 Ga (Sarkar et al., 1988).

The granulites from the Purulia-Raghunathpur area in the southern domain of the CGGC are reported to have evolved through ITD (isothermal decompression) path at ca. 1000 Ma. This was probably synchronous with the amphibolite grade migmatization. The granite plutonism, which is widespread during 1.0 Ga, is believed to have affected both granulite facies domains of the CGGC (cf. Ray Barman et al., 1994). The age difference of ca. 100 Ma in the granulites from the north and south of Ranchi shear zone perhaps indicate cooling at different crustal levels rather than to show different evolutionary history (see also Ray Barman et al., 1994).

In Saltora area, Bankura district, Manna and Sen (1974) documented a near-isobaric cooling in a suite of mafic granulites from 950 to 750°C at 8 kbar. Bhattacharya and Mukherjee (1984), on the other hand, proposed a prograde history involving breakdown of hornblende followed by cooling and retrogression to amphibolite facies conditions at 600°C/6.25 kbar. Again, Sen and Bhattacharya (1993) retrieved highest T (820–840°C at 7.5 kbar) for medium scapolite-calcite-plagioclase equilibria in wollastonite-calcite-plagioclase-garnet-quartz assemblage in calc-granulites from the same locality. They also obtained lower temperatures (600–680°C) at 6–7 kbar, estimated from Fe–Mg exchange thermometer, which is attributed to retrograde cooling that led to the formation of garnet corona at the interface of hypersthene-plagioclase (Sen and Bhattacharya, 1986). According to these authors, further cooling resulted in the stabilization of mantling pyroxene in mafic granulites. Using different petrological constraints, Sen and Bhattacharya (1993) computed the peak isobaric cooling up to 680°C and 6.8 kbar. Subsequently, infiltration of mixed $H_2O–CO_2$ at 500°C/5 kbar developed muscovite, zoisite, and calcite in different proportions in rocks of the area. Sen and his coworkers (Sen and Bhattacharya, 1986) also reported anorthosites that were emplaced during second phase of deformation along the core of doubly plunging folds.

In granulites around Dumka, a near isobaric cooling path was deduced by Ray Barman et al. (1994), based on coronal garnet at the hypersthene-plagioclase interface, similar to that described by Sen and his coworkers from Bankura area in the SE. Ray Barman et al. (1994) also documented a short decompress phase wherein garnet had developed symplectites of hypersthene + plagioclase. These authors also noted the occurrence of hypersthene-sillimanite in the enclaves of granulites from south of Ranchi and Daltonganj area in North belt, which is interpreted by them to be a product of ultra high temperature (UHT) metamorphism. Thus, the granulites from both belts, on either side of the Gondwana Formations, show similar tectonothermal history. However, relationship of the granulite facies assemblages and migmatization in the rocks of the N-belt are poorly known. Ray Barman et al. (1994) consider the isothermal decompressive path at 1.0 Ga for Purulia granulites whereas an isobaric cooling path at 1.5 Ga for Dumka granulites.

The CGGC is also characterized by more than one generation of mafic intrusives, mostly dolerite and gabbros to norites in which corona structure is often noticed at several places, especially in Daltonganj, Dumka and Purulia districts. These metaigneous rocks are generally concordant with the foliation of the host gneisses. Age-wise these metaigneous bodies may be older than 1600 Ma. Geochemically, these metabasics are different, tholeiitic in granulitic rocks while ORB (ocean ridge basalt) in non-granulitic rocks. According to Murthy (1958), the presence of these large-scale coronets in the CGGC implies a major thermal episode in the evolution of the CGGC terrain. These coronites are believed to have been emplaced at midcrustal levels during a distensional stage (Mahadevan, 2002, p. 273). There are also ultramafic enclaves in the granulitic rocks, especially at Ardra (Purulia district) and Dumka, and a possible linkage is suggested with the noritic rocks generated in the subcrustal layers (Mahadevan, ibid.).

In complexly deformed and metamorphosed rocks of the CGGC, one cannot ascertain geological relationship amongst different lithounits, especially when the complex was overprinted by more than one tectono-thermal event (i.e. orogeny). Geochronological data on the rocks of the CGGC are meager and the three age clusters of K—Ar dates have been taken to indicate three cycles of orogenies. A "Simultala orogeny" was proposed to cover ages in the range of 1246–1416 Ma, obtained in the NE part of the CGGC; the time band of 850–1086 Ma from Ranchi-Gaya areas was assigned to the Satpura orogeny; and ages of 358–420 Ma obtained in the extreme NE part of the CGGC were assigned to Monghyr orogeny (cf. Mahadevan, 2002). Newer dates by Rb-Sr systematics on granitic rocks form Bihar Mica belt and charnockitic rocks from Dumka (belt III of Mahadevan, 2002) gave ages in the range 1000–1600 Ma, pointing to the impact of Satpura orogeny. However, age data based on Rb—Sr methods are susceptible to resetting at lower grades and by post-magmatic alteration, hence they tend to record younger ages. Also, a weak Pan-African thermal event (ca. 590–595 Ma) is recorded in the fission-track dating of mica from the Bihar Mica belt (Nand Lal et al., 1976). The U—Pb data on zircons

2.5 Chhotanagpur Granite-Gneiss Complex

from massive charnockite yielded 1625 and 1515 Ma (Mesoproterozoic) while its recrystallization ages are at 1071–1178 Ma, indicating Grenville orogeny. Acharyya (2003) and Ray Barman et al. (1994) tentatively correlated the near isobaric cooling event at ca. 1500 Ma and the near-isothermal decompression at ca. 1000 Ma. This time span is recorded in the granitic rocks that are characterized by high initial Sr ratios (Ray Barman et al., 1994). Considering the results of careful field studies, it can be concluded that the charnockites are intrusive into the basement rocks, namely the khondalite, basic granulite, calc-granulite etc. (Mahadevan, 2002). The intrusive granulites with igneous texture have acquired foliation due to deformation imposed on them subsequent to their emplacement. Alternatively, it is also possible that the scattered charnockites occurring amidst high-grade gneisses and migmatites owe their origin to in-situ charnockitization brought about by dehydration melting or by influx of carbon dioxide in the pre-existing gneisses and metasediments. This means that the granulite facies rocks could be as old as the associated gneisses of the CGGC and hence Mesoproterozoic or perhaps older (Late Archaean). However, no information is available about the precursor material either of the Mesoproterozoic charnockite or of granite that intruded the basement. Ghose (1983) and Banerjee (1991) suggested that the charnockite-khondalite-granulite and associated tonalitic gneisses represent the basement complex of Archaean age. Supracrustals to this basement complex are pelitic schists, paragneisses, calc-silicates and marbles—all grouped by the stated authors under *Older metasediments* (equivalent to Singhbhum Group of 2600 Ma age). According to these authors, both basement and cover were metamorphosed up to granulite facies with emplacement of plutonic rocks, such as anorthosites, gabbros, and granitoids and also pegmatites, most probably during Grenville orogeny. Later, igneous activities included intrusion of basic and granitic rocks and extrusion of Rajmahal basalts.

Considering the paucity of reliable geochroniolgical data, Mahadevan (2002) divided the CGGC into five major divisions or belts, based on broad lithological ensemble. These east-west trending metamorphic belts from S to N are:

1. South Palamau-Gumla-Ranchi-Purulia Belt.
2. Daltonganj-(North Palamau)-Hazaribagh Belt
3. North Garhwa-Chatra-Girdih-Deogarh-Dumka Belt.
4. The Bihar Mica Belt
5. Rajgir-Kharagpur Belt.

The rocks of Belt-1 show a progressive change from greenschist facies in the south, through amphibolite facies to granulite facies near Ardra and beyond. The complete Barrovian sequence ending up with charnockitic rocks is seen only in the Purulia district of West Bengal. Charnockites and anorthosites are not found westwards.

The Belt II exposes high-grade rocks, particularly granulites that occur interbanded with granites and gneisses and form concordant bands with metasediments. On the NE extension this belt merges with the E-W- striking Bihar Mica Belt (BMB). Here the most dominant rock is migmatite gneiss amidst biotite-sillimanite gneiss with or without garnet and muscovite, suggesting temperatures in excess of muscovite-quartz stability. The muscovite-free bands are mostly garnet-sillimanite gneiss (khondalite).

The Belt III contains isolated pockets of granulite facies rocks, occurring within felsic gneisses. It is not clear whether or not the whole or some parts of the felsic gneisses ever reached granulite facies conditions.

From the preceding account, one can suspect that rocks older than 1.7 Ga may have been present in the CGGC, but no systematic dating techniques have been applied to the rocks of the CGGC. Surely, U–Pb dating of individual zircons is a powerful tool in geochronology, but the high U content affects the accuracy of $^{206}Pb/^{238}U$, and not the $^{207}Pb/^{206}Pb$ ages (Williams and Hergt, 2000). Again, age data based on Rb/Sr method are susceptible to resetting by post-magmatic alteration or shearing and hence they tend to record younger ages. If the Paleoproterozoic Singhbhum mobile belt is the result of convergence of the Singhbhum craton in the south and CGGC in the north, we should expect Archaean ages in some domains of the CGGC, which could be cratonic nucleus of the complex, similar to the Archaean nucleus of the Singhbhum craton (cf. Ghose, 1992).

The Belt IV, namely the Bihar Mica Belt (BMB), is a sequence of arenaceous and pelitic rocks, interbanded with hornblende schists or amphibolites. These rocks have been intruded by large granitic bodies followed by younger basic dykes. This writer thinks that these granitic rocks with high initial Sr ratio (Pandey et al., 1986) could be the result of crustal melting of the overriding block (CGGC, including BMB) at depth, as a result of northward subduction of the Singhbhum block (cf. Saha et al., 1987). Mica schists, minor amphibolites, micaceous quartzites, and minor calc-granulites characterize the Bihar Mica Belt (BMB). Large phacolithic granitic bodies with a rim of migmatites or injection gneisses that show lit-par-lit injection of granitic material intrude the schistose rocks. The main schistose formations that host the mica-pegmatites are feldspathized and veined by more than one generation of pegmatites (Mallik, 1993). According to Mahadevan (2002) there is a conglomeratic horizon between the CGG and the overlying BMB. If this so-called unconformity happens to be tectonic breccia, a most likely possibility, the BMB belongs to the CGGC and not a separate stratigraphic unit. The BMB rocks show mineral assemblages of upper amphibolite facies appearing as a progressive grade from the underlying granulite facies rocks of the CGGC. This together with the documentation of similarity of three fold phases in the BMB and the underlying CGGC, support the contention that the BMB is a part of the CGGC. Pb–U and Pb isochron ages of several minerals from BMB give consistent ages of ca. 900–1000 Ma (Holmes 1950; Sarkar 1968; Ghose et al., 1973). Again, the granite intrusions in BMB as well as CGGC granulite facies rocks are post-F1 and perhaps also F2 as at some places they escaped F2 overprinting. Lastly, the proposition of Kumar et al. (1985) that the Tamar-Porapahar-Khatra fault (TPKF), referred to by them as Northern shear and also as the South Purulia shear zone defining the southern boundary of the CGGC, is not tenable. This is because a metamorphic belt of CGGC also occurs to the south of this fault, which is designated as South-Palamau-Gumla-Ranchi-Purulia belt (Mahadevan, 2002, pp. 258–259). This fault appears only to delimit the Gondwana outcrops within the CGGC.

Metamorphism in the vast terrain of CGGC reached middle to upper amphibolite facies. The P-T conditions have been deduced to be above 680°C (Bhattacharya, 1988). Under these high-grade conditions the pelitic rocks may have undergone partial melting to give rise to migmatites and anatectic granites. The sporadic occurrence of andalusite and cordierite may be the outcome of low-pressure metamorphism associated with granite intrusions in some areas of the CGGC. Bhattacharya (1988) indicated that the granites of the BMB originated at depth and were emplaced as structurally controlled plutons during F2 folding.

In contrast to the high-grade rocks, there is lower grade schist- phyllite-quartzite association at Rajgir and Munger (Sarkar and Basu Mallik, 1982). In Manbazar, Purulia district there is a report of dendrite rocks interlayered with BIF, mica schists and phyllite (cf. Bhattacharya, 1988).

2.5.1 Archaean Status of CGGC Questioned

From what has been discussed already, the CGGC appears to be an old Precambrian unit and perhaps as old as, or somewhat younger than, the Singhbhum-Orissa craton. However, paucity of geochronological data and problem of resetting of ages, the CGGC does not have proven Archaean-age rocks (Sarkar, 1982, 1988). This has provoked some geologists to question the Precambrian stratigraphic status of the CGGC. On the basis of structural similarity between CGGC and Singhbhum fold belt (Bhattacharya et al., 1990; Bhattacharya and Ghosal, 1992), some workers considered the two metamorphic domains, namely the CGGC and the Singhbhum fold belt (SFB) as equivalent in age. In short, the CGGC is considered as an eastern estension of the Satputra fold belt or Central Indian Tectonic Zone. Dasgupta (2004) states that the CGGC terrain contains comparable geological milieu and tectonothermal events as in the Satpura fold belt.

Following the above view, some authors (Bhattacharya et al., 1990; Dasgupta, 2004; see also Misra, 2006) think that the boundary between the two rock-units is neither any sharp lithological boundary nor a stratigraphic break, but a continuation of deformation and metamorphism. Again, based on geochronological data, Misra (2006) proposed that crustal growth in the eastern Indian region (Singhbum-Orissa) advanced from south to north and that the CGGC is younger than the Singhbhum fold belt. According to him (Misra, loc. cit) the sedimentary precursors of the CGGC are considered to have deposited in shelf region, after the deposition of the Singhbhum Group rocks in the rifted basin north of the Singhbhum (-Orissa) craton. These considerations received further support from the available radiometric ages which show that magmatic-metamorphic events in the CGGC occurred at $\sim$2.3 Ga, 1.6–1.5 Ga, and 0.9 Ga (see Misra, 2006). Again, in view of the dominant calc-alkaline gneisses amidst the CGGC rock complex and their interleaving with the high-grade pelites and calc-magnesian metasediments, some workers proposed that the CGGC was a magmatic arc (cf. Bose, 2000).

The proposal of CGGC to be younger than or as old as the fold belt is untenable on several grounds. The CGGC is a polymetamorphic complex whereas the rocks of the Singhbhum fold belt are only once metamorphosed (monometamorphic).

This finds support from the work of Singh (1998, 2001) who showed that the CGGC records two cycles of metamorphism-deformation. The first cycle ended at 1.6 Ga during which enclaves of granulites and mafic/ultramaficl rocks are formed with granite intrusions (cf. Ghose et al., 1973). The second cycle is considered to begin with Kodarma Group (conglomerate, quartzite, metapelites, amphibolites and BIF), followed by intrusion of S-type granite which have Rb-Sr ages in the range of 1.2–1.0 Ga (Mallik, 1998; Misra and Dey, 2002). These observations do not support the proposition of time equivalency of the CGGC and SFB. Since the Singhbhum orogenic belt and its early mesoscopic folds have E-W trend, it is implied that N-S compressive forces must have been generated by N-S collision of the Singhbhum (-Orissa) craton against a crustal block to the north, which seems to be none other than the Precambrian massif of CGGC lying just north of the Archaean craton of Singhbhum. Northward subduction of the Singhbhum (-Orissa) craton under the northern block requires collision with the CGGC in all evolutionary models of Singhbhum fold belt, discussed in Chap. 6. In this process the subduction and collision resulted into thickening of the crust that must have generated granitic melt (by partial melting), emplaced as granite intrusions in the overriding plate. This overriding plate was most likely the CGGC, which shows several granitic intrusions and high-grade rocks, all showing resetting of isotopic ages. Although unproven geochronologically, the CGGC should have been a Precambrian crustal block that needed to collide with the Singhbhum (-Orissa) craton in the south to develop the Singhbhum fold belt during Proterozoic.

The proposition that the CGGC is a fold belt and not a craton raises two serious geological implications. First, it needs to explain as to how the two folds could be generated in a small time interval and how were they brought as adjacent terrains of Mesoproterozoic age. The contact between the North Singhbhum fold Belt (SFB) and CGGC is a brittle to ductile shear zone called South Purlia shear zone or Tamar Porapahar shear zone. Second, if the CGGC were also a Proterozoic fold belt like the NSFB, one need to explain the absence of a crustal block that had collided against the Singhbhum craton more than once, first to generate NSFB and then the CGGC. The author recalls that the CGGC has gneisses and granites of variable chemistry and ages, besides granulites that occur as discontinuous bands. In addition, the CGGC also contains tracts of medium-grade enclaves of metasediments and basic rocks. These characteristics are not of a fold belt, and therefore distinguish the CGGC as a typical granite-gneiss massif or a craton that collided against the Singhbhum craton to evolve the Singhbhum fold belt (NSFB) (see Chap. 6). Therefore the hypothesis of some authors that the CGGC is an eastward strike extension of the Satpura fold belt of central India, is unacceptable in view of the cratonic nature of the CGGC vis-à-vis the mobile belt of the Satpura. It would be interesting if SHRIMP U–Pb zircon geochronology could be performed on the rocks of the CGGC in a more systematic way so that a crustal evolutionary model could be attempted, as done in the Dharwar, Singhbhum and other cratonic areas of the Indian shield. With the available geological database any attempt on geodynamic evolution of CGGC will be futile.

2.6 Rajasthan Craton

The Rajasthan craton (RC) is a collage of two cratonic blocks: (1) The Banded Gneissic Complex-Berach granite (BBC), and (2) the Bundelkhand Granite massif (BKC). Therefore RC is in fact a large Rajasthan-Bundelkhand craton These two cratonic blocks are separated by a vast tract of cover rocks, besides the occurrence of the Great Boundary Fault at the eastern limit of the BBC block, making the correlation between the two cratonic areas difficult (see inset Fig. 2.8). However, the two

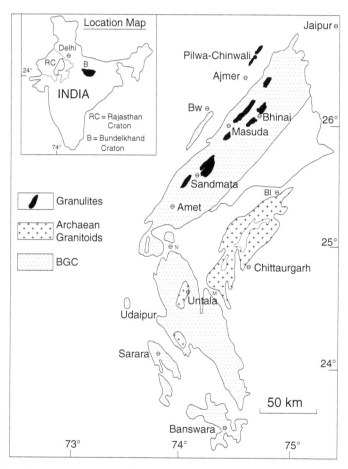

Fig. 2.8 Simplified geological map of Rajasthan craton (after Heron, 1953 and GSI, 1969), made up of Banded Gneissic Complex (BGC), Berach Granite and other Archaean granitoids. Granulite outcrops are in the BGC terrain and in the metasediments of the Delhi Supergroup (see text for details). Blank area occupied by Proterozoic fold belts and sand cover. Abbreviations: BL = Bhilwara, BW = Beawar, N = Nathdwara, M = Mangalwar. Inset shows the location of BBC (Banded gneissic complex-Berach Granite) and BKC (Bundelkhand) cratonic blocks that together constitute what is here termed the Rajasthan (-Bundelkhand) Craton, abbreviated RC

blocks have a common lithology that includes gneisses, migmatites, metavolcanic and metasedimentary rocks and a number of granitic intrusions. Both the BBC (i.e. Banded Gneissic Complex-Bearch Granite) and the BKC (i.e. Bundelkhand Granite Complex) blocks (unitedly designated Rajasthan craton, RC) have been affected by similar deformational events (Naqvi and Rogers, 1987). The two blocks also share same geodynamic settings in Proterozoic as revealed by geochemistry of their mafic magmatic rocks (Mondal and Ahmad, 2001) and same geochronological ages (Mondal, 2003), to be discussed later.

Early geological traverses and their reports were by Hacket, Oldham, La Touche, Middlemiss, Coulson and Crookshank (reviewed in Heron, 1953). The foundation of geology of Rajasthan (formerly Rajputana) is due to B.C. Gupta (1934) and Heron (1953) who gave the following broad divisions of the Precambrian rocks of Rajasthan, with ages given from recent publications (see text).

Malani Rhyolites (730–750 Ma)
       Erinpura Granite (850–900 Ma)
Delhi   {Ajabgarh Series
System {Alwar Series
-------------unconformity------------
   Raialo Series
------- -----unconformity------------
   Aravalli System
------------unconformity-------------
Banded Gneissic Complex (3.4 Ga)

2.6.1 Banded Gneissic Complex and Berach Granite

Amongst the early field geologists, Heron (1953) was the first to erect the basic framework of the Precambrian geology of Rajasthan. He recognized a gneiss-granitoid ensemble as the Archaean basement over which Proterozoic cover sequences were deposited with a profound erosional unconformity. To this basement stratigraphic unit he gave the name Banded Gneissic Complex (BGC), which also included the Bundelkhand gneiss. Pascoe (1950) renamed the Bundelkhand gneiss as the Berach granite. The geological synonymity between the crescent-shaped Berach Granite near Chittaurgarh and the Bundelkhand Granite, about 210 km away in the east (Fig. 2.8) is supported by Rb/Sr isotope data (Crawford, 1970; Sivaraman and Odom, 1982) and petrological studies (see Mondal, 2003). The BGC is predominantly a polymetamorphosed, multideformed rock-suite of tonalite-trondhjemite (TT) gneiss, migmatite, granitoids (grey granodiorite to pink granite) and subordinate amphibolite. The latter occurs as bands and enclaves/rafts of various dimensions within the gneisses, producing banded gneiss and migmatites. Besides the complex TT-amphibolite association, the BGC of Rajasthan also shows minor metasediments, mainly quartzite, frequently fuchsite bearing, low-Mg marble, mica schists and metabasic rocks (as amphibolite or greenschist) and minor ultrabasic rocks (as hornblende schists/hornblendite)—all indicating a possible greenstone remnant in the BGC terrain (cf. Upadhyaya et al., 1992). During the Archaean tectono-thermal event when the metasediments and the TT-amphibolite protoliths were recrystallized, granites (generated by partial melting of crustal rocks) were emplaced at different crustal levels, leading to stabilization of the craton around 2.5 Ga ago.

2.6.2 BGC: Antiquity Challenged and Nomenclature Changed

Gupta (1934) proposed a two-fold classification of the Banded Gneissic Complex (BGC). According to him, the BGC in the southern and central parts of Mewar is an undoubted Archaean basement underlying the Proterozoic Aravalli rocks and is designated by him as BGC-I. The migmatitic-gneissic rocks from north of Mewar have debatable basement cover relationship and were called by him as BGC-II. The basement status of the BGC-I is generally accepted on the basis of geological setting and geochronological data (see below), but the basement status of the BGC-II is controversial (cf. Bose, 1992). A brief period of confusion prevailed regarding the antiquity of the BGC-II of Gupta (1934). This was due to recognition of a structural accordance between the BGC basement and Proterozoic cover (Aravalli Supergroup) and interpretation of conglomeratic horizon (a pronounced erosion unconformity of Heron, 1953) at Morchana in Rajsamand district appearing as "tectonic inclusions". These led some geologists to refute the separate stratigraphic status of the BGC of Heron, which was conceived by them as a migmatized Aravalli (Naha and Majumdar, 1971; Naha and Halyburton, 1974). However, in view of the ambiguous erosional unconformity between the BGC and the Aravalli cover at several places and ruling out the possibility of migmatization of the Aravalli rocks due to their low-grade assemblages in contrast to the polymetamorphic character of the infracrystalline BGC (Sharma, 1988), the basement character of the BGC-II was re-established. This is further supported by the occurrence of two types of greenstone sequences in the BGC-II. The older sequence (Sequence 1), called Sawadri, is reported to have komatiite, ultramafic, chert, basic tuff and carbonates (Sinha-Roy, et al., 1998), while the younger sequence (Sequence 2), called Tanwan Group, contains greywacke, amphibolite and carbonates (Mohanty and Guha, 1995).

Another uncertainty occurred in regard to the stratigraphic status of the BGC. The metasediments in the eastern region around Bhilwara, recognized by Heron as the Aravalli system (Proterozoic), were classed as Pre-Aravalli (known as Hindoli Group by the GSI, 1993, 1998; Gupta et al., 1997, 1980; Sinha-Roy, 1985) merely on the assumption that the granitic rocks (e.g. Jahazpur granite) correlatable with the Berach granite (2.5 Ga) were intrusive into them (Raja Rao, 1971). This led Raja Rao et al. (1971) to introduce a new lithostratigraphic term, the Bhilwara Group as the basement for the younger supracrustals of the Aravalli Supergroup. Subsequently, the Bhilwara Group was elevated to the status of the *Bhilwara Supergroup* (see Gupta et al., 1980) which comprises the 2.5 Ga old Berach Granite, the

Aravalli metasediments of Heron, presently classified by the GSI as the Hindoli Group (Gupta et al., 1980), and also the large outcrop of the BGC occurring to the east of Udaipur as well as the granulite complex of Sandmata. The idea behind this grouping was the assumption by Raja Rao et al. (1970) that these metasediments were older than those occurring around Udaipur and Nathdwara. However, the structural as well as physical continuity of the Aravalli supracrustals of the type area around Udaipur into the metasedimentary belt of the Bhilwara region (Roy et al., 1981) and non-intrusive nature of the 2.5 Ga-old Berach granite, the new nomenclature of the Hindoli basement is not preferred over the BGC of Heron.

One more controversy arose when Gupta et al. (1980) of GSI classified the BGC rocks (of Heron) occurring east of the Aravalli axis into Mangalwar Complex and Sandmata Complex, and when both the complexes and the rocks of the Aravalli Sysem of Heron (1953) exposed around Bhilwara (re-named as Hindoli Group by the GSI, Gupta et al., 1980) were grouped under the Bhilwara Supergroup. *The Mangalwar Complex* is essentially a migmatitic complex while the *Sandmata Complex* is a granulite facies rock-suite, and the *Hindoli Group* as low to medium grade supracrustals. One reason that led the GSI workers to group the Sandmata Complex, Mangalwar Complex and the Hindoli Group under one Supergroup (the Bhilwara Supergroup) is the observation that the grade of metamorphism increases from Hindoli (low grade) in the east through Managalwar complex (medium grade) to high-grade of Sandmata granulites. But the present geochronological data indicate that all these rock units have not recrystallized in the same orogeny. The two complexes (Sandmata and Mangalwar) are separated by a tectonic lineament, the Delaware lineament. In a partial modification of this classification, Guha and Bhattacharya (1995) and Sinha-Roy et al. (1992) have restricted the nomenclature of the Sandmata Complex to the shear-bound high-grade granulite facies rocks within the amphibolite facies rocks that are included into the Mangalwar Complex. To combine the Sandmata complex, Mangalwar Complex and Hindoli Group under the Bhilwara Supergroup rests on the uncritical observation of increasing metamorphic grade westwards. The erection of these new Groups, Supergroups and Formation names suggested by GSI geologists is unsound and unjustified particularly when the included rock units are of different ages and show different deformational and metamorphic history. The Hindoli Group is in contact with the 2.5 Ga old Berach Granite which is regarded by GSI as intrusive into Hindoli. Other workers find field evidence for the Granite to be a basement for the Hindoli (Sharma, 1988; Roy and Jakhar, 2002). Rhyodacite of Hindoli Group gave zircon Concordia age of 1854 ± 7 Ma, indicating that it is homotaxial with a part of the Aravalli sequence (see Ramakrishnan and Vaidyanadhan, 2008, p. 275). The divisions of the basement given by GSI (Gupta et al., 1980) are therefore not based on sound Geodata.

Still another confusion in the nomenclature of the basement rock-suite occurred when Sm/Nd isotopes indicated that the biotite gneiss of the BGC east of Udaipur was coeval with amphibolites that occur as conformable bodies within the former (Macdougall et al., 1983), and that their initial isotopic ratio $\epsilon_{juv}(T) = 3.5$ for these rocks is suggestive of their derivation from a depleted mantle source. This prompted Roy (1988, p. 16) to designate these pre-Aravalli BGC rocks as *Mewar*

Gneiss, which is conceived as a heterogeneous assemblages of biotite and hornblende gneisses, amphibolites, granitic rocks, aluminous paragneisses, quartzites, marbles and pegmatites. This change of nomenclature is unwarranted and unnecessary as the rock assemblage under Mewar Gneiss is the same as those described under the BGC. Moreover, this change in name of a well-established term defies the code of stratigraphic nomenclature and respect of priority of nomenclature. The tonalitic grey gneiss, included within ~2.5 Ga old granite of the BGC terrain, later yielded younger age of 3.3 Ga (Gopalan et al., 1990). This age, supported by the 3.28 Ga single zircon age ascertained by ion probe (Wiedenbeck and Goswami, 1994) again confirms the basement nature of the BGC. Somewhat younger ages of 3.23–2.89 Ga have been obtained from single zircon grains by evaporation method ($^{207}Pb/^{206}Pb$) for these gneisses (Roy and Kroener, 1996). Revelation of Archaean ages is definitely an undeniable evidence for the nucleus of the BGC in Rajasthan craton, but it does not warrant a new name, replacing a well-established stratigraphic term of the Banded Gneissic Complex given by Heron (1953), after a dedicated fieldwork of over thirty years. The basement complex cannot obviously be wholly Archaean because later magmatic activity affecting the cover rocks had to traverse through the infracrystalline of BGC. Heron (1953) was well aware of this and similar geological phenomenon on the basis of which he stated that the BGC represented a petrological ensemble of gneisses, granites and metasediments of various ages. The 3.3 Ga old gneisses of the BGC simply represent a nucleus of the Rajasthan craton. However, some BGC rocks as at Sarara, are not Archaean in age and their disturbed dates gave scattered ages (Proterozoic) but these BGC rocks are definitely overlain by the Palaeoproterozoic Aravalli supergroup, with an erosional unconformity between them (cf. Poddar, 1965). From this and similar occurrences of BGC inliers, we cannot consider only the isotopic ages but also the geological setting of the rocks (Naha and Mohanty, 1990).

From the above discussion this writer feels that the controversies on status and nomenclature of BGC are unnecessary and unwarranted as it only caused an avoidable confusion. As a mark of respect for priority of nomenclature and conforming to the code of stratigraphic nomenclature this author retains the term Banded Gneissic Complex as given by Heron (1953), even if it is taken as a backward step (see also Roy and Jakhar, 2002). In accordance with this "follow me" tendency the Archaean geology is also shown in the regional maps incorporated in this book (see also GSI, 1993).

2.6.3 Geological Setting

The BGC including the Berach Granite occupies a large tract in the Mewar plains (Udaipur region) of south and east Rajasthan. It is skirted on the west and southwest by Proterozoic fold belts of Aravalli and Delhi Supergroups. The eastern boundary of this cratonic region is demarcated by the Vindhyan platform sediments and southern boundary is covered by Deccan Trap (Fig. 2.8). The BGC cratonic region is dominantly gneissic to migmatitic with amphibolites and metasediments of amphibolite facies, intruded by Late Archaean granites (Untala, Gingla, Berach etc.) and rare ultramafics. Amongst the gneissic rocks, grey coloured biotite gneisses are dominant with leucocratic bands as a result of which the name Banded Gneissic Complex is appropriately given by Gupta (1934) and Heron (1953). One can observe a gradational contact between the biotite gneiss (quartz-feldspar-biotite ± hornblende ± garnet) to leucogranite (quartz-feldspar) with gradual obliteration of gneissic foliation. At certain places, faint relics of gneissic foliation are seen within dominantly massive granitoid. All these features resemble those described from Finland by Sederholm (1923), and are the outcome of partial melting of the crustal rocks when they were at great depths. Intrusion of trondhjemitic veins and pegmatites are not uncommon in the banded gneisses. Retrogression of dark minerals, biotite and hornblende, into chlorite and of K-feldspar into muscovite, and of plagioclase into zoisite/epidote are common. Geochemically, the 3.3 Ga old biotite-gneiss component of the BGC from Jhamakotra, east of Udaipur, is characterized by highly enriched LREE with a relatively weak Eu anomaly (Gopalan et al., 1990). Higher silica content and LREE enrichment of these gneisses suggest evolution of the gneisses through melting of a basic (amphibolite) precursor or, through melting of earlier tonalitic/trondhjemitic felsic rock-series. The latter proposition finds support from (1) the occurrence of Al–Mg rich metasedimentary enclaves in the tonalitic gneisses (Sharma and MacRae, 1981), and (2) an earlier report of 3.5 ± 0.2 Ga Pb–Pb age of detrital zircon in Proterozoic Aravalli schists (Vinagradov et al., 1964). These Archaean gneisses are orthogneisses and the Sm/Nd as well as evaporation ages may be taken to indicate as the minimum age for crystallization of the igneous protolith of the gneisses.

According to Heron, the BGC rocks also occur as inlier within the Delhi Supergroup (Mesoproterozoic) south of Aimer in central Rajasthan. The Ana Sagar granite gneiss near Ajmer is a possible extension of the inlier, since the former rocks are dated at 2.58 Ga (Tobisch et al., 1994).

The gneiss-amphibolite association of the BGC from most part of Rajasthan, particularly from central region, shows ample evidence of partial melting that gave rise to pods and bands or layers of quartzo-feldspathic material (leucosome) within the gneisses and hornblende-biotite schist/gneiss and amphibolites. Not infrequently the amphibolites occur as enclaves within the granite-gneiss milieu. The amphibolites (greenstones) of central Rajasthan have been divided into two litho-types, namely (i) the older Sawadri Group in which the amphibolites are associated with high-grade metasediments and granulites as at Sandmata, and (ii) the Tanwan Group in which the amphibolites lack ultra basics but are associated with fuchsite-quartzite and metasediments (cf. Mohanty and Guha, 1995). However, such a division of these amphibolites from the BGC terrain is undesired because intensity of deformation and metamorphism and degree of exposures could be responsible for this association.

Amphibolites in the BGC range in size from small enclaves of irregular shape to large linear bodies showing complex fold pattern due to superposed folding. These amphibolites are older component of the BGC. Chemically, these amphibolites show similarity with basaltic andesite and are characterized by slight LREE enrichment and virtually flat HREE, to suggest their derivation from shallower mantle rock (plagioclase peridotite or spinel peridotite). Metasediments in the BGC craton are greenish quartzites (fuchsite bearing), marble, calc-silicates and ironstones, and pelitic schists. A large body of quartzite at Wahiawa (west of Mavli) is complexly folded and intruded by ca. 2828 Ma old metabasalt (amphibolite).

The BGC to the north of Nathdwara in central Rajasthan also contains granulite facies rocks comprising garnet-sillimanite gneiss, with or without cordierite, calcsilicate gneiss, enderbite-charnockite, leptynite and two-pyroxene-garnet granulite (basic granulite). This granulite-gneiss association is referred to by the name Sandmata Complex by Gupta et al. (1980). The granulites occur as isolated exposures from north of Amet in the south to Bhinai and Bandnwara in the north. Besides these rocks, granite, norite, pegmatite and quartz veins are also found as later intrusives.

It is a matter of enquiry whether these granulites are Archaean or Proterozoic in age, although B.C. Gupta (1934), a coworker of Heron, mapped these rocks under the BGC. Surely, the parent of the pelitic granulites is pelitic and the two-pyroxene granulite is obviously derived from basaltic parentage. Both these granulite types are undated, but the enderbite-charnockite occurring as intrusion in the Sandmata complex yielded 1723 Ma age by U—Pb zircon geochronology (Sarkar et al., 1989), suggesting that the granulite facies event is Mesoproterozoic (Sharma, 1995, 2003). As is described later, the metapelites and other rock units of the BGC were subjected to high temperature recrystallization (granulite facies metamorphism) as a result of underplating by basaltic magma at depth. The two-pyroxene basic granulites occurring within the Sandmata granulite complex and elsewhere in central Rajasthan, and perhaps also the norite dykes, are possible representative of this underplated/intraplated magma.

There are no age data to constrain the Archaean event of deformation and recrystallization during which the igneous tonalite rock developed the gneissic fabric and the associated sediments and basaltic supracrustals were transformed to paragneiss and amphibolite, respectively. This metamorphic event would be expressed in the development of distinctive regional metamorphic minerals, such as garnet, and also in the generation of melt phase at depth, resulting into the intrusion of granitic plutons that are recognized as Gingla granite, Untala granite etc in the BGC terrain. A two-point garnet-whole rock Sm/Nd age gave 2.45 Ga age for this mineral from a 3.3 Ga old biotite gneiss (Gopalan et al., 1990). It seems that the 2.45 Ga age of garnet is not a meaningful age, since this date is not obtained from any other mineral or rocks from the BGC. The leucocratic granite gneiss from Jagat area, showing retrograde features, yielded a zircon evaporation age of 2887 ± 5 Ma (Roy and Kroener, 1996), and an amphibolite from Mavli showing cofolding with the granite gneiss gave Sm/Nd whole-rock isochron age of 2828 ± 46 Ma (Gopalan et al., 1990). These ages may be taken to indicate the age of crystallization/emplacement of anatectic melt generated at depth during Archaean orogenesis. The main Archaean metamorphic event (pre-Aravalli) affecting the TTG-amhibolite protoliths could therefore be earlier than 2.85 Ga. This tectonothermal event was probably also responsible for the disturbance of Sm/Nd systematics in the tonalitic-trondhjemitic gneisses of the BGC (see Tobisch et al., 1994).

A second thermal event at about 2.6 Ga is also recorded in the BGC terrain as is evident from U/Pb isotopic ages for Untala and Gingla granitoids. The Untala granite is pink, medium-grained mostly massive K-feldspar bearing granite which grades on the margin into gneissic variety of granodioritic to trondhjemitic composition. The pink Untala granite is considered as the youngest phase of granite intrusion in the BGC. The Gingla granite shows variation from granodiorite to trondhjemite and even quartz-diorite. It contains a number of inclusions of amphibolite and biotite gneiss. Roy and Kroener (1996) give single zircon evaporation ages for these granites: Trondhjemite gneiss from Untala = 2666 ± 6 Ma; Low Al_2O_3 granite from Jagat = 2658 ± 2 Ma; and Leucogranitoid of Gingla = 2620 ± 5 Ma.

Outside the main BGC outcrop, there are two large bodies of granitic inlier. These are the Berach Granite near Chittaurgarh and the Ahar River Granite near Udaipur. A characteristic feature of these granites is their coarse-grained porphyritic texture, without any conspicuous foliation. The Berach granite covers an extensive area to the west of Chittaurgarh in eastern Rajasthan (Fig. 2.8). Based on identical mineral composition and physical character, Gupta (1934) correlated this granite with the Bundelkhand gneiss of central India, and claimed that the Berach Granite served as a basement rock upon which the Aravalli rocks were deposited. As discussed later, this proposition also finds support from Rb/Sr isotopic age of 2555 ± 55 Ma for the Bundelkhand gneiss and for the Berach Granite (Sivaraman and Odom, 1982; Choudhary et al., 1984; Wiedenbeck et al., 1996a, 1996b), although the two granite blocks are separated over 400 km by the Vindyhan sediments. Recent zircon evaporation ages (Roy and Kroener, 1996) also confirm the Archaean age for the granite inliers. These granitoids were interpreted by Sharma (1999) to have derived by partial melting of already existing rocks of the BGC (orthogneisses and metasediments). With the emplacement of these granitoids within and outside the BGC terrain, the Rajasthan craton was stabilized. There are a few more Late Archaean granites, for example, the Gingla Granite close to Jaisamand Lake (Dhebar lake), giving 2.8–2.6 Ga age; Bagdunda dome which is an elongate gneissic dome found within Jharol Group rocks of the Aravalli Supergroup. As a basement, the dome is rimmed by quartzite-metabasalt association of the basal Aravalli sequence.

Metamorphic studies of the BGC rocks show that they are generally recrystallized in upper amphibolite facies in most places, but polymetamorphic character of the rocks, mostly of pelitic compositions, does not allow delineation of metamorphic isograds. Sharma (1988) clearly showed that the BGC rocks suffered two metamorphic events: M1 (∼3.0 Ga) reached staurolite-kyanite zone in most parts including the Mewar region, which also shows lower grade (epidote-amphibolite facies) while M2 (∼1.6 ± 0.2) attained varying degree of metamorphism from upper amphibolite to granulite facies. However, a following shearing event retrograded these amphibolite facies rocks to lower grade mineralogy in which chlorite is the most common mineral after biotite, garnet and hornblende in different rock compositions.

Deformation in the BGC rocks is heterogenous, making correlations between different domains difficult. Structural features and deformation patterns of the BGC often resemble those in the Proterozoic cover. However, an early folding episode and related planar and linear structures distinguish the BGC terrain from the Proterozoic cover. In the northern part the deformation of BGC shows interesting features in the Khetri region where the Delhi Supergroup (Proterozoic) rests unconformably on the BGC schist-gneiss. In these gneissic rocks two prominent foliations belonging to two different episodes superimposed on the relicts of an earlier (Pre-Delhi) gneissosity/schistosity. The first deformation resulted in transposition of an already existing foliation (s-surface). Subsequent deformation of the BGC is of brittle to brittle-ductile nature, resulting in the development of regional NE-SW trending foliation as well as NW-SE striking cross faults and folds.

Moving towards south of the Khetri belt, the BGC shows large-scale folds (F3) that are coaxial with the F2 folds (developing Type 3 interference pattern of Ramsay, 1967) and control the map pattern of the BGC. Associated with the F3 are the ductile shear zones that mostly occur in conjugate sets. Later shears of brittle nature overprinted the early shears and also caused dislocation.

In central Rajasthan three fold phases (F1, F2, F3) have been identified in the BGC whose map pattern is controlled by large-scale F2 folds. The F1 and F2 folds are isoclinal, often with reclined geometry and having dominantly E-W axial plane foliation. Both F1 and F2 isoclinal folds are affected by NE-SW trending upright open folds (F3). The superposition of these later folds on coaxially folded isoclinal folds has resulted in dome and basin structures. Broad transverse warps represent the last phase (F4) of folding (see Naha et al., 1967).

Progressive deformation in the high grade BGC has been studied by Srivastava et al. (1995), particularly in Masuda-Begaliyawas, and Bandanwara area, lying to the SSW of Ajmer and NNE of Sandmata. Based on the overprinting relationship these authors recognized two groups of folds: F1 group folds comprising three sets of successively developed folds F_{1A}, F_{1B} and F_{1C}, and F2 group folds. In F_{1A} folds, the fold closure is cut across by gneissic foliation. The F_{1B} folds refolded F_{1A} folds, producing Type-3 interference pattern of Ramsay (1967). In some Type-3 interference, the early folds do not have an axial plane foliation, and they are attributed to refolding of F_{1B} by another phase of coaxial F_{1C} folds. The F2 group folds have refolded F1 axial planes and deformed the early lineation around F2 axis. Two more sets of folds have been identified by Srivastava et al. (1995). F3 folds are characterized by recumbent geometry and F4 group folds have steep axial planes and occur as transverse structures with axial planes perpendicular to the regional NE-SW or NNE-SSW striking foliation. Along the axial planes of F1 and F2 there are also shear zones in the BGC rocks. These shear zones are oblique slip type with large strike-slip component and show both dextral and sinistral sense of movement. However, slip lineation on the shear surfaces is reportedly absent (Srivastava et al., 1995).

In Sandmata area, three deformation phases have been recognized (Gupta and Rai Choudhuri, 2002) which produced three sets of closely related planar structures. Strain partitioning has been noted in the development of planar structures with varied intensity in different parts of the area. These authors also noticed that granulite facies rocks were overlying the amphibolite facies rocks without structural discontinuity, and this superposition is interpreted by them due to overfolding (see Gupta and Rai Choudhuri, 2002), indicated by the interrelationship between S1 and S2 foliation planes. Conjugate shear zones associated with F3 are recognized by these authors and by Joshi et al. (1993). From geothermometry and geobarometry, Joshi et al. (1993) gave temperatures in the range 850–600°C and pressures from about 11 to 4.5 kbar. Somewhat lower P but comparable temperatures are also obtained by Dasgupta et al. (1997) for these granulites. From these P-T data and microstrucutral criteria, Joshi et al. (1993.) deduced counter-clockwise P-T path for the Sandmata granulites. Gupta and Rai Choudhuri (2002) noticed that the shear zones showed variations in the intensity of development of s-c fabrics that have steep or vertical dips, but the shear sense indicators suggest oblique slip with dominant strike-slip component. The ductile shearing is considered responsible for the variation in the plunges of fold axis and for excavating these granulites from deeper levels to shallower depths (Sharma, 1993).

2.6.4 Evolution of the Banded Gneissic Complex

The evolution of both units of Rajasthan craton are separately described, taking first the Banded Gneissic Complex (BGC) and then Bundelkhand Granite massif. The evolution of the BGC is described below and summarized in Table 2.1 (modified after Sharma, 1999).

From the occurrence of folded enclaves of amphibolites within the 3.3 Ga old biotite gneisses it is inferred that the earliest deformation in the BGC was definitely ductile, some time during Mid to Late Archaean. These early amphibolites are comparable to type 1 amphibolite from Jagat area, described by Upadhyaya et al. (1992). These, together with the associated talc-antigorite schist as metamorphosed ultramafics along with some metasediments like the paragneisses and marble, represent an early greenstone association. Although the oldest rock so far dated in Rajasthan is the 3.3 Ga old banded gneiss (biotite gneiss-amphibolite) from Jhamakotra area (Gopalan et al., 1990), supported by the ca. 3230 and 3281 Ma single zircon ages obtained respectively by evaporation method (Roy and Kroener, 1996) and ion microprobe (Wiedenbeck and Goswami, 1994), still older crust in Rajasthan is indicated by the report of 3.5 Ga age for the detrital zircon from the cover of the Aravalli metasediments (Vinagradov et al., 1964). Another mafic activity is indicated by the 2.83 Ga old metabasalt/amphibolite (Gopalan et al., 1990) that intrudes the quartzites and biotite gneisses at Rakhiawal (Roy et al., 2000). This lithological association indicates local rifting of the BGC rocks of the Rajasthan craton during Late Archaean. The rifted basin with its rock deposits was subsequently deformed and metamorphosed during which the 2.9 and 2.6 Ga old granitoid phases of Untala and Gingla were intruded within the BGC. With the emplacement of typical Potassic granites of 2.5 Ga age, notably the Berach Granite and Ahar River Granite, the Rajasthan craton was finally stabilized (see Table 2.2).

In the Archaean craton of Rajasthan, two magmatic activities of Proterozoic age are also recorded. First is the carbonatite intrusion in the 2.5 Ga Untala granite at Newania and second is the intrusion of 1550 Ma old granites (Bose, 1992) that occur in a series of outcrops at Tekan, Kanor and several other localities within the

Table 2.2 Evolution of Rajasthan craton with summary of events, modified after Sharma (1999)

Age (Ga)	Geological events	Remarks
>3.5	1. Basaltic and tonalitic liquid extruded and intruded to form bimodal suite, making up the early crust in Rajasthan	3.5 Ga old detrital zircon from early Proterozoic Aravalli schist (Vinagradov et al., 1964)
	2. Deformation of the bimodal suite, folding and gneissic foliation of the mafic and felsic components	Evidenced by folded enclaves of old amphibolites (greenstone 1) within 3.3 Ga old biotite gneisses (Roy and Jakhar, 2002, p. 60)
3.3	Synkinematic melt generation to form second stage TT rocks from 3.5 Ga old orthogneisses and amphibolites (greenstone 1) at deeper levels of thickened crust	Occurrence of 3.3 Ga old biotite gneisses/banded gneisses as at Jagat (Gopalan et al., 1990; Upadhyaya et al., 1992)
2.9	Rifting and basin sediments subsequently deformed and metamorphosed with decompression melting of upper mantle	Folded quartzite-biotite gneiss at Rakhiawal intruded by metabasic (amphibolite 2) and dismembering of older greenstones as at Jagat (Upadhyaya et al., 1992)
	Archaean gneissic complex originated	
2.83	Crustal dialation	Emplacement of mafic dykes e.g. at Mavli (Gopalan et al., 1990)
2.65–2.5	Decompression melting of orthogneisses and paragneisses and also amphibolites at depth to produce felsic melts to intrude upwards	Granite intrusion of Gingla, Untala, Berach, and Ahar River (Roy and Kroener, 1996; Wiedenbeck et al., 1996b; Choudhary et al., 1984)
	Archaean crust of Rajasthan (stabilized)	

BGC. This Proterozoic granite is a leucogranite and perhaps related to the Delhi orogeny (see Chap. 4). The carbonatite forms a low ridge (2 × 0.5 km area) with NW-SE trend within the milieu of Untala granite. According to Chattopadhyay et al. (1992), the Newania carbonatite is a funnel-shaped body with a low E-W plunging axis. This alkaline body has a notable zone of fenitization. The sequence of carbonatite emplacement is dolomite followed by ankeritic and finally calcitic carbonatite (Pandit and Golani, 2000). Schleicher et al. (1997) gave Pb–Pb isochron age of 2273 ± 3 Ma (MSWD = 18.1%) for the dolomitic carbonatite and 1551 ± 16 Ma (MSWD = 22%) for the ankeritic carbonatite. These two ages can be seen to coincide with the opening of the Paleoproterozoic Aravalli basin and the intrusion of Mesoproterozoic Tekan granites, respectively. Earlier, Deans and Powell (1968) determined 959 ± 24 Ma age for the Newania carbonatite on the basis of Strontium isotope data. This age could represent a thermal imprint during later phase of the Delhi orogeny. Because of the Proterozoic tectono-thermal events the BGC rocks also show re-working as at Sarara, southeast of Udaipur and at other places like Anasagar, near Ajmer, where the gneissic rock is an inlier (Tobisch et al., 1994).

Besides the above-mentioned magmatic bodies of Proterozoic age, the BGC also contains ~1725 Ma old granulites, mainly at Sandmata (Gupta et al., 1980). These granulites are mainly three types: the pelitic granulites with polymetamorphic assemblage, the garnet leptynite, and the garnet-bearing two pyroxene basic granulite with coronitic texture (Sharma, 1988; see also Dasgupta et al., 1997). The basic granulite is obviously mantle derived and most probably from the melt that had underplated the Archaean crust in this part of the Indian shield. As a result, the BGC rocks overlying the mafic magma recrystallized at high T and P of the granulite facies (Sharma, 1988) in a magma-underplated crust, documented by their anticlockwise P-T path (cf. Sharma 1999, 2003). As shown in Chap. 4, the granulite rocks at depth were also intruded by 1723 Ma old charnockite-enderbite (Sarkar et al., 1989). It is for this reason that the banded gneisses (BGC) are seen as enclaves within the unfoliated granulite, charnockitic as well as basic granulite (see Srivastava, 2001). Subsequently, the granulite complex as a whole was excavated from depth to shallower level along a shear zone, which abounds the Sandmata complex (cf. Joshi et al., 1993). The enderbite-charnockites with mafic granulites are the major lithology in the Bhinai-Bandanwara area. Recently, Roy et al. (2005) dated their zircons by evaporation technique that yielded ages in the range of 1620–1675 Ma. The geochronological data as well as the nature of the occurrence of the granulite rocks at Sandmata, Bhinai, Bandanwara and other places of the BGC suggest that the granulite facies metamorphism is Mesoproterozoic, connected with the Delhi orogeny and not with the Paleoproterozoic (2100–1900 Ma) Aravalli orogeny. This is because the Aravalli orogeny predates the 1900 Ma old syntectonic Derwal granite (Choudhary et al., 1984) that intruded the Aravalli metasediments during their first phase of folding (Naha et al., 1967). Considering the age of 1723 Ma for the charnockite-enderbite intrusion (Sarkar et al., 1989) and the fragmental occurrence of the granulites in the Delhi metasediments, it is emphasized that the 1723 Ma old granulite are not reworked BGC as conceived by Roy and Jakhar (2002). They are recrystallized lower crustal rocks with minor upper mantle material in form of basic granulites and perhaps norite dykes. During Delhi orogeny these lower crustal rocks were exhumed along ductile shear zone and are now seen as fragmented or torn "pieces" in the Delhi metasediments (Fareeduddin, 1995) (see Chap. 4, for discussion).

2.6.5 Bundelkhand Granite Massif (BKC)

As discussed above, the Bundelkhand Granite massif is also a part of the Rajasthan craton in the northern Indian shield. The Bundelkhand Granite massif or the Bundelkhand Granitoid Complex (abbreviated BKC) is a semi-circular outcrop, occupying nearly 26,000 km^2 (Basu, 1986) in parts of Central Indian shield (Fig. 2.9). It contains polyphase TTG gneisses (ca. 3.5–2.7 Ga), metamorphosed volcano-sedimentary rocks, and syn- to post-tectonic granitoids (ca. 2.5–2.4 Ga). The rock-suite is intruded by numerous quartz reefs followed by dolerite dyke swarms of ca. 1700 Ma age. The dolerite dykes are possibly related to the development of

2.6 Rajasthan Craton

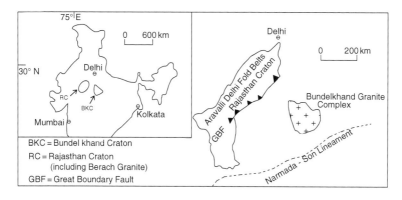

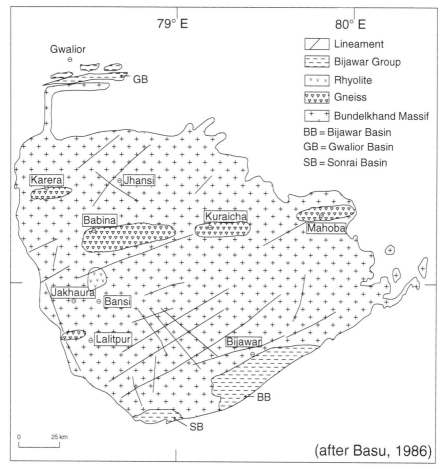

Fig. 2.9 Location and geological map of the Bundelkhand craton/massif (after Basu, 1986). BB = Bijawar basin, GB = Gwalior basin, SB = Sonrai basin

rift basins of Bijawar (or Mahakoshal Group) and Gwalior Formations. The overall tectonic trend of the BKC is E-W to ENE-WSW. The BKC is flanked along its southern and northwestern margins by the Pre-Vindhyan siliciclastic shelf sequence—the Bijawar Formation. The west, south and east margins of the BKC are covered by the Vindhyan rocks (1400 Ma and younger), while the northern border is covered under alluvium. The Bundelkhand craton is correlatable with the 2.5 Ga old Berach Granite of Rajasthan on its west but the continuity between the two granite complexes is concealed under the younger cover (including Vindhyan rift sediments), making the geological correlation only tentative.

The bulk of the Bundelkhand massif is made up of 2500–2600 Ma old granites that are coarse-grained equigranular and sometimes porphyritic to fine-grained (Sarkar et al., 1995). The granitic rocks contain numerous enclaves of schists, gneisses, banded magnetite, calc-silicates and ultramafics (Basu, 1986). Inclusions of older tonalitic gneiss are seen in the early phase of K-rich granite in the northern part. This grey to pink medium-grained gneiss is assumed as a partial melt of garnet-amphibolite parent at depth (Sharma, 1998; Sharma and Rahman, 1996). The granitoids and their gneissic varieties are calc-alkaline in nature and are characterized by highly fractionated REE patterns and depleted in HREE. They resemble the Archaean TTG suites and may thus represent relics of early crustal components in BKC. The gneisses are composed of quartz and plagioclase with accessory magnetite, apatite and zircon. They are highly deformed. The $^{207}Pb/^{206}Pb$ systematics in zircon of 19 samples of granitic gneisses, enclaves and granitoids and rhyolite from BKC massif gave Archaean ages (Mondal, 2003). These geochronological data indicate the presence of two-age components: one indicates higher age value of 3270 ± 3 Ma, which is interpreted as the "minimum" age for crystallization of the protolith of the gneiss. The other age with lower value of 2522–2563 Ma is due to Pb loss and indicates a thermal event, documented in overgrowth of old zircons in the gneisses.

Besides the gneisses, granitoids have also been studied in the BKC. The dark-coloured hornblende granitoid is calc-alkaline in nature similar to the biotite-granitoid which is a pink-coloured, coarse to medium-grained rock, occasionally with some hornblende. The leucogranitoid has very little (<10 vol.%) ferromagnesian minerals. This rock is reported with equal proportions of plagioclase quartz and K-feldspar (mostly perthite), suggesting equilibrium crystallization under the load pressure of over 12 km (cf. Tuttle and Bowen, 1958).

The age of 3.25 Ga for zircons from deformed gneissic enclave within Mahoba granitoid indicates that this rock represents the earliest phase of TTG magma and hence an early Archaean crustal component in the Bundelkhand massif. Besides, the ages of 2.8–2.9 Ga and even 2.7 Ga in gneisses indicate the presence of multiple protolith age components within the massif. It is interesting to note that the 3.3 Ga old gneisses are cofolded with amphibolites and metasedimentary rocks from Mahoba and Kuraicha. This probably indicates granite magmatism and deformation in the BKC at about 3.3 Ga ago.

The younger gneisses are restricted to E-W shears. The zircon ages of 3 suites of granitoids (Mondal, 2003) suggest that their emplacement occurred in a quick succession at ∼2.5 Ga, within a few tens of million years (Panigrahi et al., 2002).

The 2517 Ma old rhyolite from Bansi indicates felsic volcanism (anorogenic) in Proterozoic period, unrelated to an arc magmatism (cf. Mondal, 2003). The age of 2492 ± 10 Ma for the Lalitpur leucogranitoid (Mondal, 2003) is taken to indicate the time of stabilization of the Bundelkhand massif, like that of the BGC of Rajasthan. The BKC has a general trend in E-W to ENE-WSW direction, similar to the Rajasthan craton, hence indicating its correlation with the BGC and Berach granitoid on the west.

Following the granite intrusions, there occurred intrusion of giant quartz reefs in the BKC. The southern part of the massif contains striking NE-trending quartz ridges up to 200 m high with Cu mineralization (Crawford, 1970). Mondal et al. (2002) describe these giant quartz reefs to have been from residual silica left over from large-scale granite magmatism and assigned 2.3–1.9 Ga ages for them. Sharma (1998) links the development of the quartz reefs occurring along brittle-ductile shear zones to K-rich granite magmatism during the Paleoproterozoic, connecting this period to late-stage cratonization in the Bundelkhand area. The geochronological data on various felsic intrusives tightly constrain their emplacement age between 2.5 and 2.2 Ga (Crawford, 1970; Sarkar et al., 1990; Mondal, 2003). The extensive development of holocratic intrusive quartz-reefs along the brittle-ductile shear zones is a unique feature of the BKC. According to Roday et al. (1995) the quartz reefs are recrystallized or sheared quartzites. But they could also be a product of metamorphic differentiation (secretion) following the opening of cracks (shear zones) that acted as low- pressure domains within the BKC granitoids. Mineral ages of muscovite obtained from pyrophyllite of giant quartz veins suggest that hydrothermal activity was between 1480 ± 35 and 2010 ± 80 Ma (Pati et al., 1997).

A number of basic dykes of tholeiitic affinity occurs at several places in the BKC. These dykes generally truncate the early-formed quartz reefs. It is believed that the mafic magmatism occurred in two phases, in the NW-SE and NE-SW, suggesting that dyke activity in the region was episodic, which may be slightly older than the initiation of volcanism in the peripheral Gwalior and Bijawar basins. The mean ages of 2150 and 2000 Ma of the mafic dykes in the region (see Rao et al., 2005) clearly demonstrate that the dykes are younger than the granitic rocks (2500–2600 Ma) of the Bundelkhand massif and older than the volcano-sediment deposited in the rift-related basin (Gwalior traps 1830 ± 200 Ma; Crawford and Compston, 1970). The age data on mafic dykes by Mallickarjun Rao et al. (2005) constrain the timing of mafic magmatism within the Bundelkhand granite massif.

The dyke activity in the Bundelkhand massif is similar to the activity in the Rajasthan craton in the northern Indian peninsular shield. Again, both BKC and the BGC-Berach Blocks of Rajasthan craton have stabilized at 2.5 Ga ago. It must be pointed out here that the Singhbhum region in eastern India stabilized much earlier, at 3.1 Ga (Mishra et al., 1998). Zircon ages for the various lithounits of the Bundelkhand massif and BGC-Berach Granite of Rajasthan craton are summarized in Table 2.3.

The oldest granitic crustal component in the two terrains/blocks of Rajasthan and Bundelkhand has similar ages: 3300 Ma for Kuraicha gneiss of the BKC, and 3281± and 3307 ± 33 Ma (Pb–Pb and Sm/Nd ages respectively) for Mewar gneiss

Table 2.3 Geochemical and geochronological comparisons between BGC-Berach Granite and Bundelkhand massif (after Mondal, 2003)

Features	Bundelkhand massif (BKC)	BGC-Berach Granite of RC
Composition of gneiss	Tonalite, Trondhjemite, highly fractionated REE patterns with depletion of HREE, small negative Eu anomaly	Tonalite, Trondhjemite, highly fractionated REE patterns without depletion of HREE, small negative Eu anomaly
Composition of granitoid	Calc-alkaline, moderately fractionated REE pattern without HREE depletion, large Eu anomaly	Calc-alkaline, moderately fractionated REE pattern without HREE depletion, large Eu anomaly
Regional metamorphism and deformation	3.29 Ga (oldest gneissic component at Kuraicha); 2.5 Ga (youngest gneissic component at Panchwara)	3.28–3.3 Ga (oldest gneissic component S of Udaipur); 2.5 Ga (youngest gneissic component in Vali River, Udaipur district)
Emplacement of granitoid	∼2.5 Ga	∼2.5 Ga
Age of stabilization	∼2.5 Ga; no widespread evidence for much younger activity	∼2.5 Ga; evidence for the presence of Mid-Precambrian and younger activity
Mafic rocks	Mafic dyke swarms	Mafic volcanism

of Rajasthan craton (Gopalan et al., 1990; Wiedenbeck and Goswami, 1994). Furthermore, the age of 2506 ± 4 Ma for the youngest gneissic event (Vali River gneiss) in the Rajasthan craton (Wiedenbeck and Goswami, 1994) is very close to the age of the youngest gneissic event in the Bundelkhand massif (Panchwarz gneiss and Karera gneiss) (Mondal, 2003). Also, the gneisses from both these terrains share common compositional characteristics (Gopalan et al., 1990; Mondal, 2003). The Mid-Archaean gneissic components as well as the Late Archaean granitoids intruding the gneiss in both cratonic regions are also geochemically similar (Ahmad and Tarney, 1994; Sharma and Rahman, 1995).

The above discussion leads one to conclude that the two cratonic areas, namely the BGC-Berach Granite and the Bundelkhand massif, evolved as a single large protocontinent, named here as Rajasthan craton, RC, that stabilized at ∼2.5 Ga ago (see Mondal, 2003, p. 10; see also Yedekar et al., 1990).

2.7 Meghalaya Craton

The Precambrian rocks in the eastern India occur in the Shillong Plateau of the Meghalaya State and in the Mikir Hills (also called Kabri Hills) Plateau located in the State of Assam (see inset of Fig. 2.10). The Shillong-Mikir Hills Massif is here called the Meghalaya craton. It is roughly a rectangular plateau, about 50 km wide and 200 km long made up of crystalline rocks with nearly E-W, rather ENE-WSW, trend. It is bordered on the north by Brahmaputra River. The river flows nearly E-W between the Shillong and Mikir Hills Plateaus and the Main Boundary Thrust of

2.7 Meghalaya Craton

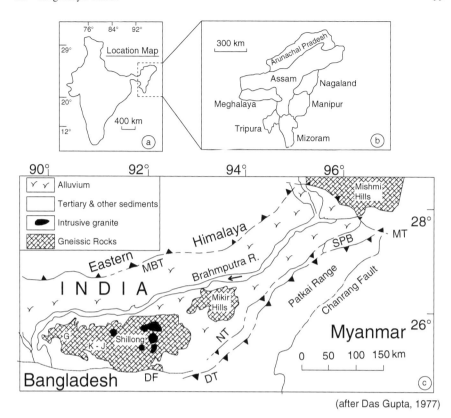

(after Das Gupta, 1977)

Fig. 2.10 Location of Meghalaya in the northeastern states of India (**a** and **b**) and Geological map (**c**) of the Shillong-Mikir Hills Plateau, herein called the Meghalaya craton, with location of Garo Hills (G), Khasi-Jantia Hills (K-J), Mikir Hills (renamed as Kabri Hills by some authors). Other abbreviations: DF = Dauki Fault, DT = Disang thrust, MBT = Main Boundary thrust, MT = Mizu thrust, NT = Naga thrust, SPB = Schuppen Belt (Zone of imbrication)

the Eastern Himalaya in the north. The southern boundary is demarcated by Dauki Fault, which also defines the northern limit of the Sylhet Trap (110–133 Ma).

Oldham (1856), Medlicott (1869), Desikachar (1974) and more recently Nandy (2001) and Dasgupta and Biswas (2000) gave a general geological account of the Shillong-Mikir Hills Plateaus, although the fringes of these plateaus are covered by Cretaceous to Eocene sediments. The oldest unit in the Meghalaya craton is the Archaean Gneissic Complex or Older Metamorphic Group (Mazumder, 1976). It consists of granite gneiss, augen gneiss and upper amphibolite to granulite facies metamorphic rocks. The latter rocks are cordierite-sillimanite gneiss, quartz-feldspar gneiss (orthogneiss) and biotite schists with or without hornblende (Rahman, 1999). In the central part of the Shillong Plateau, particularly in the Khasi Hills the biotite-quartz-sillimanite-cordierite and quartz-sillimanite rocks associated with sillimanite-corundum occur as lenses within granite- gneiss. These occupy a belt one kilometer wide with E-W strike of foliation. To the east, these rocks are cut by granite in which lenses of sillimanite rocks may be found. Recently, Bidyananda and Deomurari (2007) dated zircons from quartzo-feldspathic gneisses of the Meghalaya craton. The ^{207}Pb/^{206}Pb isotope gave 2637 ± 55 Ma for core and 2230 ± 13 Ma for rim of the mineral, indicating that the cratonic gneisses are Archaean, not unlike other cratonic regions of the Indian shield, and that the thermal overprinting on these basement rocks occurred during Proterozoic event(s). The overgrowth on detrital zircons from the cover trocks of the Shillong Group gave 1.5–1.7 Ga which is considered to be the age of metamorphism related to the Mesoproterozoic orogeny that affected both basement and cover rocks of the Indian shield (cf. Ghosh et al., 1994).

Recently, Chatterjee et al. (2007) dated monazite from high-grade metapelites of the craton. Their study provide well-constrained age of 1596 ± 15 Ma for the Garo-Golapar Hills, which is considered to date the counter-clockwise P-T path with near peak conditions of 7–8 kb/850°C. The EPMA monazite age in the Sonapahar high grade area are about 500 ± 14 Ma which nearly coincides with 408 Ma Rb-Sr dates of the porphyritic granite that intruded the Meghalaya gneiss complex. These dates correspond to Pan-African event which is widely documented in the southern granulite terrain (SGT) and in the Himalaya, besides at other places of the Indian shield.

The Gneissic Complex is unconformably overlain by the Proterozoic Shillong Group, which is mainly siliciclastic sedimentary association, now observed as phyllites and quartzites. According to Mitra (2005), the metasedimentary units (mica schists, quartzites, phyllites, slates etc.) belonging to the Shillong Group have been involved in four phases of folding. The earliest structures are very tight to isoclinal folds (F1) on bedding plane (So). These folds have high amplitude to wavelength ratio, with a penetrative axial plane cleavage (S1). These structural elements have been coaxially deformed into open to tight upright F2 folds with axial plane striking NNE, with the development of crenulation cleavage (S2). Both F1 and F2 folds appear to be buckle folds. In the more schistose rocks, NE-trending recumbent folds (F3) developed on F1 and F2 axial plane foliations. The last structures are upright conjugate folds and kink bends (F4) with axial plane striking NE, EW, and cheveron folds with NW striking axial planes. The structures of F4 thus provide evidence of longitudinal shortening of the Shillong Group of rocks (Mitra, 2005, p. 117).

The maximum time limit of the Shillong Group is given by Rb−Sr whole rock isochron age of 1150 ± 26 Ma old granite gneiss that occurs at the base of the Shillong Group (see Ghosh et al., 1991). The Millie granite (607 Ma; Chimote et al., 1988; see also Crawford, 1969) occurring as intrusive into the Shillong Group appears to suggest an event of thermal reactivation, because there is evidence that both the basement and the cover rocks of the Shillong Group are intruded by such anorogenic granites whose Rb−Sr whole-rock ages are in the range of 885–480 Ma (Ghosh et al., 1994). Both rock groups (the Gneissic complex and the cover of Shillong Group) are also intruded by basic and ultrabasic rocks, some of which belong to Late Cretaceous period (Mazumder, 1986). There is also a report of a carbonatite intrusion (Late Jurassic) from the Sung Valley of Meghalaya. Fission track ages of zircons from the adjoining Kyrdem Granite gives 1043 ± 101 Ma (Nandy, 2001). The granitic plutons occurring amidst the gneisses also extend in the direction of the regional foliation. The general alignment of these granitic bodies in the Meghalaya craton is E-W to ENE-WSW, conformable with the tectonic axis or the major lineaments of the Shillong Plateau (GSI, 1973). Rahman (1999) reports a narrow thermal aureole, about 400 m, suggesting shallow depth of the granite intrusion. The granite varies in texture from porphyritic to fine grained, consisting of quartz-microcline-micropethite and oligoclase with some biotite.

In 1964, Evans conceived that the Sylhet Trap (ca. 110 Ma old) located on the southern part of the Shillong plateau represents a part of the Rajmahal Trap (105–117 Ma; Bakshi, 1995) that is situated on the NE part of the Chhotanagpur Granite Gneiss Complex (CGGC). This led him (Evans, 1964) to believe that the Shillong Plateau has dextrally moved along the Dauki Fault for about 250 km towards east. The Dauki Fault, according to Evans (1964) is a tear fault (transcurrent or strike slip fault) that trends transverse to the strike of the deformed rocks of the Shillong-Mikir Hills Plateaus. But Murthy et al. (1969) contradicted this proposition of Evans and argued with evidence that the Duke fault has a vertical uplift to the north, causing the Shillong-Mikir Hills Plateaus as an uplifted region with northward tilting. Another view for the elevated Shillong-Mikir Hills Plateaus is proposed, according to which the upliftment occurred as a consequence of nearly N-S collision of the Indian plate with the Eurasian plate. Dasgupta and Biswas (2000) believe that the Mishmi Hills added another compressional force from the NE in aiding upliftment of the Meghalaya craton (i.e. Shillong-Mikir Hills Plateaus).

According to Dasgupta and Biswas (2000), the Dauki Fault splits towards the east, with southern part linking up with the Disang Thrust that merges further east in the "Schuppen Belt" or belt of imbricated thrusts (namely Naga, Disang and Patkai Thrusts) (Fig. 2.10). More recent study involving remote sensing and ground checks by Srinivasan (2005) revealed that the Dauki Fault is a single gravity (Normal) fault dipping towards south. However, a limited dextral slip is indicated by bending towards NE to ENE of the axial surfaces in folded Miocene sediments (Surma Series) (cf. Srinivasan, 2005). Dasgupta and Biswas (2000) pointed out that there are a number of E-W faults. Besides these, there are also N-S trending faults that are considered to have developed during the Late Jurassic-Early Cretaceous during which ultramafic-alkaline-carbonatite complexes (ca. 107 Ma old) have been emplaced (Srivastava and Sinha, 2004; Chattopadhyay and Hashimi, 1984).

The Mishmi Hills, lying mostly outside the Indian Territory, is mainly a granodioritic complex with high-grade schists and migmatites. Its geographic position has prevented any detailed study, but the predominance of granite gneisses reported from the Indian part suggests that the Mishmi Hills is similar to that of the Shillong Plateau. The Indo-Burma fold belt is thrust upon the Mishmi Hills, while the Roing Fault marks the SE end of the Mishmi Hills complex. The fault terminates the Himalayan MBT and the Siwalik belt. Whether the Mishmi Hills is a part of the Indian crust (Shillong and Mikir Hills Plateaus), or it is more closely related to any part of the two neighbouring colliding plates of Burma and South Tibet, needs to be resolved by collaborative studies.

The basement complex was block-faulted or rifted in Permian to Early Cretaceous during which Gondwana sediments were deposited. Subsequently, the composite crustal mass of Shillong and Mikir Hills Plateaus is criss-crossed by a large number of nearly rectilinear lines along which rivers and streams have cut straight valleys. The above account of the Precambrian cratonic rocks of the eastern India is general and demands intensive geological studies on modern lines, similar to that attempted on other cratonic regions and Proterozoic fold belts.

References

Acharyya, S.K. (2003). A plate tectonic model for Proterozoic crustal evolution of Central Indian Tectonic Zone. Gond. Geol. Mag., vol. **7**, pp. 9–31.

Ahmad, T. and Tarney, J. (1994). Geochemistry and petrogenesis of late Archaean Aravalli volcanics, basement enclaves and granitoids, Rajasthan. Precambrian Res. vol. **65**, pp. 1–23.

Anantha Iyer, G.V. and Vasudevan, V.N. (1979). Geochemistry of the Archaean metavolcanic rocks of Kolar and Hutti Gold Fields, Karnataka, India. J. Geol. Soc. India, vol. **20**, pp. 419–432.

Bakshi, A.K. (1995). Petrogenesis and timing of volcanism in the Rajmahal flood basalt Province, northeastern India. Chem. Geol., vol. **121**, pp. 73–90.

Balakrishnan, S., Rajamani, V. and Hanson, G. (1999). U-Pb ages for zircon and titanite from Ramagiri area, southern India: evidence for accretionary origin of the eastern Dharwar craton during the Late Archaean. J. Geol., vol. **107**, pp. 69–86.

Bandyopadhyay, B.K., Bhoskar, K.G., Ramachandra, H.M., Roy, A., Khadse, V.K., Mohan, M., Barman, T., Bishui, P.K. and Gupta, S.N. (1990). Recent geochronological studies in parts of the Precambrian of central India. Geol. Surv. India Spl. Publ., vol. **28**, pp. 199–211.

Bandyopadhyay, P.K., Chakraborti, A.K., Deo Murari, M.P. and Misra, S. (2001). 2.8 Ga old anorogenic granite—acid volcanics association from western margin of the Singhbhum-Orissa Craton, eastern India. Gondwana Res., vol. **4**, pp. 465–475.

Bandyopadhyay, B.K., Roy, A. and Huin, A.K. (1995). Structure and tectonics of a part of the central Indian shield. In: Sinha-Roy, S. and Gupta, K.R. (eds.),*Continental Crust of Northwestern and Central India*. Geol. Soc. India Mem. vol. **31**, Bangalore, pp. 433–467.

Banerjee, P.K. (1982). Stratigraphy, petrology and geochemistry of some Precambrian basic volcanic and associated rocks of Singhbhum district, Bihar, and Mayurbhanj and Kheonjhar districts, Orissa. Geol. Surv. India, Mem., vol. **111**, pp. 7–8.

Banerji, A.K. (1974). On the stratigraphy and tectonic history of the Iron-ore bearing and associated rocks of Singhbhum and adjoining areas of Bihar and Orissa. J. Geol. Soc. India, vol. **15**, pp. 150–157.

Banerji, A.K. (1991). Presidential address, Geology of the Chhotanagpur region. Indian J. Geol., vol. **63**(4). pp. 275–282.

Basu, A.K. (1986). Geology of parts of the Bundelkhand massif, central India. Geol. Surv. India Rec., vol. **117**, pt. 2, pp. 61–124.

Basu, A.R., Ray, S.L., Saha, A.K. and Sarkar, S.N. (1981). Eastern Indian 3800-million year old crust and early mantle differentiation. Science, vol. **212**, pp. 1502–1506.

Bhate, V.D. and Krishna Rao, S.V.G. (1981). Charnockitic rocks of north Bastar, Madhya Pradesh. Geol. Surv. India Spl. Publ., vol. **3**, pp. 109–113.

Bhattacharya, B.P. (1976). Metamorphism of the Precambrian rocks of the central part of Santhal Parganas district, Bihar. Quart. J. Geol. Min. Met. Soc. India, vol. **48**(4), pp. 183–196.

Bhattacharya, B.P. (1988). Sequence of deformation, metamorphism and igneous intrusions in Bihar mica belt. Geol. Soc. India Mem., vol. **8**, pp. 113–126.

Bhattacharya, A. and Bhattacharya, G. (2003). Petrotectonic study and evaluation of Bilaspur-Raigarh belt, Chhatisgarh. In: Roy, A. and Mohabey, D.M. (eds.), *Advances in Precambrian of Central India*. Gond. Geol. Mag., Spl., vol. **7**, Gondwana Geological Society, Nagpur, pp. 89–100.

Bhattacharya, D.S. and Ghosal, A. (1992). Petrofabric patterns of the Chhotanagpur Gneiss and adjoining schists of the Singhbhum Group. Indian J. Geol., vol. **64**, pp. 196–209.

Bhattacharya, D.S., Ghosal, A. and Ravi Kant, V. (1990). Chhotangapur granite gneiss in relation to the schistose rocks around Murhu, Ranchi district, Bihar: a structural approach. Proc. Indian Acad. Sci. (EPS), vol. **99**, pp. 269–277.

Bhattacharya, P.K. and Mukherjee, A.D. (1984). Petrochemistry of metamorphosed pillow and the geochemical status of the amphibolite (Proterozoic) from Sirohi district, Rajasthan, India. Geol. Mag., vol. **121**, pp. 465–473.

Bhowmik, S.K. and Dasgupta, S. (2004). Tectonometamorphic evolution of boudin-type granulites in the Central Indian Tectonic Zone and in the Aravalli-Delhi mobile belt: a synthesis and future prospectives. Geol. Surv. India Spl. Publ., vol. **84**, pp. 227–246.

Bhowmik, S. and Pal, T. (2000). Petrotectonic implication of the granulite suite north of the Sausar mobile belt in the overall tectonothermal evolution of the Central Indian mobile belt. *GSI Progress Report* (unpublished).

Bhowmik, S.K., Pal, T., Roy, A. and Pant, N.C. (1999). Evidence for Pre-Grenville high-pressure granulite metamorphism from the northern margin of the Sausar mobile belt in central India. J. Geol. Soc. India, vol. **53**, pp. 385–399.

Bhowmik, S.K. and Roy, A. (2003). Garnetiferous metabasites from the Sausar Mobile Belt: Petrology, P-T path and implications for the tectonothermal evolution of the Central Indian Tectonic Zone. J. Petrol., vol. **44**, pp. 387–420.

Bhowmik, S.K., Sarbadhikari, A.B., Spiering, B. and Raith, M.M. (2005). Mesoproterozoic reworking of Palaeoproterozoic ultrahigh temperature granulites in the Central Indian Tectonic Zone and its implications. J. Petrol., vol. **46**, pp. 1085–1119.

Bhowmik, S.K. and Spiering, B. (2004). Constraining the prograde and retrograde P-T Paths of granulites using decomposition of initially zoned garnets: an example from the Central Indian Tectonic Zone. Contrib. Mineral. Petrol., vol. **147**, pp. 581–603.

Bidyananda, M. and Deomurari, M.P. (2007). Geochronological constraints on the evolution of Megahalaya massif, northeastern India: an ion microprobe study. Curr. Sci., vol. **93**(11), pp. 1620–1623.

Bose, M.K. (1990). Growth of Precambrian continental crust—a study of the Singhbhum segment in the eastern Indian shield. In: Naqvi, S.M. (ed.), *Precambrian Continental Crust and Its Economic Resources*.Dev. Precambrian Geol., vol. **8**. Elsevier, Amsterdam, pp. 267–285.

Bose, U. (1992). A few thoughts on the evolution of the Banded Gneissic Complex in south Rajasthan and its correlation problems. In: Field guide on workshop on *Oldest Rocks—Focus on Rajasthan*. Department of Geology, M.L. Sukhadia University, Udaipur, pp. 17–23.

Bose, M.K. (2000). Mafic-ultramafic magmatism in the western Indian craton—a review. In: Dr. M.S. Krishnan Centenary Commemoration Seminar Volume Geol. Surv. India Spl. Publ., vol. **55**, 227p.

Brown, M. and Phadke, A.V. (1983). High temperature retrograde reactions in pelitic gneiss from the Precambrian Sausar metasediments of the Ramakona area, Chhindwara district, M.P. (India), definition of the exhumation P-T path and the tectonic evolution. In: Phadke, A.V. and Phansalkar, V.G. (eds.), *Professor K.V. Kelkar Volume,* Indian Society of Earth Scientists, Pune, pp. 61–96.

Chadwick, B., Vasudevan, V.N. and Hegde, G.V. (2000). The Dharwar craton, southern India, interpreted as the result of Late Archaean oblique convergence. Precamb. Res., vol. **99**, pp. 91–111.

Chadwick, B., Vasudev, V.N., Hegde, G.V. and Nutman, A.P. (2007). Structure and SHRIMP U/Pb zircon ages of granites adjacent to the Chitradurga schist belt: Implications for Neoarchaean convergence in the Dharwar craton, southern India. J. Geol. Soc. India, vol. **69**, pp. 5–24.

Chakraborty, K.L. and Mazumdar, T. (1986). Geological aspects of the banded-iron formation of Bihar and Orissa. J. Geol. Soc. India, vol. **28**, pp. 109–133.

Chatterjee, N., Crowley, J.L. and Ghosh, N.C. (2008). Geochronology of the 1.55 Ga Bengal anorthosite and Grenvillian metamorphism in the Chhotanagpur gneissic complex, eastern India. Precambrian Res., vol. **161**, pp. 303–316.

Chatterjee, N., Mazumadar, A.C., Bhattacharya, A. and Saikia, r.R. (2007). Mesoproterozoic granulites of the Shillong-Meghalaya plateau: Evidence of westward continuation of the Prydz Bay Pan-African suture into northeastern India. Precambrian Res., vol. **152**, pp. 1–26.

Chattopadhyay, S., Chattopadhyay, B. and Bapna, V.S. (1992). Geochemistry of the Newania carbonatite pluton. Indian Minerals, vol. **46**, pp. 35–46.

Chattopadhyay, N. and Hashimi, S. (1984). The Sung Valley alkaline-ultramafic-carbonatite complex. East Khasi and Jaintia Hills districts, Meghalaya. Rec. Geol. Surv. India, vol. **113**, pp. 24–33.

Chimote, J.S., Pandey, B.K., Bagchi, A.K., Basu, A.N., Gupta, J.N. and Saraswat, A.C. (1988). Rb-Sr whole-rock isochron age for the Mylliem Granite, E.K. Hills, Meghalaya. *Proceedings of 4th National Symposium on* Mass Spectrometry, IISC, Bangalore.

Choudhary, A.K., Gopalan, K. and Sastry, C.A. (1984). Present status of the geochronology of the Precambrian rocks of Rajasthan. Tectonophysics, vol. **105**, pp. 131–140.

Crawford, A.R. (1969). Reconnaissance Rb-Sr dating of Precambrian rocks of southern peninsular India. J. Geol. Soc. India, vol. **10**, pp. 117–166.

Crawford, A.R. (1970). The Precambrian geochronology of Rajasthan and Bundelkhand, northern India. Can. J. Earth Sci., vol. **7**, pp. 91–110.

Crawford, A.R. and Compston, W. (1970). The age of the Vindhyan system of peninsular India. Q. J. Geol. Soc. London, vol. **125**, pp. 351–372.

Dasgupta, S. (2004). Modelling ancient orogens: An example from North Singhbhum fold belt. Geol. Surv. India Spl. Publ., vol. **84**, pp. 33–42.

Dasgupta, A.B. and Biswas, A.K. (2000). *Geology of Assam.* Geol. Soc. India, Bangalore, 269p.

Dasgupta, S., Guha, D., Sengupta, P., Miura, H. and Ehl, J. (1997). Pressure-temperature-fluid evolutionary history of the Sandmata granulite complex, northwestern India. Precambrian Res., vol. **83**, pp. 267–290.

Deans, T. and Powell, J.L. (1968). Trace element and strontium isotopes in carbonatites, fluorites and limestones from India and Pakistan. Nature, vol. **218**, pp. 750–752.

Desikachar, S.V. (1974). A review of the tectonic and geologic history of eastern India in terms of plate tectonic theory. J. Geol. Soc. India, vol. **15**, pp. 137–149.

Dey, K.N. (1991). The older raft tonalite of Rairangapur and its bearing on Precambrian stratigraphy of the Singhbhum craton. Indian J. Geol., vol. **63**, pp. 261–274.

Drury, S.A., Harris, N.B.W., Holt, R.W., Reeves-Smith, G.J. and Wightman, R.T. (1984). Precambrian tectonics and crustal evolution in south India. J. Geol., vol. **92**, pp. 3–20.

Drury, S.A. and Holt, R.W. (1980). The tectonic framework of the south Indian craton: a reconnaissance involving landsat imagery. Tectonophysics, vol. **65**, pp. T1–T5.

Dunn, J.A. (1929). The geology of north Singhbhum including parts of Ranchi and Manbhum districts. Geol. Surv. India Mem., vol. **54**(2), pp. 1–166.

Dunn, J. and Dey, A.K. (1942). Geology and petrology of eastern Singhbhum and surrounding areas. Geol. Surv. India Mem. vol. **69**, pp. 281–456.

Evans, P. (1964). The tectonic frame work of Assam. J. Geol. Soc. India, vol. **5**, pp. 80–96.

Fareeduddin (1995). Field setting, petrochemistry and P-T regime of the deep crustal rocks to the northwest of the Aravalli-Delhi mobile belt, north-central Rajasthan. Geol. Soc. India. Mem., vol. **31**, pp. 117–139.

Fermor, L.L. (1936). An attempt at the correlation of ancient schistose formations of peninsular India. Geol. Soc. India Mem. vol. **70**, pp. 1–52.

French, J.E., Heaman, L.M., Chacko, T. and Srivastava, R.K. (2008). 1891–1883 Ma southern Bastar-Cuddapah mafic igneous events, India: a newly recognized large igneous province. Precambrian Res., vol. **160**, pp. 308–322.

Friend, C.L. and Nutman, A.P. (1991). SHRIMP U–Pb geochronology of the Closepet granite and Peninsular gneiss, Karnataka, India. J. Geol. Soc. India, vol. **38**, pp. 357–368.

GSI (1969, 1993, 1998). *Geological Map of India.* 7th ed., scale 1:2000,000. Geol. Surv. India, Hyderabad and Calcutta.

GSI and ISRO (1994). Project Vasundhara: Generalized Geological Map, scale 1:2 million. Geol. Surv. India & Indian Space Res. Orgn., Bangalore.

Geological Survey of India (1973). *Geological and Mineral Map of Arunachal Pradesh, Assam, Manipur, Meghalaya, Mizoram, Nagaland and Tripura.* 1st ed. (1:2,000,000), Publ. G.S.I., Calcutta.

References

Ghose, N.C. (1965). A note on the occurrence and origin of charnockitic rocks at Demu, district Palamau, Bihar. J. Sci. Res. Banras Hindu University, vol. **15**(2B), pp. 195–201.

Ghose, N.C. (1983). Geology, tectonics and evolution of the Chhotanagpur granite-gneiss complex, eastern India. Recent Res. Geol., vol. **10**, pp. 211–247.

Ghose, N.C. (1992). Chhotanagpur gneiss-granulite complex, eastern India: present status and future prospect. Indian J. Geol., vol. **64** (1), pp. 100–121.

Ghose, N.C., Shamkin, B.M. and Smirnov, V.N. (1973). Some geochronological observations on the Precambrians of Chhotanagpur, Bihar, India. Geol. Mag., vol. **110**, pp. 477–482.

Ghosh, J.G. (2003). 3.56 Ga tonalite in the central part of the Bastar Craton, India: Oldest Indian date. J. Asian Earth Sci., vol. **23** (3), pp. 359–364.

Ghosh, J.G. (2004). 3.5 Ga tonalite in the central part of the Bastar craton, India: oldest Indian date. J. Asian Earth Sci., vol. **23** (3), pp. 359–364.

Ghosh, S., Chakraborty, S., Bhalla, J.K., Paul, D.K., Sarkar, A., Bishui, P.K. and Gupta, S.N. (1991). Geochronology and geochemistry of granite plutons from East Khasi Hills, Meghalaya. J. Geol. Soc. India, vol. **73**, pp. 331–342.

Ghosh, S., Chakraborty, S., Paul, D.K., Bhalla, J.K., Bishui, P.K. and Gupta, S.N. (1994). New Rb-Sr isotopic ages and geochemistry of granitoid from Meghalaya and their significance in middle to late Proterozoic crustal evolution. Indian Minerals, vol. **48**, pp. 33–44.

Ghosh, D.K., Sarkar, S.N., Saha, A.K. and Ray, S.L. (1996). New insights on the early Archaean crustal evolution in eastern India: re-evaluation of Pb-Pb, Sm-Nd and Rb-Sr geochronology. Indian Minerals, vol. **50**, pp. 175–188.

Gopalan, K., Macdougall, J.D., Roy, A.B. and Murali, A. (1990). Sm-Nd evidence for 3.3 Ga old rock in Rajasthan, northwestern India. Precamb. Res., vol. **48**, pp. 287–297.

Gopalkrishnan, K. (1996). An overview of Southern Granulite Terrain, India. Geol. Soc. India Spl. Publ., vol. **55**, pp. 85–96.

Goswami, J.N., Misra, S., Wiedenbeck, M., Ray, S.L. and Saha, A.K. (1995). 3.55 Ga old zircon from Singhbhum-Orissa Iron-Ore craton, eastern India. Curr. Sci., vol. **69**, pp. 1008–1011.

Guha, D.B. and Bhattacharya, A.K. (1995). Metamorphic evolution and high grade re-working of the Sandmata Complex granulites. In: Sinha-Roy, S. and Gupta, K.R. (eds.), *Continental Crust of Northwestern and Central India*. Geol. Soc. India Mem., Bangalore, vol. **31**, pp. 163–198.

Gupta, B.C. (1934). The geology of central Mewar. Geol. Soc. India Mem., vol. **65**, pp. 107–168.

Gupta, S.N., Arora, Y.K., Mathur, R.K., Iqbaluddin, Prasad, B., Sahai, T.N. and Sharma, S.B. (1980). Lithostratigraphic Map of Aravalli Region (1:1000,000). Geol. Surv. India, Calcutta.

Gupta, S.N., Arora, Y.K., Mathur, R.K., Iqbaluddin, Prasad, B., Sahai, T.N. and Sharma, S.B. (1997). The Precambrian geology of the Aravalli region, southern Rajasthan and northern Gujarat. Mem. Geol. Surv. India, vol. **123**, 262p.

Gupta, A. and Basu, A. (2000). North Singhbhum Proterozoic mobile belt, eastern India—a review. In: Dr. M.S. Krishnan Centenary Commemoration Seminar Volume. Geol. Surv. India Spl. Publ., vol. **55**, pp. 195–226.

Gupta, P. and Rai Choudhuri, A. (2002). Tectono-metamorphic evolution of the amphibolite- and granulite facies rocks of the Sandmata complex, central Rajasthan. Indian Minerals, vol. **56**, pp. 137–160.

Hanson, G.N., Krogstad, E.J. and Rajamani, V. (1988). Tectonic setting of the Kolar Schist Belt, Karnataka, India (Abstract). Workshop on *The Deep Continental Crust of South India*, Jan. 1988, Field Excursion Guide, pp. 40–42.

Heron, A.M. (1953). The Geology of Central Rajputana. Geol. Surv. India. Mem., vol. **79**, Calcutta, 389p.

Holmes, A. (1950). Age of uraninite from pegmatite near Singar, Gaya district, India. Am. Mineral., vol. **35**, pp. 19–28.

Iyengar, S.V.P. and Murthy, Y.G.K. (1982). The evolution of the Archaean-Proterozoic crust in parts of Bihar and Orissa, eastern India. Rec. Geol. Surv. India, vol. **112**(3), pp. 1–5.

Jain, S.C., Nair, K.K.K. and Yedekar, D.B. (1995). Geology of the Son-Narmada-Tapti Lineament Zone in Central India. In: *Geoscientific Studies of the Son-Narmada-Tapti Lineament Zone*. Geol. Surv. India Spl. Publ., Nagpur, vol. **10**, pp. 1–154.

Jayananda, M., Kano, T., Harish Kumar, S.B., Mohan, A., Shadakshra Swami, N. and Mahabaleswar, B. (2003). Thermal history of the 2.5 Ga juvenile continental crust in the Kuppam-Karimangalam area, Eastern Dharwar craton, southern India. Geol. Soc. India Mem., vol. **50**, pp. 255–287.

Jayananda, M., Moyen, J.F., Peucat, J.J., Auvray, B. and Mahabaleshwar, B. (2000). Late Archaean (2550–2520 Ma) juvenile magmatism in the eastern Dharwar craton, southern India: constraints from geochronology, Nd-Sr isotopes and whole rock geochemistry. Precambrian Res., vol. **99**, pp. 225–254.

Joshi, M., Thomas, H. and Sharma, R.S. (1993). Granulite facies metamorphism in the Archaean gneiss complex from northcentral Rajasthan. Proc. Nat. Acad. Sci. India, vol. **63**(A), pp. 167–187.

Krishnan, M.S. (1935). The Dharwars of Chhotanagpur, their bearing on some problems of correlation and sedimentation. Presidential address to Section of Geology. 22nd Indian Science Congress Session, Calcutta, pp. 175–203.

Krishnana, M.S. (1982). *Geology of India and Burma*. 6th ed. CBS Publishers and Distributors, Daryaganj, New Delhi, 536p.

Krogstad, E.J., Balkrishnan, S., Hanson, G. and Rajamani, V. (1989). Plate tectonics at 2.5 Ga ago: evidence from Kolar Schist Belt, South India. Science, vol. **243**, pp. 1337–1340.

Krogstad, E.J., Hanson, G.N. and Rajamani, V. (1991). U-Pb ages of zircon and sphene for two gneiss terranes adjacent to the Kolar Schist belt, south India: Evidence for separate crustal evolution histories. J. Geology, vol. **99**, pp. 801–816.

Kumar, M.N., Das, N. and Dasgupta, S. (1985). Gold mineralization along the northern shear zone, Purulia district, West Bengal: an up-to-date appraisal. Rec. Geol. Surv. India, vol. **113**(3), pp. 25–32.

Lippolt, H.J. and Hautmann, S. (1994). Ar^{40}/Ar^{39} ages of Precambrian manganese ore mineral from Sweden, India and Morocco. Mineralium Deposita, vol. **30**, pp. 246–256.

MacDougall, J.D., Gopalan, K., Lugmair, G.W. and Roy, A.B. (1983). The Banded Gneissic Complex of Rajasthan, India: early crust from depleted mantle at 3.5 AE. EOS Trans. Am. Geophys. Union, vol. **64**(I-26), 351p.

Mahadevan, T.M. (2002). *Geology of Bihar & Jharkahnd*. Geol. Soc. India, Bangalore, 563p.

Mahadevan, T.M. (2004). Continental evolution through time: new insights from southern Indian shield (Abstract). In: International workshop on *Tectonics and Evolution of the Precambrian Southern Granulite Terrain, India and Gondwana Correlations*, Feb. 2004. National Geophys. Res. Institute, Hyderabad, pp. 37–38.

Mahalik, N.K. (1987). Geology of rocks lying between Gangpur Group and Iron-Ore Group of the horse-shoe syncline in north Orissa. Indian J. Earth Sci., vol. **14**, pp. 73–83.

Mallik, A.K. (1993). Rb-Sr geochronology of Chhotanagpur gneiss terrain and granitic plutons in Bihar Mica Belt. Geol. Surv. India Report (unpublished).

Mallik, A.K. (1998). Isotopic studies of granitic plutons of Bihar Mica Belt, Koderma. In: M.S. Krishnan Centenary Volume. Geol. Surv. India. pp. 115–116.

Manna, S.S. and Sen, S.K. (1974). Origin of garnet in basic granulites around Saltora, W. Bengal, India. Contrib. Mineral. Petrol., vol. **44**, pp. 195–218.

Mazumder, S.K. (1976). A summary of the Precambrian geology of the Khasi Hills, Meghalaya. Geol. Surv. India Misc. Publ. vol. **23**, pp. 311–324.

Mazumder, S.K. (1979). Precambrian geology of eastern India between the Ganga and Mahanadi—a review. Rec. Geol. Surv. India, vol. **110**, pp. 60–116.

Mazumder, S.K. (1986). The Precambrian framework of part of the Khasi Hills, Meghalaya. Rec. Geol. Surv. India, vol. **117** (2), pp. 1–59.

Mazumder, S.K. (1988). Crustal evolution of the Chhotanagpur gneissic complex and the Mica Belt of Bihar. In: Mukhopadhyay, D. (ed.), *Precambrian of the Eastern Indian Shield*. Geol. Soc. India Mem., Bangalore, vol. **8**, pp. 49–84.

Mazumder, S.K. (1996). Precambrian geology of peninsular eastern India. Indian Minerals, vol. **50**, pp. 139–174.

References

Mazumder, S.K. (1998). Geology of the Chhotanagpur gneiss complex—a perspective. In: M.S. Krishnan Centenary Commemorative National Seminar (Abstruct volume), Geol. Surv. India, pp. 112–114.

Medlicott, H.B. (1869). Geological sketch of the Shillong plateau in N.E. Bengal. Mem. Geol. Surv. India, vol. **7**, pt. 1, pp. 151–207.

Meen, J.K., Rogers, J.J.W. and Fullagar, P.D. (1992). Lead isotopic compositions in the Western Dharwar craton, southern India: evidence for distinct middle Archaean terrains in a Late Archaean craton. Geochim. Cosmochim. Acta, vol. **56**, pp. 2455–2470.

Mishra, V.P., Singh, P. and Dutta, N.K. (1998). Stratigraphy, structure and metamorphic history of Bastar craton. Rec. Geol. Surv. India, vol. **117**, pp. 1–26.

Misra, S. (2006). Precambrian chronostratigraphic growth of Singhbhum-Orissa craton, eastern Indian shield. J. Geol. Soc. India, vol. **67**, pp. 356–378.

Misra, S., Deomurari, M.P., Wiedenbeck, M., Goswami, J.N., Ray, S.L. and Saha, A.K. (1999). ^{207}Pb/^{206}Pb zircon ages and the evolution of the Singhbhum craton, eastern Indian shield. Gondwana Res., vol. **8**, pp. 129–142.

Misra, S. and Dey, S. (2002). Bihar Mica Belt plutons—an example of post-orogenic granite from eastern India shield. J. Geol. Soc. India, vol. **59**, pp. 363–377.

Misra, S. and Johnson, P.T. (2003). Geochronological constraint on the evolution of the Singhbhum mobile belt and late Archaean crustal growth of Eastern Indian shield. (cited in Acharyya, 2003).

Mitra, S.K. (2005). Tectonic setting of the Meghalaya Plateau and its sulphide mineralizations. J. Geol. Soc. India, vol. **56**, pp. 117–118.

Mohanty, S. (1993). Stratigraphic position of the Tirodi gneiss in the Precambrian terrain of central India: evidence from the Mansar area, Nagpur district, Maharashtra. J. Geol. Soc. India, vol. **42**, pp. 55–60.

Mohanty, M. and Guha, D.B. (1995). Lithotectonic stratigraphy of the dismembered greenstone sequence of the Mangalwar Complex around Lawa Sardargarh and Parasali areas, Rajsamand district, Rajasthan. In: Sinha-Roy, S. and Gupta, K.R. (eds.), *Continental Crust of Northwestern and Central India*. Geol. Soc. India Mem., Bangalore, vol. **31**, pp. 117–140.

Mondal, M.E.A. (2003). Are the Bundelkhand craton and the Banded Gneissic Complex parts of one large Archaean protocontinent? evidence from ion microprobe ^{207}Pb/^{206}Pb zircon ages. Newslett. Deep Continental. Stud. India, vol. **13**(2), pp. 6–10.

Mondal, M.E.A. and Ahmad, T. (2001). Bundelkhand mafic dykes, central Indian shield: implications for the role of sediment subduction in Proterozoic crustal evolution. Island Arc, vol. **10**, pp. 51–67.

Mondal, M.E.A., Goswami, J.N., Deomurari, M.P. and Sharma, K.K. (2002). Ion Microprobe ^{207}Pb/^{206}Pb ages of zircons from Bundelkhand massif, northern India: implications for crustal evolution of the Bundelkhand-Aravalli protocontinent. Precambrian Res., vol. **117**, pp. 85–100.

Moorbath, S., Taylor, P.N. and Jones, N.W. (1986). Dating the oldest terrestrial rocks—facts and fiction. Chem. Geol., vol. **57**, pp. 63–80.

Moyen, J.R., Martin, H. and Jayananda, M. (2001). Multielement geochemical modeling of crust-mantle interactions during Late Archaean crustal growth: the Closepet Granite (South India). Precambrian Res., vol. **112**, pp. 87–105.

Mukhopadhyay, D. (1976). Precambrian stratigraphy of Singhbhum—the problem and a prospect. Indian J. Earth Sci., vol. **3**, pp. 208–219.

Mukhopadhyay, D. (1984). The Singhbhum shear zone and its place in the evolution of the Precambrian mobile belt, north Singhbhum.Indian J. Earth Sci., **CEISM Volume**, pp. 205–212.

Mukhopadhyay, D. (1986). Structural pattern in the Dharwar craton. J. Geol., vol. **94**, pp. 167–186.

Mukhopadhyay, D. (2001). The Archaean nucleus of Singhbhum: the present state of knowledge. Gondwana Res., vol. **4**, pp. 307–318.

Mukhopadhyay, D., Bhattacharya, T., Chakraborty, T and Dey, A.K. (1990). Structural pattern in the Precambrian rocks of Sonua-Lotapahar region, North Singhbhum, eastern India. Proc. Indian Acad. Sci. (Earth Planet Sci.), vol. **99**, pp. 249–268.

Mukhopadhyay, D., Ghosh, A.K. and Bhattacharya, S. (1975). A reassessment of structures in the Singhbhum shear zone. Bull. Geol. Min. Met. Soc. India, vol. **48**, pp. 49–67.

Murthy, M.V.N. (1958). Coronites from India and their bearing on the origin of corona. Bull. Geol. Soc. Am., vol. **61**, pp. 23–38.

Murthy, M.V.N., Talukdar, S.C., Bhattacharya, A.C. and Chakraborty, C. (1969). The Dauki Fault of Assam. Bull. ONGC, vol. **6**(2), pp. 57–64.

Naha, K., Chaudhury, A.K. and Mukherji, P. (1967). Evolution of the Banded Gneissic Complex of central Rajasthan, India. Contrib. Mineral. Petrol., vol. **15**, pp. 191–201.

Naha, K. and Halyburton, R.V. (1974). Early Precambrian stratigraphy of central and southern Rajasthan, India. Precambrian Res., vol. **4**, pp. 55–73.

Naha, K. and Majumdar, A. (1971). Structure of the Rajnagar marble band and its bearing on the early Precambrian stratigraphy of central Rajasthan, western India. Geol. Rundsch., vol. **60**, pp. 1550–1571.

Naha, K. and Mohanty, S. (1990). Structural studies in the pre-Vindhyan rocks of Rajasthan: a summary of work of the last three decades. In: Naha, K., Ghosh, S.K. and Mukhopadhyay, D. (eds.), *Structure and Tectonics: The Indian Scene*. Proc. Indian Acad. Sci., Bangalore, vol. **99**, pp. 279–290.

Naha, K and Mukhopadhyay, D. (1990). Structural styles in the Precambrian metamorphic terranes of peninsular India. In: Naqvi, S.M. (ed.), Precambrian Continental Crust and Its Economic Resources. Dev. Precambrian Geol., vol. **8**. Elsevier, Amsterdam, pp. 157–178.

Naha, K., Srinivasan, R. and Naqvi, S.M. (1986). Structural unity in the Early Precambrian Dharwar tectonic province, peninsular India. Quat. J. Geol. Min. Met. Soc., India. Publ. Calcutta, vol. **58**, pp. 219–243.

Naik, A. (2001). A revision of stratigraphy of Chandiposh area in Eastern part of Sundergarh district, Orissa. Indian J. Geol., vol. **73**, pp. 403–432.

Nand Lal, Saini H.S., Nagpaul, K.K. and Sharma, K.K. (1976). Tectonic and cooling history of the Bihar Mica Belt, India, as revealed by fission-track analysis. Tectonophysics, vol. **34**, pp.163–180.

Nandy, D.R. (2001). Geodynamics of Northeastern India and Adjoining Region. ACB Publication, Kolkata, 209p.

Naqvi, S.M. (2005). *Geology and Evolution of the Indian Plate*. Capital Publishing Company, New Delhi, 450p.

Naqvi, S.M. and Rogers, J.J.W. (1987). *Precambrian Geology of India*. Oxford University Press, New York, 223p.

Narayanaswami, S. (1970). Tectonic setting and manifestation of the upper mantle in the Precambrian rocks of South India. *Proceedings of Second Symposium on Upper Mantle Project*, Hyderabad, pp. 377–403.

Narayanaswami, S., Chakraborty, S.C., Vemban, N.A., Shukla, K.D., Subramanyam, M.R., Venkatesh, V., Rao, G.V., Anandalwar, M.A. and Nagrajaiah, R.A. (1963). The geology and manganese ore deposits of the manganese belts in Madhya Pradesh and adjoining parts of Maharashtra. Bull. Geol. Surv. India Ser. A vol. **22**(1), 69p.

Newton, R.C. (1990). The Late Archaean high-grade terrain of south India and deep structure of Dharwar craton. In: Salisbury, M.H. and Fountain, D.M. (eds.), *Exposed Cross Section of the Continental Crust*. Kluwer Acad. Publ., Dordrecht, pp. 305–326.

Nutman, A.P., Chadwick, B., Ramakrishnan, M. and Vishwanatha, M.N. (1992). SHRIMP U-Pb ages of detrital zircon on Sargur supracrustal rocks in western Karnataka, southern India. J. Geol. Soc. India, vol. **39**, pp. 367–374.

Oldham, T.H. (1856). Remarks on the classification of the rocks of Central India resulting from the investigation of the Geological Survey. J. Asiatic Soc. Bengal, vol. **25**, pp. 224–256.

Pandey, B.K., Gupta, J.N. and Lall, Y. (1986). Whole-rock and mineral isochron ages for the granites from Bihar mica belt of Hazaribagh, India. Indian J. Earth Sci., vol. **13**(2–3), pp. 157–162.

Pandit, M.K. and Golani, P.R. (2000). Reappraisal of the petrologic status of Newania 'carbonatite'of Rajasthan, western India. J. Asian Earth Sci., vol. **22**, pp. 213–228. Elsevier.

Panigrahi, M.K., Misra, K.C., Breteam, B. and Naik, R.K. (2002). Genesis of the granitoid affiliated copper-molybdenum mineralization at Malanjkhand, central India: facts and problems. Extended Abstract. In: *Proceedings of the 11th Quadrennial IAGOD Symposium and Geocongress*, Windhoek, Namibia.

Pascoe, E.H. (1950). *A Manual of the Geology of India and Burma*. 3rd ed., pt. 1. Geol. Surv. India Publ., Calcutta, 483p.

Pascoe, E.H. (1973). A Manual of Geology of India and Burma, vol. **1**. Geol. Surv. India, Calcutta.

Pati, J.K., Raju, S., Mamgain, V.D. and Ravi S. (1997). Gold mineralization in part of Bundelkhand granite complex (BGC). J. Geol. Soc. India, vol. **50**, pp. 601–605.

Peucat, J.J., Vidal, P., Bernard-Griffiths, J. and Condie, K.C. (1989). Sr, Nd, and Pb isotopic systems in the Archaean low- to high-grade transition zone of southern India: syn-accretion vs. post-accretion granulites. J. Geol., vol. **97**, pp. 537–550.

Pichai Muthu, R. (1990). The occurrence of gabbroic anorthosites in Makrohar area, Sidhi district, Madhya Pradesh. In: *Precambrian of Central India*. Geol. Surv. India Spl. Publ., Nagpur, vol. **28**, pp. 320–331.

Pichamuthu, C.S. (1953). The Charnockite Problem. Mysore Geol. Assoc. Spl. Publ., Bangalore.

Pichamuthu, C.S. (1960). Charnockite in the making. Nature, vol. **188**, pp. 135–136.

Pichamuthu, C.S. (1961). Transformation of Peninsular gneiss to charnockite, Mysore State, India. J. Geol. Soc. India, vol. **21**, pp. 46–49.

Pichamuthu, C.S. (1965). Regional metamorphism and charnockitization in Mysore State, India, Indian Mineralogist, vol. **6**, pp. 46–49.

Poddar, B.C. (1965). Stratigraphic position of the Banded Gneissic Complex of Rajasthan. Curr. Sci., vol. **34**, pp. 433–434.

Radhakrishna, B.P. and Naqvi, S.M. (1986). Precambrian continental crust of India and its evolution. J. Geol., vol. **94**, pp. 145–166.

Radhakrishna, B.P. and Vaidyanadhan, R. (1997). *Geology of Karnataka*. 2nd ed. Geol. Soc. India, Bangalore, 353p.

Rahman, S. (1999). The Precambrian rocks of the Khasi Hills, Megahalaya, Shillong Plateau. In: Verma, P.K. (ed.), Geological Studies in the Eastern Himalayas. Pilgrims Book Pvt. Ltd., Delhi, pp. 59–65.

Raja Rao, C.S. (1970). Sequence, structure and correlation of the metasediments and genesis of the Banded Gneissic Complex of Rajasthan. Rec. Geol. Surv. India, vol. **98**(2), pp. 122–131.

Raja Rao, C.S., Poddar, B.C., Basu, K.K. and Dutta, A.K. (1971). Precambrian stratigraphy of Rajasthan: a review. Rec. Geol. Surv. India, vol. **101**(2), pp. 60–79.

Rajamani, V. (1988). The Kolar Schist Belt (Abstract), workshop on *The Deep Continental Crust of South India*, Jan. 1988, Field Excursion Guide, pp. 24–36.

Rajamani, V., Shivkumar, K., Hanson, G.N. and Shirey, S.B. (1985). Geochemistry and petrogenesis of amphibolites, Kolar Schist Belt, South India; evidence for komatiitic magma derived by low percent of melting of the mantle. J. Petrol., vol. **96**, pp. 92–123.

Ramachandra, H.M. and Roy, A. (2001). Evolution of the Bhandara-Balaghat granulite belt along the southern margin of the Sausar mobile belt of Central India. Proc. Indian Acad. Sci. (Earth Planet. Sci.), vol. **110**, pp. 351–368.

Ramakrishnan, M. (1990). Crustal development in southern Bastar, central Indian craton. Geol. Surv. India Spl. Publ., vol. 28, pp. 44–66.

Ramakrishnan, M. (1993). Tectonic evolution of the granulite terrains of southern India. Geol. Soc. India Mem., vol. **25**, pp. 35–44.

Ramakrishnan, M. (2003). Craton-mobile belt relations in Southern Granulite Terrain. Geol. Soc. India Mem., vol. **50**, pp. 1–24.

Ramakrishnan, M. and Vaidyanadhan, R. (2008). *Geology of India*. vol. **1**. Geol. Soc. India, Bangalore, 556p.

Ramiengar, A.S., Ramakrishnan, M. and Vishwanatha, M.N. (1978). Charnockite-gneiss complex relationship in southern Karanataka. J. Geol. Soc. India, vol. **19**, pp. 411–419.

Ramsay, J.G. (1967). *Folding and Fracturing of Rocks*. McGraw-Hill, New York, 586p.

Rao, Mallickarjun J., Pooranchandra Rao, G.V.S., Widdowson, M. and Kelley, S.P. (2005). Evolution of Proterozoic mafic dyke swarms of the Bundelkhand granite massif, central India. Curr. Sci., vol. **88**(3), pp. 502–506.

Ravindra Kumar, G.R. and Raghavan, V. (1992). The incipient charnockites in transition zone, khondalite zone and granulite zones of South India: controlling factors and contrasting mechanisms. J. Geol. Soc. India, vol. **39**, pp. 293–302.

Ray Barman, T., Bishui, P.K., Mukhopdhyay, K. and Ray, J.N. (1994). Rb-Sr geochronology of the high-grade rocks from Purulia, West Bengal and Jamua-Dumka sector, Bihar. Indian Minerals, vol. **48**(1 and 2), pp. 45–60.

Roday, P.P., Diwan, P. and Singh, S. (1995). A kinematic model of emplacement of quartz reefs and subsequent deformation patterns in the central Indian Bundelkhand batholith. Proc. Indian Acad. Sci. (Earth Planet. Sci.), vol. **104**, pp. 465–488.

Rogers, A.J., Kolb, J., Meyer, F.M. and Armstrong, R.A. (2007). Tectono-magmatic evolution of the Hutti-Maski greenstone belt, India: constraints using geochemical and geochronological data. J. Asian Earth Sci., vol. **31**, pp. 55–70.

Roy, A.K. (1977). Structural and metamorphic evolution of the Bengal anorthosite and associated rocks. J. Geol. Soc. India, vol. **18**(5), pp. 203–233.

Roy, A. (1983). Structure and tectonics of the cratonic areas of north Karnataka. In: Sinha-Roy, S. (ed.), Structure and Tectonics of the Precambrian Rocks, Recent Researches in Geology, vol. **10**, Hindustan Publ. Co., New Delhi, pp. 91–96.

Roy, A.B. (1988). Stratigraphy and tectonic framework of the Aravalli mountain range. In: Roy, A.B. (ed.), Precambrian of the Aravalli Mountain Range. Geol. Soc. India Mem., Bangalore, vol. **7**, pp. 3–32.

Roy, A., Hanuman Prashad, M. and Bhowmik, S.K. (2001). Recognition of Pre-Grenville and Grenville tectonothermal events in the Central Indian Tectonic Zone: implications on Rodinia crustal assembly. Gondwana Res., vol. **4**, pp. 755–757.

Roy, A.B. and Jakhar, S.R. (2002). *Geology of Rajasthan (Northwest India): Precambrian to Recent*. Scientific Publishers (India), Jodhpur, 421p.

Roy, A., Kagami, H., Yoshida, M., Roy, A., Bandyopadhyay, B.K., Chattopadhyay, A., Khan, A.S., Huin, A.K. and Pal, T. (2006). Rb-Sr and Sm-Nd dating of different metamorphic events from the Sausar Belt, central India: implications for Proterozoic crustal evolution. J. Asian Earth Sci., vol. **26**, pp. 61–76.

Roy, A.B., Kataria, P., Kumar, S. and Laul, V. (2000). Tectonic study of the Archaean greenstone association from Rakhiawal, east of Udaipur, southern Rajasthan. In: Gyani, K.C. and Kataria, P. (eds.), *Tectonomagmatism, Geochemistry and Metamorphism of Precambrian Terrains*. Proc. Nat. Sem., Udaipur, Publ. Dept. Geology, Udaipur, pp. 143–157.

Roy, A.B. and Kroener, A. (1996). Single zircon evaporation ages constraining the growth of the Archaean Aravalli craton, northwestern Indian shield. Geol. Mag., vol. **133**(3), pp. 333–342.

Roy, A.B., Kroener, A., Bhattacharya, P.K. and Rathore, S. (2005). Metamorphic evolution and zircon geochronology of early Proterozoic granulites in the Aravalli Mountains of northwestern India. Geol. Mag., vol. **142**(3), pp. 287–302.

Roy, A., Sarkar, A., Jayakumar, S., Aggrawal, S.K. and Ebihara, M. (2002a). Mid-Proterozoic plume-related thermal event in eastern Indian craton: evidence from trace elements, REE geochemistry and Sr-Nd isotope systematics of basic-ultrabasic intrusives from Dalma volcanic belt. Gondwana Res., vol. **5**, pp. 133–146.

Roy, A., Sarkar, A., Jayakumar, S., Aggrawal, S.K. and Ebihara, M. (2002b). Sm-Nd age and mantle source characteristics of the Dhanjori volcanic rocks, eastern India. Geochem. J., vol. **36**, pp. 503–518.

Roy, A.B., Somani, M.K. and Sharma, N.K. (1981). Aravalli—pre-Aravalli relationship—a study from Bhinder, south-central Rajasthan. Indian J. Earth Sci., vol. **8**, pp. 119–130.

Saha, A.K. (1972). The petrogenetic and structural evolution of the Singhbhum granitic complex, eastern India. *Proceedings of the 24th International Geological Congress*, pp. 147–155.

Saha, A.K. (1979). Geochemistry of Archaean granites of the Indian shield. J. Geol. Soc. India, vol. **20**, pp. 375–392.

Saha, A.K. (1994). Crustal evolution of Singhbhum-North Orissa, Eastern India. Geol. Soc. India Mem., vol. **27**, p. 341.

Saha, A.K., Ghosh, A., Dasgupta, D., Mukhopadhyay, K. and Ray, S.L. (1984). Studies on crustal evolution of the Singhbhum-Orissa Iron-Ore craton. In: Saha, A.K. (ed.), Crustal Evolution and Metallogenesis in Selected Areas of the Indian Shield. Indian Soc. Earth Sci., Monograph Volume, Calcutta, pp. 1–74.

Saha, A.K. and Ray, S.L. (1984a). Early-Middle Archaean crustal evolution of the Singhbhum-Orissa craton. Indian J. Earth Sci., CEISM Seminar Volume, pp. 1–18.

Saha, A.K. and Ray, S.L. (1984b). The structural and geochemical evolution of the Singhbhum granite batholithic complex, India. Tectonophysics, vol. **105**, pp. 163–176.

Saha, A.K., Sarkar, S.S. and Ray, S.S. (1987). Petrochemical evolution of the Bihar Mica Belt granites, eastern India. Indian J. Earth Sci., vol. **14**(1), pp. 22–45.

Santosh, M., Yokoyama, K. and Acharyya, S.K. (2004). Geochronolgy and tectonic evolution of Karimnagar and Bhopalpatnam granulite belts, central India. Gondwana Res., vol. **7**, pp. 501–518.

Sarangi, S. and Mohanty, S. (1998). Structural studies in the Chhotanagpur gneissic complex near Gomoh, Dhanbad district, Bihar. Indian J. Geol., vol. **70**(1 and 2), pp. 73–80.

Sarkar, S.N. (1968). *Precambrian Stratigraphy and Geochronology of Peninsular India: A Synopsis*. Dhanbad Publications, Dhanbad, pp. 1–33.

Sarkar, A.N. (1982). Structural and petrological evolution of the Precambrian rocks in western Singhbhum, Bihar. Mem. Geol. Surv. India, Publ. GSI Calcutta, vol. **113**, pp. 1–97.

Sarkar, A.N. (1988). Tectonic evolution of the Chhotanagpur plateau and the Gondwana basin in east India: an interpretation based on supra-subduction geological processes. In: Mukhopadhyay, D. (ed.), Precambrian of the Eastern Indian Shield, Mem Geol. Soc. India, Bangalore, vol. **8**, pp. 127–146.

Sarkar, A.N. and Basu Mallik, S. (1982). Study of palaeogeodynamics in the Rajgir metasedimentary belt: stress systems, crustal shortening and deformation pattern. Rec. Geol. Surv. India, vol. **112**(3). pp. 25–32.

Sarkar, S.S., Chatterjee, A., Mandy, S. and Saha, A.K. (1988). Classification of the granites of Bihar Mica Belt, eastern India, using stepwise multigroup discriminant analysis and cluster analysis. Indian J. Earth Sci., vol. **15**, pp. 189–200.

Sarkar, G., Corfu, F., Paul, D.K., McNaughton, N.J., Gupta, S.N. and Bishnui, P.K. (1993). Early Archaean crust in Bastar craton, central India—a geochemical and isotopic study. Precambrian Res., vol. **62**, pp. 127–137.

Sarkar, S.N., Ghosh, D.K. and Lambert, R.J.St. (1985). Rubidium-Strontium and Lead isotopic studies of the Soda Granites from Mosabani, Singhbhum Copper Belt, India. Indian J. Earth Sci., vol. **13**, pp. 101–116.

Sarkar, S.N., Gopalan, K. and Trivedi, J.R. (1981). New data on the geochronology of the Precambrians of Bastar-Drug, central India. Indian J. Earth Sci., vol. **8**(2), pp. 131–151.

Sarkar, A.N. and Jha, B.N. (1985). Structure, metamorphism and granite evolution of the Chhotanagpur granite gneiss complex around Ranchi. Rec. Geol. Surv. India, vol. **113**(3), pp. 1–12.

Sarkar, G., Paul, D.K., de Laeter, J.R., McNaughton, N.J. and Mishra, V.P. (1990). A geochemical and Pb, Sr isotopic study of the evolution of granite gneisses from the Bastar craton, central India. J. Geol. Soc. India, vol. **35**, pp. 480–496.

Sarkar, A., Paul, D.K. and Potts, P.J. (1995). Geochronology and geochemistry of the Mid-Archaean trondhjemitic gneisses from the Bundelkhand craton, central India. In: Saha, A.K. (ed.), *Recent Researches in Geology and Geophysics of the Precambrians (RRG)*, vol. **16**, pp. 76–92, Geol. Soc. India, Bangalore.

Sarkar, G., Ray Barman, T. and Corfu, F. (1989). Timing of continental arc-type magmatism in northwest India : evidence from U-Pb zircon geochronology. J. Geol., vol. **97**, pp. 607–612.

Sarkar, S.N. and Saha, A.K. (1962). A revision of the Precambrian stratigraphy and tectonics of Singhbhum and adjoining region. Quart. J. Geol. Min. Met. Soc. India, vol. **34**, pp. 97–136.

Sarkar, S.N. and Saha, A.K. (1977). The present status of the Precambrian stratigraphy, tectonics and geochronology of Singhbhum-Keonjhar-Mayurbhanj region, eastern India. Indian J. Earth Sci., **(S. Ray Volume)**, pp. 37–65.

Sarkar, S.N. and Saha, A.K. (1983). Structure and tectonics of the Singhbhum-Orissa Iron-Ore Craton, eastern India. In: Mukhopadhyay, D. (ed.), *Recent Researches in Geology* (Structure and Tectonics of the Precambrian rocks). Hindustan Publishing Corpn., Delhi, India, vol. **10**, pp. 1–25.

Sarkar, S.N., Saha, A.K., Boelrijk, N.A.L.M. and Hebeda, E.H. (1979). New data on the geochronology of the Older Metamorphic Group and the Singhbhum Granite of Singhbhum-Keonjhar-Mayurbhanj region, eastern India. Indian J. Earth Sci., vol. **6**, pp. 32–51.

Sarkar, S.N., Saha, A.K. and Miller, J.A. (1969). Geochronology of the Precambrian rocks of Singhbhum and adjacent regions, eastern India. Geol. Mag., vol. **106**, pp. 15–45.

Sarkar, A., Sarkar, G., Paul, D.K. and Mitra, N.D. (1990b). Precambrian geochronology of the central Indian shield—a review. Geol. Surv. India Spl. Publ., vol. **28**, pp. 453–482.

Sarkar, S.N., Trivedi, J.R. and Gopalan, K. (1986). Rb-Sr whole rock and mineral isochron age of the Tirodi gneiss, Sausar Group, Bhandara district, central India. J. Geol. Soc. India, vol. **27**, pp. 30–37.

Schleicher, H., Todt, W., Viladkar, S.G. and Schmidt, F. (1997). Pb-Pb age determinations on the Newania and Savattur carbonatites of India; evidence for multiple stage histories. Chem. Geol., vol. **140**, pp. 261–273.

Sederholm, J.J. (1923). On migmatites and associated Pre-Cambrian rocks of south-western Finland, Part 1: The Pellinge Region. Bull. Comm. Geol. Finlande, vol. **58**, 153p.

Sen, S. (1967). Charnockites of Manbhum and the charnockite problem. J. Geol. Soc. India, vol. **8**, pp. 8–17.

Sen, S.K. and Bhattacharya, A. (1986). Fluid induced metamorphic changes in the Bengal anorthosite around Saltora, West Bengal. Indian J. Earth Sci., vol. **13**, pp. 45–68.

Sen, S.K. and Bhattacharya, A. (1993). Post-peak pressure-temperature-fluid history of the granulites around Saltora, West Bengal. Proc. Nat. Acad. Sci. India, vol. **63**(A), Pt. 1, pp. 281–306.

Bandyopadhyay, P.K. and Sengupta, S. (1984). The Chakradharpur granite gneiss complex of west Singhbhum, Bihar. In: *Monograph of Crustal Evolution*. Indian J. Earth Sci., pp. 75–90.

Sengupta, S., Bandyopadhyay, P.K. and van den Hul, H.J. (1983). Geochemistry of the Chakradharpur Granite Gneiss Complex—a Precambrian trondhjemite body from west of Singhbhum, eastern India. Precambrian Res., vol. **23**, pp. 57–78.

Sengupta, S., Corfu, F., McNutt, R.H. and Paul, D.K. (1996). Mesoarchaean crustal history of the eastern Indian craton: Sm-Nd and U-Pb isotopic evidence. Precambrian Res., vol. **77**, pp. 17–22.

Sengupta, S. and Mukhopadhyay, P.K. (2000). Sequence of Precambrian events in the eastern Indian craton. Geol. Surv. India Spl. Publ., vol. **57**, pp. 49–56.

Sengupta, S., Paul, D.K., de Laeter, J.R., McNaughton, N.J., Bandyopadhyay, B.K. and de Smeth, J.B. (1991). Mid-Archaean evolution of the eastern Indian craton: geochemical and isotopic evidence from Bonai pluton. Precambrian Res., vol. **49**, pp. 23–37.

Sharma, R.S. (1988). Patterns of metamorphism in the Precambrian rocks of the Aravalli Mountain belt. Geol. Soc. India Mem., vol. **7**, pp. 33–75.

Sharma, R.S. (1993). Deep continental crust; an example from Rajasthan, NW India. Newslett. DCS-DST, vol. **3**(1), Govt. of India, New Delhi, pp. 6–10.

Sharma, R.S. (1995). An evolutionary model for the Precambrian crust of Rajasthan: some petrological and geochronological considerations. Geol. Soc. India Mem., vol. **31**, pp. 91–115.

Sharma, K.K. (1998). Geological evolution and crustal growth of Bundelkhand craton and its relict in the surrounding regions, north Indian shield. In: Paliwal, B.S. (ed.), *The Indian Precambrian, Scientific Publishers,* Jodhpur, pp. 33–43.

Sharma, R.S. (1999). Crustal development in Rajasthan craton. Indian J. Geol., vol. **71**, pp. 65–80.

Sharma, R.S. (2003). Evolution of Proterozoic fold belts of India: a case of the Aravalli Mountain belt of Rajasthan, NW India. Geol. Soc. India Mem., vol. **52**, pp. 145–162.

Sharma, M., Basu, A.R. and Ray, S.L. (1994). Sm-Nd isotopic and geochemical study of the Archaean tonalite-amphibolite association from the eastern Indian craton. Contrib. Mineral. Petrol., vol. **117**, pp. 45–55.

Sharma, R.S. and MacRae, N.D. (1981). Paragenetic relations in gedrite-cordierite-staurolite-biotite-sillimanite-kyanite gneiss at Ajitpura, Rajasthan, India. Contrib Mineral Petrol, vol. **78**, pp. 46–60.

Sharma, K.K. and Rahman, A. (1995). Occurrence and petrogenesis of the Loda Pahar trondhjemitic gneiss from Bundelkhand craton, northern Indian shield—Geochemistry, petrogenesis and tectonomagmatic environments. DST Newslett., vol. **6**, New Delhi, pp. 12–19.

Sharma, K.K. and Rahman, A. (1996). Mafic dykes in Bundelkhand granitoids and mafic volcanics in Supra-batholithic volcano-sedimentaries (Bijawar). DST Newsletter, vol. **15**, New Delhi, pp. 17–19.

Singh, S.P. (1998). Precambrian stratigraphy of Bihar—an overview. In: Paliwal, P.S. (ed.), Indian Precambrian. Scientific Publishers, Jodhpur, pp. 376–408.

Singh, S.P. (2001). Early Precambrian stratigraphy of the Chhotanagpur Province. In: Singh, S.P. (ed.), *Precambrian Crustal Evolution and Mineralization in India*. South Asian Association of Economic Geologist, Patna (Chapter), pp. 57–75.

Singh, Y., and Chabria, T. (1999). Late Archaean-Early Proterozoic Rb-Sr isochron age of granites from Kawadgaon, Bastar district, Madhya Pradesh. J. Geol. Soc. India, vol. **54**, pp. 405–409.

Singh, Y., and Chabria, T. (2002). Early Proterozoic ^{87}Rb/^{86}Sr model ages of pegmatitic muscovite from rare metal-bearing granite-pegmatite system of Kawadgaon, Bastar craton, central India. Gondwana Res., vol. **5**, pp. 889–893.

Singh, A.P., Mishra, D.C., Vijaya Kumar, V. and Vyaghreswara Rao, M.B.S. (2003). Gravity-magnetic signatures and crustal architecture along Kuppam-Palani geotransect, south India. Geol. Soc. India Mem., vol. **50**, pp. 139–163.

Singh, R.N., Thorpe, R. and Kristic, D. (2001). Galena Pb isotope data of base metal occurrences in the Heasatu-Belabathan belt, eastern Precambrian shield, Bihar. J. Geol. Soc. India, vol. **57**, pp. 535–538.

Sinha, K.K., Krishna Rao, N., Shah, V.L. and Sunil Kumar, T.S. (1997). Stratigraphic succession of Precambrian of Singhbhum—evidence from quartz-pebble conglomerate. J. Geol. Soc. India, vol. **49**, pp. 577–588.

Sinha-Roy, S. (1985). Granite-greenstone sequence and geotectonic development of SE Rajasthan. In: Proceedings of Symposium Megastructures and plate tectonics and their role as a guide to ore mineralization. Bull. Geol. Min. Met. Soc. India, vol. **53**, pp. 115–123.

Sinha-Roy, S., Guha, D.B. and Bhattacharya, A.K. (1992). Polymetamorphic granulite facies pelitic gneisses of the Precambrian Sandmata complex, Rajasthan. Indian Minerals, vol. **48**(1), pp. 1–12.

Sinha-Roy, S., Malhotra, G. and Mohanty, M. (1998). Geology of Rajasthan. Geol. Soc. India, Bangalore, 278p.

Sivaraman, T.V. and Odom, A.L. (1982). Zircon geochronology of Berach granite of Chittorgarh, Rajasthan. J. Geol. Soc. India, vol. **23**, pp. 575–577.

Solanki, J.N., Sen, B., Soni, M.K., Tomar, N.S. and Pant, N.C. (2003). Granulites from southeast of Wardha, Sidhi district, Madhya Pradesh in NW extension of Chhotanagpur Gneissic Complex: Petrography and geothermobarometric Estimation. Gond. Geol. Mag., vol. **7**, pp. 297–311.

Srinagesh, D. and Rai, S.S. (1996). Teleseismic tomographic evidence for continental crust and upper mantle in south Indian Archaean terrains. Phys. Earth Planet. Int. vol. **97**, pp. 27–41.

Srinivasan, V. (2005). The Dauki Fault in northeast India: through remote sensing. J. Geol. Soc. India, vol. **66**, pp. 413–426.

Srivastava, D.C. (2001). Deformation pattern in the Precambrian basement around Masuda, central Rajasthan. J. Geol. Soc. India, vol. **57**, pp. 197–222.

Srivastava, R.K., Hall, R.P., Verma, R. and Singh, R.K. (1996). Contrasting Precambrian mafic dykes of the Bastar craton, central India: Petrological and geochemical characteristics. J. Geol. Soc. India, vol. **48**, pp. 537–546.

Srivastava, R.K. and Singh, R.K. (2003). The Paleoproterozoic dolerite dyke swarm of the southern Bastar craton, central-east India: evidence for the Columbia supercontinent. Geol. Soc. India Mem., vol. **52**, pp. 163–177.

Srivastava, R.K. and Sinha, A.K. (2004). Early Cretaceous Sung Valley ultramafic-alkaline-carbonatite complex, Shillong Plateau, northeastern India: petrological and genetic significance. Mineral Petrol., vol. **80**, pp. 241–263.

Srivastava, D.C., Yadav, A.K., Nag, S. and Pradhan, A.K. (1995). Deformation style of the Banded Gneissic Complex in Rajasthan: a critical evaluation. Geol. Soc. India Mem., vol. **31**, pp. 199–216.

Stein, H.J., Hannah, J.L., Zimmerman, A., Markey, R.J., Sarkar, S.C. and Pal, A.B. (2004). A 2.5 Ga porphyry Cu-Mo-Au deposit at Malanjkhand, central India: implications for Late Archaean continental assembly. Precambrian Res., vol. **134**, pp. 189–226.

Straczek, J.A. and others (1956). Manganese ore deposits of Madhya Pradesh, India. In: *Symposium Sobre Yacimientos de Manganeso*. Proc. 20th Intern. Geol. Congr., Mexico, vol. **IV,** Tomo, pp. 63–96.

Subba Rao, M.V., Narayana, B.L., Divakar Rao, V. and Reddy, G.L.N. (1999). Petrogenesis of the protolith for the Tirodi gneiss by A-type granite magmatism: geochemical evidence. Curr. Sci., vol. **76**, pp. 1258–1264.

Swami Nath, J. and Ramakrishnan, M. (1981). The early Precambrian supracrustals of southern Karnataka. Geol. Surv. India. Mem., vol. **112**, p. 350.

Swami Nath, J. and Ramakrishnan, M. (1990). Early Precambrian supracrustals of southern Karnataka. A present classification and correlation. Geol. Soc. India Mem., vol. **19**, pp. 145–163.

Tobisch, O.T., Collerson, K.D., Bhattacharya, T. and Mukhopadhyay, D. (1994). Structural relationship and Sr-Nd isotope systematics of polymetamorphic granitic gneisses and granitic rocks from central Rajasthan, India—implications for the evolution of the Aravalli craton. Precambrian Res., vol. **65**, pp. 319–339.

Tuttle, O.F. and Bowen, N.L. (1958). Origin of granite in the light of experimental studies in the system $NaAlSi_3O_8–KAlSi_3O_8–SiO_2–H_2O$. Geol. Soc. Am. Mem., vol. **74**, 153p.

Upadhyaya, R., Sharma, B.L Jr., Sharma, B.L. Sr. and Roy, A.B. (1992). Remnants of greenstone sequence from the Archaean rocks of Rajasthan. Current Sci., vol. **63**(2), pp. 87–92.

Vasudev, V.N., Chadwick, B., Nutman, A.P. and Hegde, G.V. (2000). Rapid development of the Late Archaean Hutti Schist belt, northern Karnataka: Implications for new field data and SHRIMP U/Pb ages. J. Geol. Soc. India, vol. **55**, pp. 529–540.

Vemban, N.A., Subramanian, K.S., Gopalkrishnan, K. and Venkata Rao, V. (1977). Major faults, dislocations/lineaments of Tamil Nadu. Geol. Surv. India Misc. Publ. vol. **10**, pp. 53–56.

Vinagradov, A.P., Tugarinov, A.I., Zhykov, C.I., Stanikova, N.I., Bibikova, E.V. and Khore, K. (1964). Geochronology of Indian Precambrians. *Reports of 22nd International Geological Congress*. vol. **10**, New Delhi, pp. 553–567.

Wanjari, N. and Ahmad, T. (2007). Geochemistry of granitoid and associated mafic enclaves in Kalapathri area of Amgaon gneiss complex, central India. Gond. Geol. Mag. Spl., vol. **10**, pp. 55–64.

Wanjari, N., Asthana, D. and Divakar Rao, V. (2005). Remnants of early continental crust in the Amgaon gneisses, central India: Geochemical evidence. Gond. Res. (Gondwana Newsletter Section), vol. **8**. pp. 589–595.

Wiedenbeck, M. and Goswami, J.N. (1994). An ion-probe single zircon $^{207}Pb/^{206}Pb$ age from the Mewar Gneiss at Jhamakotra, Rajasthan. Geochem. Cosmochem. Acta, vol. **58**, pp. 2135–2141.

Wiedenbeck, M. and Goswami, J.N. and Roy, A.B. (1996a). An ion microprobe study of single zircons from Amet granite, Rajasthan. J. Geol. Soc. India, vol. **48**, pp. 127–137.

Wiedenbeck, M. and Goswami, J.N., Roy, A.B. (1996b). Stabilization of the Aravalli craton of northwestern India at 2.5 Ga: an ion microprobe zircon study. Chem. Geol., vol. **129**, pp. 325–340.

Williams, I.S. and Hergt, G.T. (2000). U-Pb dating of Tasmania dolerites: a cautionary tale of SHRIMP analysis of high-U zircon. In: Woodland, J.D., Hergt, J.M., Noble, W.P. (eds.), Beyond 2000: New Frontiers in Isotopic Geosciences, Lorne, 2000. Abstract and Proceedings, pp. 185–188.

Yedekar, D.B., Jain, S.C., Nair, K.K.K. and Dutta, K.K. (1990). The central Indian collision suture. In: Precambrian of Central India. Geol. Surv. India Spl. Publ., Nagpur, vol. **28**, pp. 1–43.

Yedekar, D.B., Reddy, P.R. and Divakar Rao, V. (2000). Tectonic evolution of the CIS zone, Central Indian shield. In: Roy, A. (ed.), *Deep Continental Studies in Central India*. Geol. Surv. India Spl. Publ., Calcutta, vol. **63**, pp. 163–169.

Chapter 3
The Young and Old Fold Belts

3.1 Introduction

With the basic concept of plate tectonics and mountain building processes discussed in Chap. 1, we now attempt to understand the evolution of the fold belts or mobile belts or orogenic belts in light of the plate tectonics theory. We believe that orogenic belts, being fold-thrust belts, formed at convergent margins where the adjacent plates had moved toward each other and collided. These margins are called subduction zones and are classified as either oceanic or sub-continental, depending upon whether the crust of the overriding plate above the subduction zone is oceanic or continental. Oceanic subduction zones are marked by a trench and by an arcuate chain of volcanoes on the overriding plate, called island arcs or ocean island arcs. Subcontinental convergent margins are also marked by an oceanic trench and by an arcuate chain of volcanoes that are built on the continental crust. These margins are called continental arcs or Andean-type continental margins. The Andes is the simple orogenic belt that formed when oceanic lithosphere (Nazca plate) subducted eastward under a continental margin of South America. This mountain belt is a 900 km long along the western margin of South America. The prominent features are the active Peru-Chile trench and subduction zone, the active volcanic arc on land, and an active fold-thrust-belt along the eastern margin of the mountain chain. The Nazca plate is subducting beneath South America, along much of the length of the Andes. The Andes have a long Pre-Mesozoic to Recent history (Dewey and Bird, 1970).

Continental mountain ranges can also develop as the result of a collision between an island arc and a continent, but such mountain belts should be smaller than those formed by continent-continent collision. Arc-continent collision zones are relatively rare, as they usually represent an intermediate step in the closure of an ocean and hence are short lived.

Collision fold belts typically develop when subduction to a continental margin brings a continent that makes up an integral part of the subducting oceanic lithosphere. The result is the collision of a continent with another continent. By studying the young collision orogenic belts, we can discover the relationship between orogenic structures and related plate tectonic activity. The younger orogenic belt preserves most of the sedimentation and deformation history.

3.2 Himalaya as an Example of the Young Fold Belt

We now look at the characteristics of an orogenic belt that formed by continent-continent collision and the Himalayan mountain range is the best example to understand the events of the plate tectonics in the development of the youngest collision fold belt on the earth. In the evolution of this mountain belt, like other collision fold belts, the geological record of collision is always preceded by Andean-type orogenesis. The Himalayan orogenic belt forms a part of the Alpine-Himalayan mountain chain that stretches from Spain in the west to Indonesia in the east. The Himalaya is among the youngest mountains of the world formed in the framework of plate tectonics theory and a classical example of continent-continent collision. The Himalaya offers an unusually good opportunity for understanding the complex interplay of thermal and mechanical processes that lead to the distribution of strain pattern and magmatic and metamorphic rocks that characterize the different parts of the fold belt. It is here that one can recognize the colliding plates, subduction zone (s), Arc etc. The evolution of Himalaya would thus help us in organizing our observations and developing our ability to apply plate tectonic in the evolution of the Proterozoic fold belts of India described in equal details in this book. Before we take up the evolution of the Himalayan mountain belt we should know the geological setting of this young orogenic belt of India.

3.2.1 Geological Setting

The Himalayan orogenic belt in Indian sector is 250–300 km wide and about 3000 km long from Nagaland State of India in the east to Kashmir in the western India and beyond up to central Afghanistan. The eastern boundary of the Himalaya is taken as the major bend or syntaxis, the Namche-Barva syntaxis, although doubted by some geologists. A similar syntaxial bend is seen in the western part of the Himalaya and is known as the Kashmir syntaxis or Nanga Parbat syntaxis. The Himalayan belt comprises a series of lithologic and tectonic units that run parallel to the mountain belt, maintaining a constant character for long distances (Fig. 3.1). This figure is a compilation after Auden (1937), Heim and Gansser (1939), GSI (1962), Gansser (1964) and Valdiya (1977).

The main geological features of the Himalaya (Fig. 3.1) from north to south are:

1. Tibetan or Asian Block
2. Trans-Himalayan Batholith (Early Cretaceous to Early Oligocene)
3. Indus-Tsangpo Suture Zone (Early to Middle Cretaceous)
4. The Tethys (Tibetan) Sedimentary Zone (Late Precambrian to Early Eocene)

    ——-South Tethyan Detachment Fault, STDS (Trans Himadri Fault of Valdiya, 1989)——-

5. Higher Himalayan Crystalline (HHC) (Late Precambrian with Old and Young Granites incl. Miocene)

    ————————-Main Central Thrust (MCT)————-

3.2 Himalaya as an Example of the Young Fold Belt

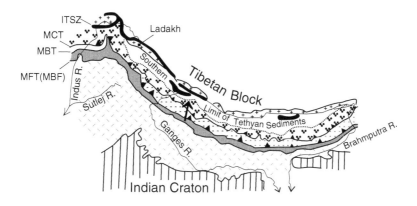

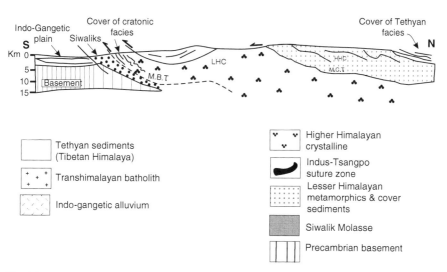

Fig. 3.1 Generalized geological map of the Himalaya showing the principal geological divisions (litho-tectonic units), based on the compilation map of GSI (1962), Gansser (1964), and Valdiya (1977). Bottom figure is a crustal section showing the relationship of different shear zones. Abbreviations: HHC = Higher Himalayan Crystallines, ITSZ = Indus-Tsangpo shear zone, MBT = Main Boundary Thrust, MCT = Main Central Thrust, MFT (MBF) = Main Frontal thrust (Main Boundary fault)

6. Lesser Himalayan Crystalline (LHC) and Cover Sediments (Proterozoic to Palaeozoic/Caledonian Granitoids)

——————————— Main Boundary Thrust (MBT)———————————

7. Subhimalaya or Outer Himalaya (Cenozoic Molasse: Siwaliks)

——————————— Himalayan Frontal Fault ———————————

8. Indian Plate (northern margin covered with alluvium)

The major units of Himalaya occur mainly in the Indian sector where the Himalaya is divided as Lesser or Lower Himalaya (elevation of 1500–3000 m) and Higher Himalaya with elevations from more than 3000 to 8000 m. The Lesser Himalayan rocks (LHC) consist of both Precambrian metamorphic and cover sediments (weak to unmetamorphosed) and are thrust over the sub-Himalayan rocks along Main Boundary Thrust (MBT), whereas the Higher Himalayan Crystallines (HHC) have been thrust over the LHC rocks along the Main Central Thrust (MCT) which is a zone of ductile deformation of several kilometers thick (bottom sketch of Fig. 3.1). The metasediments at higher levels of the HHC unit are intruded by granites of Miocene age, which originated by melting of the pelitic schists in upper crustal conditions when the HHC unit was exhumed along the south-vergent, north-dipping Main Central Thrust (MCT). The southward thrusting of the Himalayan rocks commenced mostly in Middle and Late Tertiary and some movements continue even today (see Auden, 1937).

The Himalaya is also divided into three geographic divisions from west to east as Western, Central and Eastern Himalaya. Western Himalaya encompasses mountain ranges (Ladakh, Zanskar, Dauladhar, Pir Panjal etc.) of the State of Jammu and Kashmir, Himachal Pradesh, Uttarakhand (Kumaun and Garhwal Hills) and the adjoining Karakoram, Kohistan, Kailas, Mansarovar ranges. Central Himalaya essentially covers Nepal. The mountain ranges of Sikkim, Darjeeling, Bhutan and Arunachal Pradesh are included in the Eastern Himalaya.

The Himalaya is bordered on the south by the Indian shield covered under the molasse and alluvium of Indus and Ganges river system. Prior to the occurrence in the Himalaya, all of the HHC and the Lesser Himalayan crystallines were once a part of the northern Indian cratonic rocks, mostly of Rajasthan-Bundelkhand craton, Chhotanagpur Gneiss Complex and the Megahalaya craton. These rocks were a variety of Neoarchaean to Proterozoic gneiss, granite and older metasediments which were involved, along with their sedimentary cover in the Himalayan orogeny and were variously deformed and metamorphosed. The sedimentary successions of Eocambrian, Paleozoic and Mesozoic periods were extensively developed in various parts of the Himalaya, notably in Kashmir, Chamba, Zanskar-Spiti and Kinnaur-Kumaun areas of the Western Himalaya. The Lesser Himalaya in Kumaun has also several Proterozoic sedimentary basins, namely the Krol belt, Shali belt (Deoban-Garhwal) and Chamoli-Tejam belt. Most of these sediments were deposited in intracratonic and pericratonic basins (corresponding to the Purana platform basins of the Indian shield) when India was a member of the Eastern Gondwana located in southern hemisphere. The present configuration of these basins is due to the Tertiary Himalayan orogeny and subsequent erosion along various structural levels (Bhargava and Bassi, 1999).

The Indus-Tsangpo (Zangpo) Suture Zone (ITSZ) is considered to mark the southern junction of the Indian and Asian plates in the Himalaya (Gansser, 1964) along which the Neo-Tethys oceanic lithosphere was subducted as early as 140 Ma ago (Searle et al., 1987; Ahmad et al., 2008). The suture is a few kilometers narrow zone chiefly of ophiolite mélange that dips steeply south. The ophiolites are not continuous and at places are replaced by sediments typical of fore-arc environment.

The ITSZ continues westward into Pakistan where it is known as Main Mantle Thrust (MMT). Rocks within the suture are mostly Cretaceous or Early Tertiary in age. Geological evidence shows that all of the Himalayan rocks to the south of the suture were once part of the Indian plate and not derived from the Asian plate. Immediately to the north of the ITSZ is a Trans-Himalayan Batholith for almost entire length of the Himalaya. The batholith is a magmatic arc and formed as a result of melting of the Neo-Tethys oceanic floor which subducted beneath the Asian plate, prior to India-Asia collision. The Batholith received different names, Kohistan batholith in Pakistan, Ladakh batholith in Indian Territory; Kailas and Gandese pluton in southern Tibet; while in Arunachal Himalaya it is called Lohit batholith. The island arc rocks of Ladakh in India (and of Kohistan arc in Pakistan) are Cretaceous to Early Tertiary in age, ranging from 140 to 60 Ma. These rocks invade and overlie the Precambrian-Early Mesozoic continental crystalline rocks. However, the volcanic rocks in Kohistan arc appear to be part of an intra-oceanic island arc.

To the north of the ITSZ is the Southern Tibetan plate with cover of Palaeozoic and Mesozoic sediments that are intruded by the Cretaceous-Eocene (Powell and Conaghan, 1975) Trans-Himalayan granite batholith, the Kangdse Granite, formed in response to the northward underthrusting of the Tethyan Ocean, prior to India-Tibet (Asia) collision. Its equivalent unit in the Himalayan sector of Pakistan is the Kohistan arc that is formed within the Tethys in Mid-Cretaceous times, ultimately to become squeezed between the converging Indian and Asian Plates at about 100 Ma ago. It should be mentioned that the Kohistan arc is separated from the Karakoram Granite batholith to the north by a suture called Bangong-Nuijiang (Anduo) suture. Further north, the Northern Tibetan Block (Quangtong) is separated from the Southern Tibetan Block by an E-W suture, called the Kakoxili suture (Chang et al., 1986). The Northern Tibetan Block is apparently equivalent to the Central Pamir block as it is characterized by Precambrian basement overlain by a Silurian-Permian sedimentary sequence of Gondwana affinity, extensive Cretaceous calc-alkaline volcanics and a thick Tertiary sequence.

Before we discuss plate tectonic evolution of the Himalaya as a single fold belt, we give below the geological characteristics based on agreed observations and geochronological data.

1. The Himalayan lithounits are bounded by thrusts/shear zones, running along the length of the Mountain belt.
2. The Himalayan thrusts are north dipping.
3. The Higher Himalayan Crystallines (HHC or GHS) are characterized all along by inverted metamorphism wherein higher grade metamorphism is in higher structural levels and not in the deeper nappes.
4. Metamorphic isograds are nearly parallel to the regional foliation in the Himalayan Metamorphic Belt (HMB).
5. SW-vergent folds and NNE-SSW stretching lineations are widespread in the (HMB).
6. Higher Himalayan Crystallines (HHC) were thrust along the Main Central Thrust (MCT) on the Lesser Himalayan rocks (LHC). Erosion of the hanging

wall rocks (i.e. HHC) left them as Klippe (erosional remants), while exposures of footwall by erosion of the HHC show the LHC as tectonic windows at several places.
7. Despite being overridden by HHC, the LHC rocks do not show higher pressure assemblages!
8. The basement to the Himalayan sediments is Archaean to Proterozoic metamorphic rocks, similar in nature to the rocks of the northern Indian continental crust.
9. Although protolith ages of the Himalayan metamorphic rocks of Higher Himalayan Crystallines (HHC) and Lesser Himalayan Crystallines (LHC) are Precambrian, their mineral ages are Tertiary.
10. Leucogranite of 18–20 Ma age occurs all along the length of the HHC in the upper levels, near the contact of the HHC with the Tethyan sedimentary zone along Trans-Himadri Fault.
11. Subduction polarity in the Himalaya is similar to that in the Alps, but convexity of the Himalaya is just opposite to that of the Alps and so is the metamorphism; Alps shows normal Barrovian zones while the Himalaya shows inverted metamorphism.

With this basic information, we now describe evolution of the Himalayan fold belt.

3.2.2 Evolution of Himalaya

The Himalaya formed in response to the collision of India with Asia. The collision was brought about by the northward migration of the Indian plate at about 180 Ma ago. As the intervening oceanic crust of Tethys Sea was subducted below the southern margin of the Asian plate, India's journey started northward for about 2500 km. Before the evolution of the Himalaya or main continent-continent collision of Indian and Tibetan plate, there was a successive collision of terrains (or microcontinental blocks), derived from the northern margin of the Gondwanaland. This is evidenced by the presence of many E-W trending ophiolite belts that showed southward younging (ascertained by radiometric dating). Impressed by this feature, Chang and Cheng (1973) suggested that there was a successive collision of several crustal blocks after the progressive closure of several small oceans during Phanerozoic (Fig. 3.2). First, the Northern Tibet welded to Eurasia along Kokoxili suture (KS) at about 140 Ma. It was followed by a step-back in the subduction zone to the south of North Tibet. This underthrusting brought the Southern Tibet plate into juxtaposition with the North Tibet and eventually welded the North Tibet and South Tibet along the Bangong-Nujiang suture (BNS). A further step-back in underthrusting preceded the main collision event. India collided with Southern Tibet along the Indus-Tsangpo suture zone (ITSZ). The successive collision of these blocks and southward stepping of north-dipping suture zone led Allegre et al. (1984) to propose the evolutionary history of the Himalaya, depicted in Fig. 3.2. It is obvious

3.2 Himalaya as an Example of the Young Fold Belt

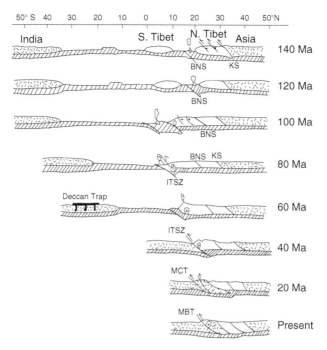

Fig. 3.2 Sequence of events in the evolution of the Himalaya (redrawn after Allegre et al., 1984). Abbreviatons: BNS = Bongong-Nujiang suture, ITSZ = Indus-Tsangpo suture zone, KS = Kokoxili suture, MCT = Main Central Thrust, MBT = Main Boundary Thrust

from this that the Himalaya contains at least two microcontinents (North Tibet and South Tibet) that represent suspect terrains.

Evolution of the Himalaya began when India, after separation from the Gondwana assembly, migrated northward to collide with the accreted Asian plate. A large area of intervening oceanic plate of ca. 2500 km—the site of Tethys which separated India and Asia—was destroyed by subduction under the Asian plate. The subducting ocean floor basalt (and some thin veneer of sediments that had deposited on the ocean floor) partially melted to produce andesitic/basaltic lava with granitoids crystallizing below it as calc-alkaline magmatic arc. This arc first developed as an oceanic Island arc, exposed as Kohistan-Ladakh arc to which the southern margin of the Asian plate (Lhasa or Tibetan Block) accreted along what is now the Shyok suture. The north-moving Indian plate also accreted to the Arc here and its eastern extension along Indus-Tsangpo suture zone (ITSZ). The collision is considered to have occurred obliquely in the NW, and the first impingement of Indian plate was in Kohistan-Ladakh sector (65 Ma ago), followed by anticlock-wise rotation of Indian plate until final collision with South Tibet (Lhasa Block of Asia) at 50 Ma ago (Dewey et al., 1989). Accretion further east developed Andean-type batholith or continental-type magmatic arc (Sharma, 1987), called the Trans-Himalyan Batholith (100–60 Ma old). This island arc ourcrops almost continuously along the southern margin of the Asian plate for a distance of about 3000 km from Pakistan to Nagaland

State of India (see Fig. 3.1). The magmatic-arc plutonic complex from Kailas mountain area is dated by U–Pb method on sphene at 107 ± 1.4 Ma with a lower intercept of 33.8 ± 7.3 Ma (Miller et al., 2001). The magmatic arc, although undeformed, experienced an episodic uplift after the Indian-Asian collision, possibly due to the underthrusting of Indian lithosphere and also because of rapid exhumation of ultra high pressure rocks of Tso-Morari dome (see further). With continued subduction, a mixture of pillow lavas, dunites, peridotites and serpentinites as hydrated ultramafic rocks transported from their birth place arrived to the final position within a convergent plate boundary where this mixture increased in volume by several hydration reactions (giving rise to serpentinite). Here, the mixture along with Cretaceous-Miocene flysch (deposited on the basaltic crust of the Neo-Tethys) was subjected to load pressure of the overlying Asian plate and developed glaucophane in basalt and greywacke compositions under high pressure and low temperature, giving rise to blue schists. The heterogenous mixture of crystallized and recrystallized rocks and marine sediments (as accretion wedge) piled up due to "bulldozing effect" by the leading edge of the Asian plate. This amalgamated pile called *ophiolite complex* was emplaced as scrapped off material of the subducting plate along weak zones (sutures) in the overlying accreted Asian plate, due to combined effect of tectonics and buoyancy. By this emplacement, the ophiolite complex occurs as tectonic fragments on both sides of the Island arc and is familiarly known as ophiolite mélange zone. The mélange on the east formed the Shyok ophiolite zone while on the west the ophiolite mélange zone became the Indus ophiolite complex that is believed to be the site of the trench zone and hence called the Indus-Tsangpo Suture Zone (ITSZ). The geology of the ITSZ ophiolite complex is comparable to that of the Shyok suture, but the two occurrences differ somewhat in their geochemistry (Ahmad et al., 2005). The ITSZ is demarcated by an ophiolite complex that is a remnant of the subducting oceanic crust, thrusted up at some stage onto downgoing continental crust of India. The ophiolites here are not well preserved complexes as elsewhere, but are dismembered lithologic sequences emplaced as tectonic slices or mélanges. The ophiolite complex not only occurs as tectonic slices or isolated blocks, but also appears as pebbles within continental detrital series (e.g. Shergol conglomerate).The emplacement of the ophiolite mélanges is believed to have raised the Trans-Himalayan Batholith at about 60 Ma ago and build the lofty mountain of the Himalaya. The occurrence of ophiolite is considered to demarcate the contact of Indian and Asian plates hence the ITSZ denotes the join between these two continental blocks. It is a matter of debate as to how the ophiolite complex was emplaced. According to one view, the ophiolites were emplaced onto the continental margin during arrested subduction of the buoyant Indian plate. To attain isostasy the Indian plate rose up, lifting on its margin a torn off slice of overriding plate as ophiolite complex. An alternative model proposes that the ophiolite complex was emplaced onto an active Indian plate by a process called obduction. Here, a part of the subducting oceanic plate detached along a pre-existing fault and was shoved onto the Indian continental margin as the Indian plate continued to subduct. Many ophiolites can thus be tectonically emplaced not only on continental crust but also over island arc, normally with a thin layer of amphibolite at the base. The radiometric age data on the glaucophane yielded 100 ± 20 to 67 ± 12 Ma (Sengor, 1989), implying that the blueschist facies metamorphism occurred much before the main Himalayan collision.

With further subduction, India came close to the accreted Asian plate and continent-continent collision of the India and Asia commenced in the Late Eocene (about 45 Ma ago). Collision along the ITSZ occurred between 45 and 55 Ma, as evidenced by paleomagnetic data (Klootwijk et al., 1992), paleontological criteria of similarity of fauna and fossils between India and Asia after collision (Sahni et al., 1981), the end-phase of subduction-related magmatism (Searle et al., 1987), stratigraphic record of the termination of Neo-Tethys ocean floor crust (Beck et al., 1995), and onset of purely terrigenous molasses along the ITSZ (Powell and Conaghan, 1973) as well as the last marine sediment in the Himalayan region as nummilitc limestone of Early Eocene age in the Lesser Himalaya (Subathu Formation) and in the northern parts of the Tethyan Himalaya, marking the regression of Tethys sea. Estimates showed that northward movement of India continued for at least a further 1500 km after the initial collision. This movement seems to have been accommodated by the crustal shortening within the colliding plates and the effects are observed in the development of folds and thrust in the Proterozoic and Lower Palaeozoic sequences of the northern Indian plate. It is estimated that the N-S compression generated due to collision of Indian-Asian plates gave rise to crustal shortening of 300–700 km within the colliding plates (cf. Kearey and Vine, 1996; see also Saklani, 1993). Structures attributed to this deformation include folds, thrust belts, strike-slip faults, and extensional rift system. The N-S compression produced immense crustal thickness in the Proterozoic and Lower Palaeozoic cover sequences of the northern Indian plate margin during underthrusting of the Indian plate. This crustal thickening was followed by thermal relaxation (Himalayan metamorphism, 32–30 Ma ago) so that the sediments and the basement rocks recrystallized (cf. Dewey and Burke, 1973), developing Barrovian regional metamorphism up to kyanite isograd in these structurally duplicated rocks, attaining maximum pressure of 8–11 kbar at temperatures up to 650°C (Searle et al., 1987).

While India began to underthrust the Asian plate, the subducting lithosphere may have anchored some continental pieces of this northern Indian margin, which recrystallized at greater depths, corresponding to the P-T stability of coesite and diamond. The metamorphosed thick crust disturbed the isostasy in this part of the Earth. The lithostatic pressure of 30 km (equivalent to 10 kbar) was due to the load of the ophiolite-bounded magmatic arc exerting on these "Indian" rocks below. As a result, the HHC metamorphic rocks exhumed only as mid crustal rocks, leaving the Indian lower crust at its own place in the colliding plate. This explains why no rocks typical of lower crust have been exhumed south of ITSZ, despite metamorphic assemblages indicating upper mantle depths (Sharma, 2008). At the time when the HHC rocks were undergoing regional metamorphism, the Lesser Himalayan rocks near the leading edge of the Indian plate also recrystallized up to garnet grade with normal metamorphic progression It is completely ruled out that the inverted metamorphism found all along the Himalayan Metamorphic Belt (HMB) was due to reverse geothermal gradient in the deep crust, for there is no physical explanation for reversal of the isotherms in the deep crust/mantle, particularly when regional metamorphic belts are commonly characterized by normal Barrovian metamorphism. The HMB rocks of the Himalaya are, unlike most regional metamorphic terrains, associated with continental collisions, where normal progressive metamorphism is found from deeper tectonic levels (showing high grade rocks) to successively higher structural levels (showing lower grade rocks). It implies that the inversion of metamorphic isograds in the Himalayan Metamorphic Belt (HMB) occurred later for which different explanations have been advanced, as discussed below. To restore isostasy, there occurred crustal extension of the deformed and crystallized continental rocks of the Indian plate margin. The Higher Himalayan Crystalline rocks were exhumed along the south directed, north-dipping Main Central Thrust (MCT), as evidenced by NE-SW trending mineral stretching and S-C fabrics in HHC rocks. As a result, the HHC were thrust upon the Lesser Himalayan rocks made of Proterozoic to Lower Cenozoic sedimentary cover of the Indian continent. Cartlos et al. (2004) state that the MCT shear zone in the Sikkim Himalaya was active during 22–10 Ma and show that the average monazite ages decrease toward structurally lower levels. They also showed that the peak metamorphism in the Lesser Himalaya occurred at 610°C/7.5 kb and 12–11 Ma and that of the MCT zone at 525°C/6 kb at 14–12 Ma. Some geologists consider it a possibility that the SW-directed stacking of the HHC was contemporaneous with the NE-directed stacking of the Tethyan Sedimentary zone. During this exhumation, the HHC rocks, particularly its pelitic schists, underwent decompression melting, forming 18–22 Ma old leucogranites that occur in the higher levels of the HHC near its contact with the Tethyan sedimentary zone. It is the dehydration melting reaction of muscovite that gave rise to the formation of migmatite and sillimanite isograd (4–6 kbar/750 $\pm$ 50°C) much after the Barrovian index mineral kyanite in the HHC rocks. The sillimanite zone thus formed occurs at the top of the inverted Barrovian isograds. Harris et al. (2004) documented decompression melting of kyanite zone metapelites in western Sikkim at 23 Ma to generate leucogranites.

Both HHC and Lesser Himalayan metamorphics exhibit *inverted metamorphism*, as stated earlier. Various models or explanations have been suggested for this inverted metamorphism. These can be grouped into three categories: (i) recumbent folding of the isograds; (ii) syn- to post-metamorphic thrusting; and (iii) shear heating. Some even suggested that ductile deformation caused translation of material along ductile shears, which consequently caused inverted metamorphism (Jain and Manickavasagam, 1993; Jain et al., 2002). In this model, translation of material occurs by grain-scale processes across a zone. This translation is accommodated by the distributed ductile shears. An analogue for this regional non-coaxial deformation could be a card deck where small amounts of slip between individual cards is likely to accommodate much larger displacements across the whole deck. This ductile deformation is assumed to translate material from deeper crustal levels whereby higher metamorphic grades overlie shallower rocks of lower grade, causing inverted metamorphism. The displacements on ductile zone will re-align pre-existing markers, e.g. lithological contacts, isograds etc, into approximate parallelism with the shear zone boundary. In this model of ductile shearing, the isograds need not be synchronous with the shearing and it is a question whether the entire Himalayan Metamorphic Belt represents a vast intracontinental ductile shear zone. Since recumbent folding is uncommon in the Himalayan metamorphic belt, unlike the Alps, inverted metamorphism cannot be attributed to a huge recumbent fold. In the model of post metamorphic thrusting, the higher temperature slab is assumed to thrust over the lower temperature slab (Le Fort, 1975, 1989). To preserve inverted metamorphism, exhumation (uplift and erosion) would have to be extremely rapid. The model of shear heating invokes generation of frictional heat during thrusting of one rock slab over the other, whereby this heat increases T in the lower slab in the immediate vicinity whilst upper slab cooling. Preservation of this inverted metamorphism requires rapid exhumation. Also, if the shear zone is wide, as in the MCT, thermal effects are unlikely to occur in the broad zone of 5–10 km, because wide ductile shear zone contains abundant micas and tourmalines which on shearing would liberate volatile phase (fluid fluxing) and hence unlikely to allow frictional heating. Joshi and Rai, (2003), after analyzing the P-T data for the shear zone transects across the MCT, demonstrated that the P-T distribution is chaotic and suggested post-metamorphic fractal displacements in the MCT zone as the cause of inverted metamorphism. The cause of the inverted metamorphism is still not completely clear and it is very likely that different models may be appropriate in different parts of the Himalaya. Dasgupta et al. (2004) investigated Lesser Himalaya of Sikkim for which P-T conditions are found to increase from 500°C/5 kb at garnet isograd to 715°C/7.5 kb at the sillimanite-K feldspar isograd, contrary to the common belief that P decreases progressively toward higher structural levels.

Recently, Sharma (2005) linked the inversion of metamorphic isograds with the southern convexity of the Himalaya; the two regional phenomena appear to be linked for this mountain belt. It is suggested by Sharma (loc. cit.) that the Indian plate during final collision was too buoyant to be carried down further (England and McKenzie, 1982). Consequently, the northward push of the Indian plate resulted in southward movement or thrusting along the Main Boundary thrust (MBT) of the deformed-metamorphosed rocks that eventually rested over the cold rock of the colliding Indian plate. This formed the Lesser Himalayan Crystallines (LHC) with their (meta-) sediments. This southward thrusting of the LHC reduced the drag load at the tip of the down-going Indian plate and the Indian plate was raised and simultaneously retreated during the ongoing collision. Sharma linked this with flipping up of the Indian plate whereby causing both reversal of the curvature of the Himalayan arc and southward movement of the hotter and deeper rocks. This happened when thrusting of the Higher Himalayan Crystallines (HHC) occurred over the lower grade rocks of the Himalayan Metamorphic Belt (HMB) and resulted in the inversion of metamorphic grade. In this event, the Indian plate once again rose and retreated, whilst the northward push continued. While the HHC moved along the MCT, the result was a decrease of P and hence decompression melting of the continental crust, which produced the 18–20 Ma old leucogranite that occur all along the HHC, close to the major structural discontinuity between HHC and the lower grade metasediments of the Late Proterozoic Haimantas toward the north (Sharma, 2005). This mechanism is shown diagrammatically in Fig. 3.3a–d, with thrusting

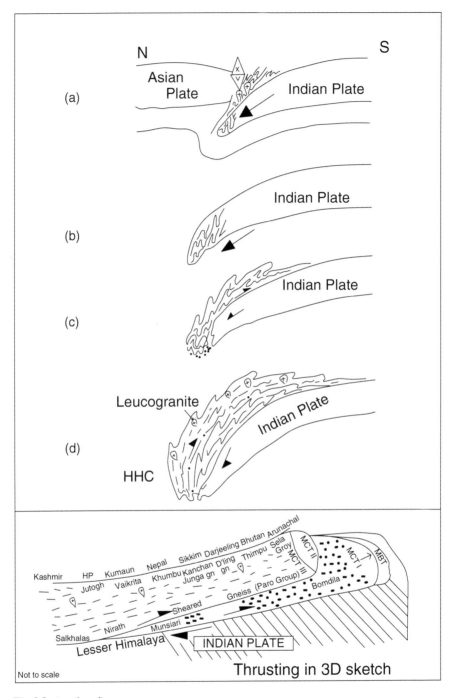

Fig. 3.3 (continued)

mechanism in a three-dimensional sketch (in bottom of Fig. 3.3). In this model, Sharma presumes that the metamorphism was normal but the reversal occurred due to successive thrusting of the sheets wherein the high grade one was overlain by the next lower grade one which in turn was overridden by still lower grade and so on, linking this inversion with the southern convexity of this lofty mountain belt.

To explain the inversion of isograds and exhumation of HHC, a newer mechanism called *channel flow model* has been proposed (see Grujic, 2006). According to this model, the HHC channel between MCT and STDS attained critical low viscosity at depth and under focused denudation and gradient in lithostatic pressure (due to high elevation of Tibetan plateau against low elevation southwards) the HHC rocks yielded to lateral flow and hence extrusion from mid-crustal depths. The lateral flow is the consequence of pure shear or Poisuille (pipe) flow in the middle and Cuette flow (simple shear) at the margins of the channel, thereby causing thinning of the channel, inversion of the isograds and exhumation of the mid-crustal rocks of the Higher Himalaya. According to this model, the inverted metamorphic isograds are not primary features but are tectonic artifacts that reflect post-metamorphic thrusting or imbrication. The HHC between SW-vergent, north-dipping Main Central Thrust (MCT) and noth-dipping South Tibet Detachment System (STDS), juxtaposed the HHC against the weakly metamorphosed Tethyan Sedimentary zone. The Himalayan inverted metamorphic rocks experienced a rapid decompression. As a result, dehydration melting of muscovite in pelitic schists occurred in the higher levels (4–6 kbar, $700 \pm 50°C$) of the HHC in the age range of 15–22 Ma, giving rise to the leucogranite and migmatite in the upper levels of the HHC near the North-Himadri Fault, i.e. South Tibet Detachment System (STDS).

In this model of channel flow, how much is the role of isostasy and erosion (including that by antecedent river system) is unknown, although both play a significant role in exhumation of the deep-crustal rocks, possibly without retaining their high pressure/high temperature mineralogy.

As an alternative to the channel flow model, Sharma (2007) advanced the *slab breakoff model* wherein it is considered that termination of the collision occurred

---

Fig. 3.3 (continued) A plausible model (after Sharma, 2005) for the inverted metamorphism in the Himalayan metamorphic belt (HMB), shown in stages: (**a**) Indian-Asian plate collision, but Indian plate was too buoyant to be carried down further, (**b**) Plastic deformation and recrystallization of the leading edge of the Indian plate, being at greater depth and hence at higher P & T, (**c**) Northward push of the subducted Indian plate resulted into southward thrusting of the plastically deformed and recrystallized rocks over the Precambrian rocks (Lesser Himalayan/basement) and overlying sediments. Emplacement of these rocks reduced drag load at the "tip" of the subducted continental crust, retreating and raising up the Indian plate, and this flip also caused southern convexity of the thrust belt all along the Himalayan length, (**d**) Continued northward push of the Indian plate resulted into renewed southward thrusting of still higher grade Higher Himalayan Crystalline (HHC) rocks during Oligocene-Miocene. Upthrusting of HHC caused decompression melting (later solidified as the Leucogranite), occasional folding of isograds and parallelism of "S" and "C" fabrics in the HHC rocks. Bottom figure shows thrusting in 3-dimension, differentiating MCT into I and II

with the detachement of the Indian lithospheric slab whereby the HHC wedge destabilized, exhuming the HHC rocks either in pulses or in a single event when the ultra high pressure rocks were rapidly excavated from deeper levels.

As stated earlier, the fragments of the continental crust anchored by the lithosphere were dragged down to 100–130 km depth and developed ultra high pressure (UHP) mineralogy with coesite in garnet (Mukherjee and Sachan, 2001) or in silicate rocks and diamond in graphite-bearing rocks. It is conceived by this writer that end stages of continent-continent collision experienced slab detachment. As a result, the UHP rocks, including the eclogites (formed from basaltic parent), which had been subducted deep at the coesite and diamond stability between 55 and 45 Ma, had a rapid return due to rebound effect of the slab breakoff and resided as "nodules" in the rocks near the ITSZ. The not-so-deep subducted continental rock returned earlier and got emplaced as a distinct unit, called the Tso-Morari Crystalline (TMC) that occur as a dome between low-grade Tethyan Sedimentary zone and ITSZ. During its ascent the TMC may have possibly elevated the Tethyan sedimentary zone. Because of its early arrival and location, the TMC became the host for the UHP fragments coming later from greater depths. By this model of slab breakoff, we can understand the geological setting of the Tso-Morari Crystallines (TMC) that was once a part of the HHC but during subduction process changed its mineralogy with high pressure assemblages (Sharma, 2008). Due to its limited subduction (reflected in its mineralogy) it disengaged first from the downgoing slab and ascended rapidly to mid crustal depths and emplaced amidst Tethyan zone near ITSZ. The proposed mechanism finds support from the occurrence of post-orogenic, undeformed basic rocks (being products of decompression melting of the upwarped asthenosphere as a consequence of slab detachment) intruding the Himalayan metamorphics (Sharma, 1962) and from the deep mantle anomalies recently discovered by seismic tomography at the base of lithosphere in West Pakistan (van der Voo et al., 1999).

The geological events for the Himalayan evolution are enumerated below

The paradigm of slab breakoff is an efficient mechanism for exhumation of the Himalayan rocks and is independent of the parameters of the channel flow model that can be challenged on several accounts Sharma (2007, p. 40) as given below.

1. Continuity of lithologic units in the HHC (GHS) for over 1000 km along strike (Gansser, 1964) indicates lack of internal stratigraphic disturbance.
2. The Model is beset with uncertainties:

    (a) Whether HHC represents a *complete section of mid crust*

    or

    An *extruded segment* of a cooling channel?

    (b) Whether the HHC bounding faults associated with *late-stage* exhumation of

    or

    Formed at depth beneath the Tibetan Plateau?

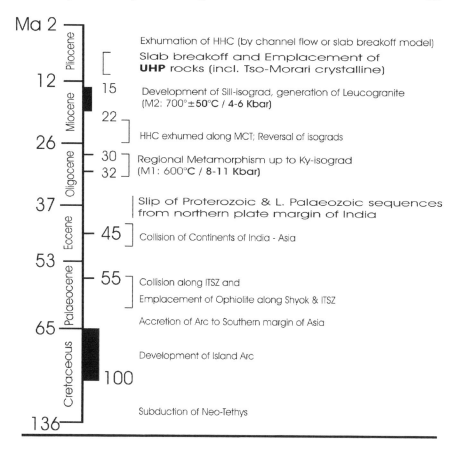

Geological Events during Himalayan Orogeny

(c) Whether the fabric within HHC are related to *flow during channeling*/extrusion

or

Pre-date the Himalayan event?

3. The Model presumes that the SW-directed stacking of the HHC (GHS) was contemporaneous with the NE-directed stacking of the Tethyan Sedimentary Zone. This is invalidated by the available geochronological data which suggest that STDS was active in the interval of 17–14 Ma whereas the majority of HHC-Leucogranite, formed during slip on MCT, were emplaced between 24 and 19 Ma (Harrison et al., 1999).

4. Geochronology of metamorphic monazites from MCT in Central Nepal indicates that MCT shear zone was active at 6 Ma (Harrison et al., 1997) while ductile deformation seems to have terminated during the Early Miocene (Schelling and Arita, 1991).

5. Denudation playing a predominant role in the Model, implies that insufficient denudation is responsible for the absence of (Lower crust) granulites. If granulites are anhydrous residue of deep crustal melting, why channel flow did not operate?.
6. The Model requires rapid and large magnitude of denudation for the minor effect that decompression has on melting of the source composition.
7. The Model requires generation of multiple anatectic phases via decompression, which seems difficult and does not fulfill the requirement (by the Model) of definite timing linking Slip with STD and Anatexis.
8. Omission of parts of the Tethyan section across STDS at the top of HHC slab shows that some fault movements clearly post-date peak metamorphism (Steck et al., 1993).
9. The exhumation of HHC rocks metamorphosed first (in M1) up to kyanite grade at P of 8–11 kbar and T below anatexis of crustal rocks (Searle et al., 1992) could not have occurred by channel flow but requires another process.

The recently discovered crustal thickening beneath the Himalaya is attributed to on-going underthrusting of India. It has been proposed that the thickening was accomplished by delamination of the Indian crust from the lithospheric mantle along a brittle-ductile transition at the level of Moho. This means that some process of "mantle peeling" has taken place beneath the Asian plate so that Indian crust could be placed directly beneath the Tibetan (Asian) plate. This is revealed by the seismic tomographic analysis of Roecker (1982) and by the seismic studies of Hirn et al. (1995), both showing that no Indian lithosphere is present at depth north of the ITSZ. Therefore, another model is proposed by Molnar (1984) in which only limited underthrusting of the Indian plate is postulated and no crust-mantle decollement is required. In consideration of this proposition and that of McKenzie (1969) that the Indian crust was too buoyant to be subducted deeper under the Asian plate, Butler (1986) suggested that crustal thickening was accomplished by thrusting along a detachment climbing from the Moho. According to this model the footwall section of the crust would be depressed below the level at which the granulites of the lower crust undergo a phase change to eclogite. The increased density of the basal eclogite could then decrease the buoyancy of the crust and allow limited subduction to take place. Bouguer anomalies are found increasingly negative northwards from India, suggesting that the crust thickens in this direction. A detailed imaging of crustal structure across the Himalaya and Tibet shows crustal thickness of ca. 35 km of the Indian shield, and this increases to 55 km under the Himalaya and to 70 km beneath Tibet. Furthermore, the Moho topography is not smooth and exhibits a number of breaks or steps and at places overlapping of Moho at depth. This suggests that crustal thickness did not occur by simple underplating and that India did not descend smoothly and coherently beneath Asia; rather it took place in response to intracontinental thrusting affecting both the crust and the upper mantle.

A combined interpretation of satellite images and focal mechanism solution of earthquake has revealed that the pattern of faulting in the Himalayan region is such that thrust faulting is restricted to a relatively narrow belt north of the Himalayan

3.2 Himalaya as an Example of the Young Fold Belt

Frontal Faults. Strike-slip faulting is dominant in a region some 1500 km wide to the north of the Himalaya and eastward into Indo-China. The MBT and MCT thrusts can be cited as two most spectacular examples of thrust faulting. The high-grade crystallines (HHC) which behaved like a basement to the Tethyan sediments were also transported on the Indian continental margin, along MCT, to rest on the Lesser Himalayan low grade metasedimentaries and seen further south to cover even the Eocene rocks. It needs to be pointed out that there is a serious confusion in the use of the term LHC. The LHC are believed by most workers as Lesser Himalayan Crystallines equivalent of HHC which are geographically located in the Lesser Himalaya but are equivalent to HHC.

Two major strike faults are remarkable on the west and east end of the Indian plate. At its western end, the Indus suture is terminated by a large sinistral strike fault, the Quetta-Chaman fault that continues into the Indian Ocean as Owen Transform Fault. At the eastern end of the Indus-Tsangpo Suture there is a large dextral strike fault, the Saquang Fault, which continues through Burma to connect with the Indonesian subduction zone. This fault perhaps also continues into the Indian Ocean as the now largely inactive Ninety-East Ridge Transform. The Indian plate, therefore, may be regarded as a two-pronged wedge that has driven northwards into the Tibetan plate between two large transform faults. An analogy has been drawn between the pattern of faulting in Asia when it is indented by a rigid indenter of India and the lines of failures (slip lines) developed in a plastic medium experimented by Tapponier and Molnar (1976). Alternative to this Indentation model, there is another model, known as Continuum model that places greater emphasis on the role of forces arising from crustal thickening (Tapponier et al., 1982). Both indentation model and continuum model have been challenged by England and Molnar (1990) who suggested that the major sinistral strike-slip faults are a consequence of the presence of a wide deforming zone between India, China and Eurasia (see also, Lyon-Caen and Molnar, 1983). If the faults rotate in a clockwise sense, as suggested by recent palaeomagnetic measurements, their sinistral motion arises from their location in a broad region undergoing dextral shear in the N-S direction. This model explains the eastward motion of Tibet relative to Asia and India, and its northward movement relative to SE China (Fig. 3.4).

The Himalaya is undergoing rapid uplift at rates between 0.5 and 0.4 mm per year. It is, therefore, experiencing rapid erosion, with deposition of these clastic sediments in a foredeep or foreland basin. These Siwalik molasses reach a thickness of as much as 6–8 km near the Himalayan front, and the coarseness of the basin fill decreases away from the mountain front. Conglomerates pass into sandstones and shales, extending southwards into the Ganga basin where they are bounded by the Main Frontal thrust (MFT). In some foredeep, the clasts in the conglomerates and sandstones reveal an unroofing sequence in which stratigraphically younger deposits contain debris from successively deeper levels in the mountain.

Geophysical evidence suggests that in the Himalaya, oceanic portions of the Indian plate are still being underthrust at the Indonesian arc (Zhao et al., 1993). In the west of India, oceanic crust of the northwest Indian Ocean is still subducting beneath the Makran region of southwest Pakistan and Iran (see Moores and Twiss, 1995).

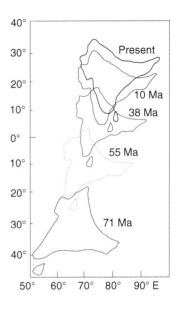

Fig. 3.4 Northward drift of India with respect to Asia from 71 Ma to the present, determined from magnetic lineations in the Indian and Atlantic oceans (redrawn from Molnar and Tapponier, 1975)

3.3 Old Fold Belts

After having discussed the plate tectonics theory and its application to the evolution of the Himalayan orogenic belt, we now look at the characteristics of the older fold belts, mainly Proterozoic fold belts of India. Here we consider the plate tectonic events that may have led to their development. The record of the older plate tectonic activity is expected to be found in continental rocks and especially in the Proterozoic fold belts. Here we need to recognize the plates that converged, their direction of movement, the collision suture, as well as the magmatic arc as relict arc on the flanking continent. For the nature of the Proterozoic plate tectonic events, we need to rely on the palaeomagnetic evidence for the plate movement. However, we do not have palaeomagnetic data in any of these mobile belts and their flanking cratonic areas. Therefore, we have to search for other "petrotectonic indicators", such as blueschist and calc-alkaline rocks that suggest subduction, which might have occurred in the concerned fold belts. We may fail to recognize ophiolite in the Proterozoic fold belts for some reasons such as thermal overprinting, although ophiolites and blueschists have been reported from Middle to Lower Proterozoic terrains. Blueschists are characteristics of consuming margins, and ophiolites are typical in collision belts. The lack of blueschists and ophiolites presents problems in any attempt to apply plate tectonics models to the Middle to Early Proterozoic terrains. Their absence in Proterozoic fold belts implies either that the plate tectonics of Palaeozoic type has not operated or that some other petrotectonic indicators have taken the place of blueschists and ophiolites. If the Precambrian geotherm was higher, the metamorphic conditions in the subduction zone might have been a high-P, low-T to form rocks such as kyanite ± talc schists that occupied the tectonic position of blueschists in the Proterozoic fold belts. It is also possible that the deformed ultramafic complexes that occur in some ancient fold belts represent the subducted oceanic crust during Precambrian times. Despite these problems, mantle convection cannot be denied for the Precambrian times and some form of plate tectonics has to be conceived as pleaded by Kroener (1981).

In considering the nature of global tectonic activity in Precambrian times, Kroener (1981) suggested three approaches. First, a strictly uniformitarian approach can be taken in which Precambrian tectonics originated by the same mechanism of plate tectonics actively operated in Phanerozoic times. Second, a modified uniformitarian approach can be postulated in which plate tectonic processes in the Precambrian were somewhat different from present because the physical conditions affecting the crust and mantle have changed throughout geological time (Hargraves, 1981). Third, a completely different tectonic mechanism can be invoked for Precambrian times. Only the first two of these approaches will be considered, as there are no cogent reasons for not doing so.

The oldest orogenic system that originated in plate tectonic time is the Pan African-Baikalian-Brazilian system present in East Asia and throughout much of Africa and South America. It dates back to Neoproterozoic (1000–550 Ma) and has many features that also resemble those of Mesozoic-Cenozoic orogenic regions. Long-distance correlations are not possible but it is suggested that around ca. 1000 Ma ago, a supercontinent formed by the collision of formerly separate continents to develop the Grenville orogenic belt and its correlatives. At this time, North America and East Gondwanaland (Australia and Antarctica) may have been joined. In the latest Precambrian time, this supercontinent (called Rodinia by some geologists) broke up with the separation of Laurentia, and another supercontinent formed, because of the Pan African-Baikalian-Brazilian orogeny. The Canadian geologist, Paul Hoffman suggests that one supercontinent called Gondwanaland may have developed by turning inside out (or extraversion) of a previous supercontinent. This extraversion resulted in the rifting of Western North America away from Antarctica-Australia and the opening of the Protopacific (Panthalasa) Ocean. At the beginning of Palaeozoic times, the plates were rearranged into a supercontinent. In Cambrian times, this supercontinent fragmented with the opening of the first proto-Atlantic ocean. This closed and the supercontinent reassembled in Late Silurian times, producing the *Caledonian orogeny*. In Early and Middle Devonian times, transcurrent movements between the Northern and Southern parts of the supercontinent produced the *Acadian orogeny*. A second phase of opening in the Late Devonian created the second proto-Atlantic, which closed in Middle Carboniferous times, producing the *Variscan orogeny* and the re-assembly of the plates into the supercontinent, the Pangae.

In the early Palaeozoic, orogeny during Middle to Late Ordovician is called Taconian in the USA, Grampian in the Canada and Fennarkian in the Scandinavia. A Middle Palaeozoic orogeny during Silurian-Devonian is known as the Acadian orogeny in North America and Caledonian orogeny in Great Britain and Scandinavia. A third orogeny in Late Carboniferous (Pennsylvanian)-Permian is referred to as the Alleghanian orogeny in the USA and Hercynian-Variscan in southern Europe. The timing of these orogenies is different at different places along the strike of the Appalachian-Caledonian system. This belt is probably the best example of an orogen that conforms to the composite orogenic model, evolved in the plate tectonic times. The presence in it of foredeep sediments, fold-and-thrust belts, a core with nappes and ophiolites, as in Newfoundland and Scandinavia, are interesting features that need to be looked in also for Proterozoic fold belts of India.

3.4 Proterozoic Fold Belts of India

As stated earlier, the Proterozoic fold belts are of special interest from the geotectonic point of view because it is believed that Proterozoic plate tectonics/ convergence were responsible for the evolution of these fold belts (Condie, 1982).

In India the Proterozoic fold belts are well developed in Rajasthan (western India), in Singhbhum-Orissa (eastern India), Madhya Pradesh and Maharashtra (central India), Eastern Ghats and Southern Granulite Terrain. All the fold belts are peripheral to the cratonic areas made up of Archaean gneiss-amphibolite-migmatite-granite association that served as basement for the supracrustal rocks forming the fold belts. These basement rocks also crop out amidst the deformed and metamorphosed supracrustal rocks of the fold belts, suggesting that both cover and basement were folded together in the mountain building orogeny.

For the Proterozoic mountain building process (orogenesis), there could be Phanerozoic style of plate tectonics in which lithospheric plates from far off distances approached towards each other because of subduction of the intervening oceanic crust (Black et al., 1979). In this mechanism it is very important to identify the colliding plates, a suture zone along which subduction occurred, a magmatic arc which formed of melts generated from subducting plate, and other features like ophiolite belt—a remnant ocean floor. This belt is the most direct evidence of a collision suture, but often unrecognizable in the Proterozoic fold belts. Some sutures, on the other hand, are characterized by mylonitic or ductile shear zone. Without collating other evidence, these sutures do not necessarily imply the presence of a subduction. These sutures are likely to form in ensialic orogenesis, slightly different from that by plate tectonics where orogeny is associated with ocean floor subduction. In the ensialic orogenesis, the rift basin is floored by continental rocks and not by oceanic crust. The structures in older terrains on either side of the fold belt are virtually unaffected (albeit overprinted) through the younger mobile belt. On the other hand, the sialic blocks, welded together by plate tectonics process, show markedly different stratigraphy or tectonic history and have a discontinuity in the orientation or style of structures. Nearly all the Proterozoic fold belts of India seem to have evolved by the ensialic orogenesis that is described below.

3.5 Ensialic Orogenesis Model

The ensialic orogenesis is a modified version of plate tectonic theory, appears suitably applicable for the evolution of the Proterozoic fold belts of the Indian shield. The ensialic orogenic model initially starts with segmentation of once large sialic block due to crustal thinning by ductile stretching or rifting. Extension and thinning of the continental crust eventually causes the floor of the rift valley to drop below sea level. This depositional basin then receives sediments that are laid upon sialic continental basement, and nowhere upon mafic oceanic crust. The volcanic lava or basic flows, generated by decompression melting of the upper mantle are likely to interrupt the sedimentation processes. Subsequently, these supracrustals (sediment + lava) along with their basement are compressed, deformed, and metamorphosed and even partially melted at depth. The whole rock complex is finally raised in an orogenic or mobile belt that is accreted to the converging sialic blocks. The ensialic orogenesis involves only a reworking of the older continental material and is thus incompatible with the crustal accretion from far off distances as in the plate tectonic mechanism. However, in the ensialic orogenesis, a limited subduction is envisaged which is called Ampferer subduction (abbreviated A-subduction). Here the mantle part of the lithosphere is conceived to detach from the crust and the crust eventually undergoes heating from the mantle magma to result in metamorphism of both basement and cover rocks of the region. Thus, it seems that in the ensialic orogenesis model the structures in older terrains on either side of the fold belt are virtually unaffected (albeit overprinting) through the younger mobile belt. Sharma (2003) explained the reversal of the movement direction of the crustal blocks that had drifted away from the site of continental rifting. He suggested that sediment load and decay of the underlying "plume" or mantle diapir and consequent change in vertical component of stress are the main cause for reversal of the movement direction of the separating crustal blocks. Geological evidence in nearly all Proterozoic fold belts of India suggests that they owe their origin to ensialic orogenesis model that is a modified plate tectonic process in Phanerozoic time. It is believed that the stable Archaean cratons were subdivided by mobile belts in which deformation is almost wholly ensialic.

Toward the end of Archaean, 2.5 Ga ago, major global changes seem to have occurred, and one of the significant changes was demise of Archaean processes, reflected in global stabilization of continental areas or cratons. The Archaean crust subsequently underwent ductile stretching or rifting because of drag effects of sub-lithospheric convection currents. This resulted in the formation of "geosynclinal" basins whose floor was made of continental crustal rocks or oceanic crust (true geosyncline). These geosynclinal basins became the site for deposition of submarine basalts (poured out as a result of decompression melting of upper mantle) and of shallow to deep-water facies sediments and clastics derived in part from the pre-existing terraines. These supracrustal rocks of the basin were subsequently deformed by convergence of opposing continental blocks/lithospheric plates. This deformation resulted into thickening of the crust that led to heating (radiogenic and even by heat from igneous intrusions) and hence metamorphism of the rocks which ultimately were raised into lofty mountains or fold belts. Thermal modelling of crustal overthickening tectonics (England and Thompson, 1984) has indicated that increased heat production and insulation are the most important factors for the perturbed geotherm, whereas the effects of erosion and isostasy are important in terminating metamorphism.

Heating of a crustal segment could also take place by mere lithospheric thinning and magma underplating/interplating. Because of the magma addition below or within the crust, the affected rocks are heated to very high temperatures, designated as ultra high temperature (UHT) metamorphism, resulting in the formation of granulite facies rocks. These rocks are found to show anticlockwise (ACW) trajectory in the P-T space, unlike the rocks of geosyncline that document clockwise P-T path. There is no reasonable thermal scenario for creating anticlockwise P-T path with simple burial. Such paths clearly appear to require magmatic heating early in their evolution. This heating can be brought about by (a) accretion of large volumes of mantle-derived magmas in the lower/middle crust (Harley, 1988), and (b) lithospheric thinning or crustal extension due to which basic magma generated by decompression melting of mantle would move upwards to underplate the lower crust (Sandiford and Powell, 1986). Such mechanisms are proposed for the Eastern Ghats belt (Dasgupta and Sengupta, 2002), discussed later. Admittedly, extreme mantle thinning is capable of generating granulite facies conditions but excessive crustal lithosphere thinning would give rise to such rapid heat loss from the crust that granulite facies conditions cannot be attained. It means that granulite metamorphic conditions can only be generated by extreme lithosphere thinning with minimal crustal thinning. Such a scenario is most realistically generated by asymmetric lithosphere extension (Sandiford and Powell, 1986).

The anticlockwise P-T path for most granulite facies rocks is often beset with a problem of their excavation from deep crustal levels. Some granulite terrains with ACW path show isothermal decompression which implies a rapid erosion or gravitational spreading/stretching of tectonically thickened crust (cf. Mohan et al., 1997). Difficulties arise, however, when the retrograde path of the ACW is an isobaric cooling. This implies that these granulites had their residence at great depths and in order to be exhumed they must be involved in another orogeny or upthrusted along a shear zone. This controversy of the Eastern Ghats belt has been discussed by Bhattacharya and Gupta (2001). In order to resolve a type of tectonic setting during metamorphism, the P-T data, obtained from calibration of geothermobarometric models and from petrogenetic grids based on experimental mineralogy and textural relationship, are considered in conjunction with geochronological data. However, these critical data are not always available for all the Proterozoic fold belts of India, believed to have evolved through the ensialic orogenesis.

References

Ahmad, T., Harris, N.B.W., Islam, R., Khanna, P.P., Sachan, H.K. and Mukherjee, B.K. (2005). Contrasting mafic magnetism in the Shyok and Indus suture zones: geochemical constraints. Himalayan Geol., vol. 26(1), pp. 33–40.

Ahmad, T., Sivaprabha, S., Balkrishnan, S., Thanh, N.X., Itaya, T., Sachan, S.K., Mukhopadhyay, D.K. and Khanna, P.P. (2008). Geochemical-isotopic characteristics and K-Ar ages of magmatic rocks from Hundar Valley, Shyok Suture zone, Ladakh. Himalayan J. Sci., vol. 5 (Special Volume on 23rd Himalaya-Karakoram-tibet Workshops, Leh, India), Extended Abstract, p. 18.

Allegre, C.J. et al. (1984). Structure and evolution of the Himalaya and Tibet orogenic belt. Nature, vol. 307, pp. 17–22.

References

Auden, J.B. (1937). The structure of the Himalayas in Garhwal. Rec. Geol. Surv. India, vol. 71, pp. 407–433.

Beck, R. and others (1995). Stratigraphic evidence for an early collision between northwest India and Asia. Nature, vol. 373, pp. 55–58.

Bhargava, P.N. and Bassi, U.K. (1999). Western Himalaya Tethyan Eocambrian-Mesozoic basin and events. Gondwana Res. Group Mem., vol. 6, pp. 51–59.

Bhattacharya, A. and Gupta, S. (2001). A reappraisal of polymetamorphism in the Eastern Ghats mobile belt—a view from north of the Godavari rift. Proc. Indian Acad. Sci. (Earth Planet. Sci.), vol. 110(4), pp. 369–383.

Black, R., Caby, R., Moussine-Pouchkine, A. et al. (1979). Evidence for late Precambrian plate tectonics in West Africa. Nature, vol. 278, pp. 223–227.

Butler, R.W.H. (1986). Thrust tectonics, deep structure and crustal subduction in the Alps and Himalayas. J. Geol. Soc. Lond., vol. 143, pp. 857–873.

Chang, C. and Cheng, H. (1973). Some tectonic features of the Mt. Jolmo Lungma area, southern Tibet. Sci. Sinica, vol. 16, pp. 257–265.

Chang, C. et al. (1986). Preliminary conclusions of the Royal Society and Acadmica Sinica 1985, geotraverse of Tibet. Nature, vol. 323, pp. 501–507.

Cartlos, E.J., Dubey, C.S., Harrison, T.M. and Edwards, M.A. (2004). Late Miocene movement within the Himalayan Main Central Thrust shear zone, Sikkim, northeast India. J. Metamorphic Geol., vol. 22(3), pp. 207–226.

Condie, K.C. (1982). *Plate Tectonics and Crustal Evolution*. Pergamon Press, New York.

Dasgupta, S. and Sengupta, P. (2002). Ultrahigh temperature metamorphism in the Eastern Ghats mobile belt, India: evidence from high Mg–Al granulites. Proc. Indian Acad. Sci., Part A, vol. 68(1), pp. 21–34.

Dasgupta, S., Ganguly, J. and Neogi, S. (2004). Inverted metamorphic sequence in the Sikkim Himalaya: crystallization history, P-T gradient, and implications. J. Metamorphic Geol., vol. 22(3), pp. 207–226.

Dewey, J.F. and Burke, K.C.A. (1973). Tibetan, Variscan, and Precambrian basement reactivation: products of continental collision. J. Geol., vol. 81, pp. 683–692.

Dewey, J.F. and Bird, J.M. (1970). Mountain belts and the new global tectonics. J. Geophys. Res., vol. 75, pp. 2625–2647.

Dewey, J.F., Cade, S. and Patman, W.C. (1989). Tectonic evolution of the India-Eurasia collision zone. Eclog. Geol. Helv., vol. 82, pp. 717–734.

England, P.C. and McKenzie, D.P. (1982). A thin viscous sheet model for continental deformation. Geophys. J. Roy. Astr. Soc., vol. 70, pp. 295–321.

England, P.C. and Thompson, A.B. (1984). Pressure-temperature-time paths of regional metamorphism, I—heat transfer during the evolution of regions of thickened continental crust. J. Petrol., vol. 25, pp. 894–928.

England, P.C. and Molnar, P. (1990). Late Cenozoic uplift of mountain ranges and global climate change; chicken or egg? Nature, vol. 346, pp. 29–34.

Gansser, A. (1964). *Geology of the Himalayas*. Wiley-Interscience, London, 289p.

Grujic, D. (2006). Channel flow and continental collision tectonics: an overview. In: Law, R.D., Searle, M.P. and Godin, L. (eds.), *Channel Flow, Ductile Extrusion and Exhumation in Continental Collision Zones*. Geol. Soc. Lond. Spl. Publ. 268, London, pp. 25–37.

GSI. (1962). *Geological Map of India, Scale 1:2000,000*. Geol. Surv. India, Calcutta.

Hargraves, R.B. (1981). Precambrian tectonic style: a liberal uniformitarian interpretation. In: Kroner, A. (ed.), *Precambrian Plate Tectonics*. Elsevier, Amsterdam, pp. 21–56.

Harley, S. (1988). Granulite P-T paths: constraints and implications for granulite genesis. In: Vielzeuf, D. (ed.), *Granulite and Their Problems*. Terra Cognita, European Union of Geosciences, Wien, vol. 8(3), pp. 267–268.

Harris, N.B.W., Caddick, M., Kosler, J., Goswami, S., Vance, D. and Tindle, A.G. (2004). The pressure-temperature-time path of migmatites from Sikkim Himalaya. J. Metamorphic Geol., vol. 22, pp. 249–264.

Harrison, T.M., Lovera, O.M. and Grove, M. (1997). New insights into the origin of two contrasting Himalayan granite belts. Geology, vol. 25, pp. 899–902.

Harrison, T.M., Grove, M., McKeegan, K.D., Coath, C.D., Lovera, O.M., Lefort, P. (1999). Origin and episodic emplacement of the Manaslu intrusive complex, Central Himalaya. J. Petrol., vol. 40, pp. 3–19.

Heim, A. and Gansser, A. (1939). The Central Himalayas: geological observations of the Swiss expedition of 1936. Mem. Soc. Helv. Sci. Nat., vol. 73, pp. 1–245.

Hirn, A., Jiang, M., Sapin, M. et al. (1995). Seismic anisotropy as an indicator of mantle flow beneath the Himalaya and Tibet. Nature, vol. 375, 571–574.

Hirn, A., Nercessian, A., Sapin, M. et al. (1984). Lhasa block and bordering sutures—a continuation of a 500-km Moho traverse through Tibet. Nature, vol. 307, pp. 25–27.

Jain, A.K. and Manickavasagam, R.M. (1993). Inverted metamorphism in the intracontinental ductile shear zone during Himalayan collision tectonics. Geology, vol. 21, pp. 407–410.

Jain, A.K., Singh, S. and Manickavasagam, R.M. (2002). Himalayan collision tectonics. Gond. Res. Group Mem., vol. 7, pp. 1–114.

Joshi, M. and Rai, J.K. (2003). An appraisal of models for inverted Himalayan metamorphism. Mem. Geol. Soc. India, vol. 52, pp. 359–380.

Kearey, P. and Vine, F.J. (1996). *Global Tectonics*. 2nd ed. Blackwell Science, Cambridge, 333p.

Klootwijiik, C., Gee, J.S., Peirce, J.W., Smith, G.M. and Mcfadden, P.L. (1992). An early India-Asia contact: paleomagnetic constraints from Nentyeast Ridge, ODP Leg 121. Geology, vol. 20, pp. 395–398.

Kroener, A. (1981). Precambrian plate tectonics. In: Kroener, A. (ed.) *Precambrian Plate Tectonics*. Elsevier, Amsterdam, pp. 57–90.

Le Fort, P. (1975). Himalaya: the colliding range. Present knowledge of the continental arc. Am. J. Sci., vol. 275, pp. 1–44.

Le Fort, P. (1989). The Himalayan orogenic segment. In: Sengor, A.M.C. (ed.), *Tectonic Evolution of the Tethys Region*. Kluwer, Boston.

Lyon-Caen, H. and Molnar, P. (1983). Constraints on the structure of the Himalaya from an analysis of gravity anomalies and a flexural model of the lithosphere. J. Geophys. Res., vol. 88, pp. 8171–8191.

McKenzie, D.P. (1969). Speculation on the consequences and causes of plate tectonics. Geophys. J. Roy. Astr. Soc., vol. 18, pp. 1–32.

Miller, C., Thoeni, M., Frank, W., Grasemann, B., Kletzli, U., Guntli, P. and Braganits, E. (2001). The early Plaeozoic magmatic event in the northwest Himalaya, India: source, setting and age of emplacement. Geol. Mag., vol. 138(3), pp. 237–251.

Mohan, A., Tripathi, P. and Motoyashi, Y. (1997). Reaction history of sapphirine granulites and decompression P-T path in a granulite complex from Eastern Ghats. Proc. Indian Acad. Sci. (Earth Planet. Sci.), vol. 106, pp. 105–130

Molnar, P. (1984). Structure and tectonic of the Himalaya. Ann. Rev. Earth Planet. Sci., vol. 12, pp. 489–518.

Molnar, P. and Tapponier, P. (1975). Cenozoic tectonics of Asia: effects of a continental collision. Science, vol. 189, pp. 419–426.

Moores, E. and Twiss, R.J. (1995). *Tectonics*. W.H. Freeman Company, New York, 415p.

Mukherjee, B.K. and Sachan, H.K. (2001). Discovery of coesite from Indian Himalaya: a record of ultra-high pressure metamorphism in Indian continental crust. Curr. Sci., vol. 81(10), pp. 1358–1361.

Powell, C.Mc.A. and Conaghan, P.J. (1973). Plate tectonics and the Himalaya. Earth Planet. Sci. Lett., vol. 20, pp. 1–12.

Powell, C.Mc.A. and Conaghan, P.J. (1975). Tectonic models of the Tibetan plateau. Geology, vol. 20, pp. 727–731.

Roecker, S.V. (1982). The velocity structure of the Pamir-Hindu Kush Region: possible evidence of subducted crust. J. Geophys. Res., vol. 87, pp. 945–959.

References

Sahni, A., Bhatia, S.B., Hartenberger, J.L., Jaeger, J.J., Kumar, K., Suder, J. and Vianey-Laude, M. (1981). Vertebrates from the Subathu formation and comments on the biogeography of the Indian subcontinent during the early Paleogene. Bull. Soc. Geol., vol. 23, France, pp. 689–695.

Saklani, P.S. (1993). *Geology of Lower Himalaya (Garhwal)*. International Books, Delhi, 240p.

Sandiford, M. and Powell, R. (1986). Deep crustal metamorphism during continental extension: modern and ancient examples. Earth Planet. Sci. Lett., vol. 79, p.151–158.

Schelling, D. and Arita, K. (1991). Thrust tectonics, crustal shortening, and structure of the far-eastern Nepal Himalaya. Tectonics, vol. 10, pp. 851–862.

Searle, M.P. and others (1987). The closing of Tethys and the tectonics of the Himalaya. Geol. Soc. Am. Bull., vol. 98, pp. 678–701.

Searle, M.P., Waters, D.J., Rex, D.C. and Wilson, R.N. (1992). Pressure, temperature and time constraints on Himalayan metamorphism from eastern Kashmir and western Zanskar. J. Geol. Soc. Lond., vol. 149, pp. 753–773.

Sengor, A.M.C. (1989). *Tectonic Evolution of the Tethyan Region*. Kluwer, Boston.

Sharma, K.K. (1987). Crustal growth and two-stage India-Eurasia collision in Ladakh. Tectonophysics, vol. 134, pp. 17–28.

Sharma, R.S. (1962). On the occurrence of an olivine-dolerite laccolith at Ranikhet, dist. Almora (U.P.). J. Sci. Res., B.H.U., vol. 12(2), Varanasi, pp. 231–237.

Sharma, R.S. (2003). Evolution of Proterozoic fold belts of India: a case of the Aravalli mountain belt of Rajasthan, NW India. Geol. Soc. India Mem., vol. 52, pp. 145–162.

Sharma, R.S. (2005). Metamorphic history of the Himalayan rocks and a new model for the inverted metamorphism. Himalayan Geol., vol. 26(1), pp. 19–24.

Sharma, R.S. (2007). Tracking Himalayan rocks in subduction-collision zones and their exhumation: a paradigm for UHP crustal rocks sans lower crust. Himalayan Geol., vol. 28(3), pp. 39–41.

Sharma, R.S. (2008). Geohistory of Tso-Morari crystalline, Eastern Ladakh, India: a plausible model for ultra-high pressure rocks in the Himalaya. Himalayan J. Sci., vol. 5(7), pp. 139–140.

Steck, A., Spring, L., Vinnay, J.C. and others (1993). Geological transect across the Northwestern Himalaya in Eastern Ladakh and Lahul (a model for the continental collision of Indian and Asia). Eclog. Geol. Helv., vol. 86, pp. 219–263.

Tapponier, P. and Molnar, P. (1976). Slip-line field theory and large-scale continental tectonics. Nature, vol. 264, 319–324.

Tapponier, P., Peltzer, G., Le Dain, A.Y., Armijo, R. and Cobbling, P. (1982). Propagating extrusion tectonics in Asia: new insights from simple experiments with plasticene. Geology, vol. 10, pp. 611–616.

Valdiya, K.S. (1977). Structural setup of the Kumaun Lesser Himalaya. In: *Eclogie Geologie de l' Himalaya*, C.N.R.S., Paris Coll. Intern., vol. 268, pp. 301–318.

Valdiya, K.S. (1989). Trans-Himadri Intracrustal Fault and basement upwarps south of Indus-Tsangpo Suture Zone. Geol. Soc. Am. Spl. Publ., vol. 232, pp. 153–168.

Van der Voo, R., Spakman, W. and Bijwaard, H. (1999). Tethyan subducted slabs under India. Earth Planet. Sci. Lett., vol. 171, pp. 7–20.

Zhao, W., Nelson, K.D. and Project INDEPTH team. (1993). Deep seismic reflection evidence for continental underthrusting beneath southern Tibet. Nature, vol. 366, pp. 557–559.

Chapter 4
Aravalli Mountain Belt

4.1 Introduction: Geological Setting

The Aravalli mountain of Rajasthan, western India, has NE-SW orographic trend and runs for over 700 km through the States of Gujarat, Rajasthan, Haryana and Delhi. This belt is made up of Proterozoic supracrustal rocks of the Aravalli and Delhi Supergroups, which are distinct from each other in their deformational and metamorphic history. These Proterozoic supracrustals were deposited on the Archaean basement gneiss complex, called Banded Gneissic Complex (BGC) which is predominantly a polymetamorphosed, multideformed rock-suite of tonalite-trondhjemite gneiss, amphibolite, migmatite and granitoid, most of which are Archaean (3.4–2.6 Ga old) (see Chap. 3, Sect. 3.5). To the west of the BGC occurs the Delhi fold belt while the Aravalli fold belt made up of the Aravalli Supergroup is located to the east of the Delhi fold belt and also amidst the BGC terrain (Fig. 4.1). The fundamental contributions on the Aravalli and Delhi fold belts, besides their basement BGC, and their constituent rocks are by B.C. Gupta (1934) and Heron (1953) who gave crisp geological account of field relationship along with impressive geological maps that were subsequently modified by later workers but the claim of the original author(s) is retained in all the maps presented here in a simplified form. Subsequent geological studies on the Precambrian rocks of Rajasthan are by Gupta et al. (1980, 1997), Sinha-Roy et al.(1998) and Roy and Jakhar (2002).

4.2 Aravalli Fold Belt

It is an arcuate-shape mobile belt, having NE trend in the northern part and NW-SE trend in the southern part. According to Heron (1953), the Aravalli fold belt runs through Bhilwara up to Sawar in the north while its southern extension is near Champaner in Gujarat. The width of the fold belt in the north is about 40 km and it increases up to 150 km toward the south, especially north of Banswara district of Rajasthan.

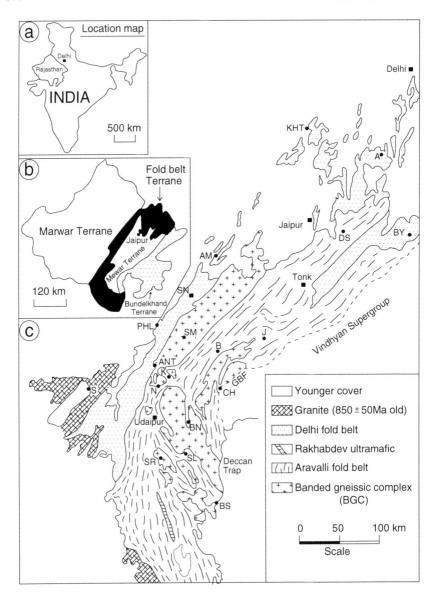

Fig. 4.1 Generalized geological map of the Proterozoic fold belts (Aravalli and Delhi) and cratonic rocks (BGC) in Rajasthan (after GSI, 1969, 1993). Inset (**a**) is the location map and (**b**) shows the major Crustal Terranes in Rajasthan (after Sinha-Roy, 2000). Abbreviations: A = Alwar; AM = Ajmer; ANT = Antalia; B = Bhilwara; BN = Bhinder; BS = Banswara; BY = Bayana; CH = Chittaurgarh; DS = Dausa; GBF = Great Boundary Fault; J = Jahazpur; K = Kankroli; KHT = Khetri; N = Nathdwara; PHL = Phulad; S = Sirohi; SL = Salumbar; SM = Sandmata; SN = Sendra; SR = Sarara

The Aravalli Supergroup occurs in three sectors, without any line of separation. These are: (1) the NE Bhilwara sector, consisting of amphibolite facies schists, gneisses and migmatites; (2) the Central Udaipur sector, consisting of greenschist, conglomerate, quartzite, carbonate and metapelites—all showing low-grade metamorphism; (3) the Southern sector, consisting of metapelite, mafic/ultramafic rocks and minor carbonates that are low grade in the northern part and high-grade in the southern and eastern part, especially in the Salumbar area (Mohanty and Naha, 1986). It is the Udaipur sector (2) in which one finds a complete succession and is known as the Type area of the Aravalli Supergroup. Here, the Aravalli rocks in their eastern part are shelf facies rocks while deep-sea or distal facies (Jharol Group) occurs in the west (Gupta et al., 1980). The shelf facies succession starts with basic lavas at the base followed by a thick sequence of granite-derived sediments (coarse clastics) and carbonate. The presence of high-Mg basalt in the mafic lava reported by Ahmad and Tarney (1994) suggests extension of the lithoshphere and decompression melting of the underlying upper mantle. The deep-water facies in the west is dominantly made of metapelite, carbonates and minor quartzites. The contact between the two facies is marked by a linear ultramafic body, which is almost completely altered to serpentinite. The Aravalli rocks from the type area are metamorphosed under greenschist facies conditions.

The Bhilwara sector on its west has a shear contact against high-grade gneisses and granulites of Sandmata Complex. This shear is called the Delwara Dislocation Zone (Sinha-Roy and Malhotra, 1989). The eastern margin of the Bhilwara sector shows a low-grade belt of the Aravalli rocks (Gwalior Series of Heron, 1953), indicating that the progressive metamorphism in the Aravalli sequence is from east to west. Sharma (1999, 2003) has taken this as an evidence for a west-directed subduction during the Aravalli orogeny. This author does not find any convincing argument to support the suggestion of Roy and Rathore (1999) that the high-grade Bhilwara belt is a pop-up sequence during the phase of Aravalli basin inversion, or that the Bhilwara belt was exhumed along the abounding shear zone, subsequently proposed by Roy and Jakhar (2002, p. 145). The contact of the Palaeoproterozoic Aravalli fold belt with the Mesoproterozoic fold belt on the west in Udaipur sector is nearly straight. This western contact is variably described as an angular unconformity (Heron, 1953), sheared unconformity (Naha et al., 1984) or a suture zone (Sinha-Roy, 1988; Sugden and Windley, 1984). Mukhopadhyay (1989) and Gupta and Bose (2000) have indicated that the Aravalli and Delhi Supergroups were never in contact and that the southward extending basement rocks separated the Aravalli fold belt from the South Delhi fold belt. This inference has great geodynamic significance, implying that these Proterozoic cover rocks occurring in the two fold belts had deposited unconformably on the Archaean basement complex, familiarly known as the Banded Gneissic Complex (BGC). The supracrustal rocks of the Aravalli fold belt, now complexly folded and metamorphosed are clastic sediments with minor chemogenic and organogenic assemblages plus basic volcanics—overlying the BGC with the first order unconformity (Heron, 1953).

4.2.1 Aravalli Stratigraphy: Controversies

The Aravalli stratigraphy has been drastically modified subsequent to Heron's (1953) work, which described these rocks unconformably resting on the BGC under the name Aravalli System. According to Heron, the Aravalli System has local amygdaloids and tuffs, quartzites, grits and conglomerates, overlain by impure limestones, quartzites, phyllites and schists. The sequence occurs in several distinct belts separated by intervening outcrops of the BGC. This sequence of Heron (Heron, op. cit.) has been modified on several occasions, especially by the publications of the Geological Survey of India (GSI). Following the code of stratigraphic nomenclature of the IUGS, the Aravalli System (including the Raialo Series of Heron) was designated by the lithostratigraphic term, the Aravalli Supergroup (Gupta et al., 1980; see also Roy et al., 1984). Poddar and Mathur (1965) classified the Aravalli sequence into two divisions: the lower shelf facies and upper deep-water facies. In 1977, the GSI proposed a two fold classification of the Aravalli Supergroup—the lower Udaipur Group and the upper Jharol Group with distinct lithological attributes. Subsequently, Gupta et al. (1980) classified the Aravalli Supergroup into 9 Groups (Debari, Udaipur, Kankroli, Bari Lake, Jharol, Dovla, Nathdwara, Lunavada and Champaner Groups). Recognizing an unconformity between the two Groups, Roy (1988) also classified the Aravalli sequence into the Lower Aravalli Group and Upper Aravalli Group, which correspond respectively to the shelf facies and deep-water facies of Poddar and Mathur (loc. cit). In 1993, Roy et al. revised their two-fold classification into three-fold one, viz. the Lower, Middle and Upper Aravalli Groups, each of which was subdivided into two or more formations (see Roy et al., 1993). During the same time, Sinha-Roy et al. (1993b) also subdivided the Aravalli Supergroup into three Groups, namely the lower Delwara Group, the middle Debary Group (both shelf facies) and the upper Jharol Group (deep-water facies). A few more attempts have been made in dividing the Aravalli Supergroup (see Gupta, 2004, pp. 174–176). The different stratigraphic schemes contradict one another because of the difference of opinion of geologists with regard to the age, aerial extent of the Groups/Formations and because of basement-cover relationship. Not only this, the contradictions have been further complicated by using the same geographical name for the rock-suites that are dissimilar in their lithocontents. The foregoing discussion allows us to accept the following stratigraphy of the Aravalli supergroup:

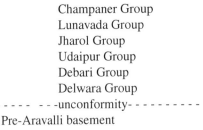

Champaner Group
Lunavada Group
Jharol Group
Udaipur Group
Debari Group
Delwara Group
- - - - - - - - -unconformity- - - - - - - - - - -
Pre-Aravalli basement

The Aravalli rocks of Heron exposed in and around the 2.5 Ga old Berach Granite, notably in the Hindoli and Jahazpur areas, were considered older than the Granite, taken to be intrusive into these low- to medium-grade metasediments (graywackes, pelites), volcanics (felsic and mafic), volcaniclastics and BIF. These rocks were considered as Pre-Aravalli and named Hindoli Group by geologists of the GSI (see Gupta et al., 1980), although the Berach Granite is undoubtedly a component of the Archaean basement (BGC) in Rajasthan, as revealed from field observation by Heron and confirmed by geochronological data on zircons (Sivaraman and Odom, 1982; see also Crawford, 1970). It is possible that the Berach Granite, like the granite-gneiss bodies of Ahar River, Udaisagar, Jaisamand etc. occurring amidst the Aravalli Supergroup rocks as the basement inliers (see Roy, 1988), has been marginally re-mobilised to appear as intrusive into the Aravalli metasediments (Naha and Halyburton, 1977; Sharma, 1977, 1988).

From the above description it seems that the proposed schemes of stratigraphic succession subsequent to Heron's (1953) work are mainly due to inference drawn from misreading of field relations.

Another instance of misreading of field relations is in regard to the linearly occurring ultramafic bodies of Rakhabdev, described by some workers as an ophiolite belt demarcating a suture zone. In satellite imageries this elongated ultramafic body, known as the Rakhabdev lineament, occurs between shelf and distal facies of the Aravalli rocks, conformable in the Jharol belt between Gogunda and Jharol (Fig. 4.1) and cross-cutting at some other places, especially to the west of Rakhabdev. This ultramafic body, which is now seen as serpentinite body, shows intrusive relationship documented by the local occurrence of skarn in dolomitic limestone that surrounds this ultrabasic body in the region around Kherwara and Rakhabdev (Roy and Jakhar, 2002, p. 106). The intrusive relationship is also evident when the ultramafic body cuts across the fold trend in the Aravalli rocks (see Roy and Jakhar, 2002, Fig. 5.31, p. 107). The chemistry of these mafic/ultramafic rocks is also without signatures of subduction zone (see Abu-Hamatteh et al., 1994). These relationships are compelling evidence to reject the hypothesis that the Rakhabdev ultramafic bodies are dismembered bodies of obducted ophiolitic complexes (cf. Sinha-Roy, 1984; Sugden et al., 1990).

In spite of these controversies, the Aravalli Supergroup shows two contrasting lithofacies: (1) the shelf facies made up of shale-sand-carbonate assemblage, and (2) deep-water facies of carbonate-free shale-arenite association in which ultramafic rocks (serpentinites) have been emplaced sometime in the evolutionary history of the Aravalli fold belt. Regardless of these different divisions, the various lithofacies/Group/ Formations have been deformed and metamorphosed together during the Aravalli orogeny (see later).

Lunavada Group in the south of Aravalli fold belt consists of greywacke, phyllite and quartzite. It shows structural discordance with the underlying rock Groups and is devoid of certain deformational phases of the Aravalli Supergroup rocks sensu stricto, but reported to contain deformation of Satpura orogeny (Mamtani et al., 1999). The Champaner Group in Gujarat State is surrounded by Godhra Granite (950 Ma old; Gopalan et al., 1979), making its geological relationship with the main Aravalli rocks unclear. The Champaner Group and Lunavada Group have E-W structural trend, thus deflected from the NE-SW trend of the Aravalli sequence in the north. This deflection is attributed to tectonic movements along the Son-Narmada Lineament (cf. Mamtani et al., 1999).

4.2.2 Geochronology: Time of Aravalli Orogeny

The upper and lower age limits of the Aravalli supracrustals, which rest on the BGC with a profound unconformity (Heron, 1953), are not precisely known. This is in view of a large scatter of isotopic data for the Aravalli rocks (Macdougall et al., 1983) and the disputed origin of the Derwal Granite, which was emplaced during the first-deformation phase of the Aravalli rocks (Naha et al., 1967). The whole rock Rb–Sr isotopic analysis of this granite yielded an age of 1900 ± 60 Ma (Choudhary et al., 1984). This age can apparently be taken as the minimum age for the Aravalli rocks. However, Pb isotope studies of the sedimentary Pb–Zn–Cu deposits hosted by rocks of the Aravalli Supergroup yielded modal ages of 1800 and 1700 Ma (Deb et al., 1989). The younger age of 1700 Ma is from the type area of the Aravalli rocks to the south of Udaipur while older age of 1800 Ma is from the Aravalli rocks of the Bhilwara belt to the west of Berach granite. It can thus be inferred that the minimum lower limit of the Aravalli rocks is about 1700 Ma. This finds support from the Sm–Nd systematics for the metavolcanics, associated with the Aravalli metasediments, which yielded an age of about 1700 Ma (Volpe and Macdougall, 1990). From what has been discussed above it can be suggested that the time span of the Aravalli Supergroup and their stratabound ore deposits had a probable range from about 1900 to 1700 Ma.

The 2150 Ma Pb–Pb age of the Delwara volcanics occurring with the conglomeratic horizon in the type Aravalli indicates the time of rifting of the Aravalli basin, but the time of closure of the Aravalli basin is uncertain. However, the age of 1900 Ma of the Derwal Granite (Choudhary et al., 1984) emplaced during syn-F1 deformation of the Aravalli rocks could be taken as the closing time of the Aravalli orogeny.

The next succession overlying the Aravalli Series of Heron is the Raialo Series, which Heron (1953) found to show unconformable relationship with his Aravalli System. The rocks constituting Raialos are marble, metapelite and quartzite. This angular relationship was interpreted by some geologists, mainly Naha and his coworkers (Naha and Halyburton, 1974), as a result of superposed deformation, causing angularity between the axial-plane foliation and the limbs of the cofolded Raialo Series/Aravalli system which was first deciphered in the area around Udaipur. In northern Rajasthan, the Raialo Series is found to underlie the Delhi System of Heron (1953). In either case, if the Raialo Series of Heron has an independent stratigraphic status, it must be placed between Aravalli and Delhi Supergroups in Rajasthan. But it is emphasized that there is no record of a separate Raialo orogeny, unlike the Aravalli orogeny and Delhi orogeny, which gave rise to the Aravalli fold belt and Delhi fold belt rocks in Rajasthan The marble-schist-quartzite association, occasionally with black shales and BIF occurs unconformably over the basement gneisses were classed Raialo Series. These belts occur in varying sizes as synclinal keels within the BGC (Mangalwar Complex and Hindoli Groups of the Bhilwara Supergroup). Sinha-Roy et al. (1998) described them as Aravalli-equivalent deposited in pull-apart basins with the Mangalwar complex. Important belts are Jahazpur, Sawar, Pur-Banera, Rajpura-Dariba, Rampura-Agucha and Bhindar. Most of these belts are mineralized with lead, zinc, copper etc.

4.3 Delhi Fold Belt

This fold belt (excluding the Sirohi belt) forms the orographic axis of the Aravalli Mountain. This Mesoproterozoic fold belt extends nearly straight with NE-SW trend from Gujarat (Deri-Ambaji) in the south to Delhi and further in the north. This fold belt, showing unconformable relationship with the BGC, has straight boundaries against its eastern and western margins (Fig. 4.1). The eastern boundaries of the fold belt against the pre-Delhi rocks are highly tectonized, which is also evident from the truncation of the Aravalli rocks near Gogunda. From south of Khetri belt to Anatolia the Delhi fold belt contacts the gneiss-granulitic rocks of the Sandmata complex, but further south the fold belt contacts the metasediments of the Aravalli Supergroup. The Delhi fold belt (DFB) is made up of Delhi Supergroup intruded by 850 $\pm$50 Ma old granite The Delhi Supergroup sequence comprises lower arenaceous rocks (the Alwar Series of Heron, 1953) and upper calcareous-argillaceous metasediments with basic and ultrabasic rocks (Ajabgarh Series of Heron, ibid.).

The rocks of the Delhi System of Heron (1953) form a broad synclinorium, called the Main Delhi Synclinorium, which fans out in the NE-SW from a narrow zone in the central part (Fig. 4.1). The Delhi System was later designated as Delhi Supergroup, following the code of stratigraphic nomenclature (Gupta et al., 1980). Solely on the basis of the ages of granites occurring in the Delhi Supergroup, the Delhi fold belt has been divided into two sectors (Sinha-Roy, 1984), with Ajmer region as the dividing line. These are: (1) the Northern Delhi fold belt (NDFB), containing granites of 1750 Ma or slightly younger ages, and (2) the Southern Delhi fold belt (SDFB) characterized by intrusive granites of 850 $\pm$ 50 Ma, in central and southern Rajasthan and northern Gujarat. The South Delhi fold belt is in fact the Main Delhi synclinorium of Heron (1953) who correlated his Delhi system rocks from the Main Delhi Synclinorium with Delhi belts of the NE region purely on the lithological similarity. The SDFB is developed in Ajmer-Beawar tract with extension in NW Gujarat, while the NDFB is developed in Khetri-Alwar-Ajabgarh-Bayana in northern Rajasthan. The NDFB contains a folded sequence of graben-filled clastics (Alwar Series of Heron, 1953) and volcani-clastics and carbonates (Ajabgarh Series of Heron, loc. cit.). The SDFB has dominating mafic and felsic volcanics in the western part, named Barotiya-Sendra belt (Bose et al., 1990; see also Gupta et al., 1988), while the eastern part (Bhim-Rajgarh belt) is mainly clastic sediments and carbonates.

4.3.1 Delhi Stratigraphy: Controversies

Although Heron's classification of the Delhi Supergroup (Delhi system of Heron) is broadly acceptable by subsequent workers, controversies arose in regard to the physical continuity of the above-stated two sectors and their involvement in the same orogeny (Gupta et al., 1980). Sinha–Roy (1984) divided the Delhi rocks into an older NDFB and younger SDFB, essentially on the basis of Rb–Sr ages of granites reported by Choudhary et al. (1984). On the basis of these isotopic ages, Sinha-Roy (1984) suggested the diachronous evolution of the Delhi fold belt. Accepting the proposition of the older and younger Delhi fold belts, Fareeduddin et al. (1995) even recognized a link between the N and S belts and designated it as Ajmer Formation. According to Roy and Jakhar (2002), this formation is a time transgressive ensemble that includes: (i) the pre-Delhi inlier east of Beawar (Heron, 1953); (ii) the possible inlier of Anasagar granite (Mukhopadhyay et al., 2000; Tobisch et al., 1994); and (iii) the high-grade Delhi metasediments that surround the Anasagar Granite. This concept of older NDFB and younger SDFB is contradicted from detailed field studies of this "link region" by Sengupta (1984). Considering the structural data and younging direction, the Rajgarh quartzites of northern region stratigraphically overlie the carbonate association of Shyamgarh-Bhim belt in the south. Geologically, this implies that the rocks in the north are younger, and not older, than those occurring in the south. This stratigraphic relationship established by Sengupta (1984) rejects the division of the Delhi basins into an older NDFB and a younger SDFB. To emphasize, the SDFB is also found to continue up to Sambhar, NW of Ajmer (Fareeduddin et al., 1995). Thus, the division of N and S Delhi fold is dismissed and deserves no further consideration while discussing the evolution of the Delhi fold belt.

In 1988 Sinha-Roy recognized three subdivisions in the Delhi fold belt of the northeastern sector, namely the Bayana sub-basin, the Alwar sub-basin, and the Khetri sub-basin. Based on isotopic ages around 1700 Ma and slightly younger dates on the granites from the northeastern sector, and accepting their intrusive character, it was suggested that the Delhi Supergroup is older than Mesoproterozoic (Choudhary et al., 1984; see also Crawford, 1970; Gopalan et al., 1979). If these Mesoproterozoic granites served as the basement for the Delhi sediments, as shown by Gangopadhyay and Sen (1972), the controversy of diachronous origin of the Delhi fold belts does not arise. In this situation, one needs to account for the absence of 850 ± 50 Ma old granite intrusions that are so widespread in the South Delhi Fold Belt (SDFB). It is almost certain that in all these basins there must have been a basement, Pre-Delhi Granite of Mesoproterozoic age or the BGC of Archaean age for the deposition of the Delhi sediments. Furthermore, one cannot also expect a similar rock sequence in these different Delhi sub-basins, which also show different basement-cover relationships. The abrupt break in stratigraphy or in the depositional environment in the NDFB is a possibility due to scanty outcrops.

The Delhi rocks of the South Delhi fold belt are regionally metamorphosed to staurolite-kyanite grade in the east and sillimanite-muscovite grade in the west. The isograds are nearly parallel to the compositional layering, indicating syntectonic metamorphism (Sharma, 1988). The Delhi rocks also show thermal overprinting by 850 ± 50 Ma old intrusions of Erinpura Granite which have produced assemblages up to pyroxene-hornfels facies. Further south in the area between Balaram and Mawal in northern Gujarat, the Delhi fold belt reached high temperature facies (granulite facies or pyroxene hornfels facies) in which pelitic granulites (garnet-sillimanite-cordierite-spinel ±hypersthene) and calc-granulites (wollastonite-scapolite-diopside-calcite-plagioclase) have intrusive relationship with the gabbro-norite (Desai et al., 1978). This high-temperature facies appears to be the result of thermal overprinting on the already regionally metamorphosed Delhi (Pre-Delhi BGC) rocks.

In the North Delhi fold belt, which lies to north of Ajmer, there are no granites of Erinpura Granite age. The granites of this northern fold belt are mainly in the age range of 1500–1700 Ma (Choudhary et al., 1984). The metasediments of this belt show medium grade metamorphism (amphibolite facies), but granulite facies rocks have been recently reported from Chinwali-Pilwa-Arath areas of this belt (Fareeduddin, 1995), which are similar to the Sandmata granulites occurring amidst the BGC (Sharma, 1988).

4.3.2 Geochronology: Time of Delhi Orogeny

The Delhi rocks, like the Aravalli rocks are also poorly constrained in regard to their upper and lower age limits. Ages in the range of 1750–1850 Ma have been obtained for granite-gneisses occurring in the domain of the Delhi fold belt. Zircons from Anasagar granite-gneiss near Ajmer yielded 1849 ± 9 Ma (Mukhopadhyay et al., 2000) and Rb–Sr whole-rock isochronal age for the Jasrapur granite gneiss in the Khetri belt, northern Rajasthan, gave 1844 Ma (cited in Roy and Jakhar, 2002; see also Gupta et al., 1980). These granitoids, along with the Median Inlier of Beawar, represent in all probabilities remobilized basement rocks of the BGC which was rifted to form the depositional basin for the Delhi rocks. There are also granites in the Northern Delhi fold belt that gave Rb–Sr whole rock isochron age around 1750 Ma. These granites were presumed to be intrusive into the Delhi metasediments but they could be components of basement rocks for the Delhi rocks (cf. Gangopadhyay and Das, 1974). Biju-Shekhar et al. (2003) dated monazites and zircons from granitoids of North Delhi Fold Belt (NDFB), and gave ages between 1.71 and 1.78 Ga which they interpreted as crystallization age of the magma. These accessory minerals from associated metasediments gave ages between 1.7 and 1.9 or more. The older ages are interpreted by Biju-Shekhar et al. (2003) as the ages of the provenance rocks. Recent monazite age from cordierite-anthophyllite rocks of NDFB (Mahendragarh area of Haryana), yielded ages of 952–945 Ma (Pant et al., 2008), which are considered by them as age of metamorphism. If so, the metamorphism in the NDFB is possibly related to widespread granite activity of 900 ± 50 Ma in the SDFB. These dates clearly indicate, as this write thinks, that both NDFB and SDFB had Mesoproterzoic granite intrusions in the late stage of Delhi orogeny which should have occurred between 1.6 and 1.0 Ga (see below).

Accordingly, the 950 Ma event in the metasediments from NDFB (Pant et al., 2008) and 900–1000 Ma old granite intrusions of SDFB can be interpreted as cooling ages of the rocks of the Delhi fold belt, recrystallized metasediments and the emplaced granite plutons.

The 1700 Ma age obtained for the detrital zircons from pelitic granulites that occur in Parbatshar-Pushkar region, along the western margin of the Delhi basin indicate the period of Delhi sedimentation The time of rifting for Delhi basin is also supported by the 1700 Ma age of the metabasalt from Ranakpur (see Volpe and Macdougall, 1990) where they are seen to have been cofolded with the Delhi metasediments and also isogradically metamorphosed with them (Sharma, unpublished data). These metabasalts with relic pillow structures indicate synchronous volcanism in the rifted Delhi basin formed by crustal extension. It remains to be explained as to why the model ages from Saladipura in North Delhi fold belt and from Deri-Ambaji in the South Delhi Fold belt gave contrasting ages of 1800 and 1100 Ma, respectively (cf. Deb et al., 1989, 2001).

The closure of the Delhi basin should have been much earlier than the 1012 ± 78 Ma (Sm–Nd) old undeformed diorites that intruded the Delhi metasediments of Ranakpur in SDFB (see Volpe and Macdougall, 1990). This Post-Delhi age is also reflected in the Pb–Pb model ages of 1102 and 1087 Ma for the basemetal deposits at Deri-Ambaji and Barotiya, respectively, in the south Delhi fold belt (Deb et al., 1989). A more plausible age for the closing stage(s) of the Delhi basin can be guessed from the granite activities in the age range of 1480 and 1350 Ma (Gopalan et al., 1979; Choudhary et al., 1984), which may be related to main tectono-thermal event during the Delhi orogeny. A single zircon date of ca. 1450 Ma (cited in Roy and Jakhar, 2002, p. 149) from a complexly folded outcrop of biotite gneiss from Rampura-Agucha area of the BGC terrain is also indicative of this tectonothermal event related to the Delhi orogeny.

Mineral isochron (Sm–Nd) ages of the diorites yielded 838 ± 36, 835 ± 43 and 791 ± 43 Ma (Volpe and Macdougall, 1990), which is attributed to ca 800 Ma old tectono-magmatic event recognized by Volpe and Macdougall (loc. cit.). In the southern Delhi fold belt profuse granite activity at 900 ± 50 Ma is recorded, especially around Erinpura and Abu, southern Delhi fold belt (cf. Choudhary et al., 1984). Recently, Pandit et al. (2003) have obtained U–Pb zircon (TIMS) age of 967.8 ± 1.2 Ma for the Chang pluton near Sendra, which is very close to the whole rock Rb–Sr determination of 966 ± 250 Ma by Tobisch et al. (1994). From these post-Delhi dates on the plutonic intrusives, Pandit et al. (2003) suggest that the Chang and Sendra granites are coeval with diorite intrusions at Ranakpur.

The lower age limit of the Delhi rocks is about 1750 Ma. This age limit renders the Delhi sedimentation younger than the lower age limit for the Aravalli rocks by about 200 Ma only. This age difference of 200 ± 50 Ma between the Aravalli and Delhi sedimentation requires a more or less continuous orogenic cycle during which the Aravalli rocks seem to have been involved earlier than the Delhi supracrustals. This is because the rocks of both Aravalli and Delhi Supergroups document mineral assemblages of only one regional metamorphism (single orogeny) (Sharma, 1977, 1988). If so, the orogeny affecting the Aravalli rocks must have started earlier. This is supported by the first deformation phase (AF1) that occurs in the Aravalli rocks (and their basement BGC) but absent in the Delhi Supergroup (Naha et al., 1984). Since the regional metamorphism of the Aravalli rocks is syntectonic, as deduced from integrated petrological and fabric studies (Sharma, 1988), the age of this metamorphism should be near the age of the 1900 Ma old syntectonic Derwal granite intruding the Aravalli rocks during AF1, much before the Delhi supracrustals were metamorphosed. It should be stated that the regional metamorphism of the Delhi rocks was later at about 1450 Ma ago, and much before the 1012 Ma old diorite intrusion in the amphibolite and calcgneisses of the Delhi Supergroup (Volpe and Macdougall, 1990).

A strong metamorphic re-equilibration in the Delhi rocks is documented at about 850 ± 50 Ma by the Rb–Sr isotopic data on the Erinpura Granite (Crawford, 1970; Choudhary et al., 1984) and by the Sm–Nd mineral isochron in the metabasic and diorite rocks (Volpe and Macdougall, 1990). This thermal event is also reflected in the Rb–Sr isotope of muscovite from 2800 Ma old Untala granite (Choudhary et al., 1984). But this thermal event is yet to be documented in the 2600 Ma old Berach Granite by a more detailed geochronological work, although a thermal event at 710 Ma is revealed from the lower intercept of the Discordia for Berach Granite (Sivaraman and Odom, 1982, Fig. 2). This younger event apparently corresponds to another igneous activity, namely the Malani volcanism-plutonism that occurred at about 750 Ma ago to the west of the Aravalli ranges (Torsvik et al., 2001).

West of Udaipur the Delhi Supergroup rocks abut against the rocks of the Aravalli Supergroup. Their contact relationship is unclear—whether it is a structural discordance (Gupta et al., 1980) or major tectonic suture (Sugden and Windley, 1984) or a sheared unconformity (Naha et al., 1984). Petrological studies by the author have revealed that the Aravalli rocks in the immediate contact with the Delhis and further east have greenschist facies assemblages while the Delhi metasediments on the west show amphibolite facies metamorphism (Sharma, 1988). This is perhaps one of the most critical problems of these Proterozoic supracrustals in that the older Aravalli sequence is lower grade than the younger Delhi sequence in immediate contact. A detailed petrological study also revealed that each of the Proterozoic sequences have been involved in only one orogenic cycle, unlike the polymetamorphic BGC rocks, and the mineral assemblages in both the supracrustal sequences are the outcome of a single regional recrystallization, disregarding the thermal overprinting by the Erinpura granite intrusions (Sharma, 1988).

4.4 Deformation of the Proterozoic Fold Belts

Both Aravalli and Delhi fold belts, constituting the Aravalli Mountain in Rajasthan, are made of Proterozoic volcano-sedimentary rocks which were laid upon the basement BGC during Palaeoproterozoic and Mesoproterozoic times.

Structural studies by Naha and his coworkers deciphered two main fold phases in the Precambrian rocks of Rajasthan: an early isoclinal and reclined folding (F1) on W-WNW plunging axes, and a late upright folding (F2) on NNE-striking axial planes. In some terrains the F1 folds are coaxially refolded by F1a folding and both are superposed by F2 folding. The early F1 folds found in the rocks of the Aravalli

Supergroup (and their basement BGC) are not documented in the rocks of the Delhi Supergroup in which first folding (DF1) is isoclinal on NE-striking axial planes, thus coinciding with the second folding (AF2) in the Aravalli (and BGC) rocks. The Delhi rocks also show coaxial superposed folding with DF2 folds on upright axial planes. In subsequent studies Naha (1983) and Naha and Mohanty (1990) also established two additional deformation phases, which produced conjugate/kink folds: DF3 and DF4 in the Delhi rocks and AF3 and AF4 in the Aravalli rocks. The AF3 folds are reclined and developed by a vertical load (compression) on steep, fissile rocks of the Aravalli Supergroup. The corresponding DF3 conjugate folds in the Delhi rocks, also developed by vertical compression, are coaxial with DF1 and DF2. The AF4 conjugate and kink folds are the result of longitudinal shortening along NE-SW horizontal directions. The corresponding DF4 kink folds in the Delhi rocks are also the result of horizontal squeezing in the NE-SW and perhaps N-S directions.

The BGC, as a crystalline basement to the Proterozoic cover, shows a more complex deformation history, although it is cofolded with its Proterozoic supracrustal sequences. The Pre-Aravalli (Archaean) features are documented by rootless folds and by angularity of the Archaean planar structures with the prevalent gneissosity (Naha and Roy, 1983; Sharma, 1977; Mukhopadhyay and Dasgupta, 1978). A more recent structural study by Srivastava et al. (1995) reveals that the early folds (Pre-Aravalli) in the gneisses are non-cylindrical and virtually coplanar.

The deformational history in the Precambrian rocks of Rajasthan is summarized in Fig. 4.2.

4.5 Geohistory of Granulites

The granulite or granulite facies rocks are completely absent in the Aravalli Supergroup, but they occur as shear-bound rocks in the BGC around Sandmata and as fragments in the Delhi Supergroup near the contact with the BGC to the north of Ajmer. At Sandmata and around it, the granulites occur as discontinuous oval-shaped outcrops (Sharma and Joshi, 1984; Sharma, 1988). The granulite complex is a suite of charnockite, pelitic granulite, leptynite, two-pyroxene-garnet granulite and cordierite-garnet-sillimanite-kyanite gneiss. Norite dykes intrude this granulite complex and the whole suite has a sheared contact with the BGC (Sharma et al., 1987). The recently discovered granulites from the Delhi metasediments occur as fragments, but have petrological similarities with the Sandmata granulites (cf. Fareeduddin, 1995).

The pelitic granulites show mineral assemblages of two periods of metamorphism wherein kyanite predates the sillimanite that shows different orientation and crosscutting relationship with the former. The sillimanite prisms are seen to change into kyanite needles whose significance will be discussed later. The garnet-bearing two pyroxene granulites or basic garnet granulites structurally underlie the pelitic granulites. The presence of garnet-clinopyroxene-quartz corona at the interface of hypersthene and plagioclase suggests that after emplacement this basic

4.5 Geohistory of Granulites

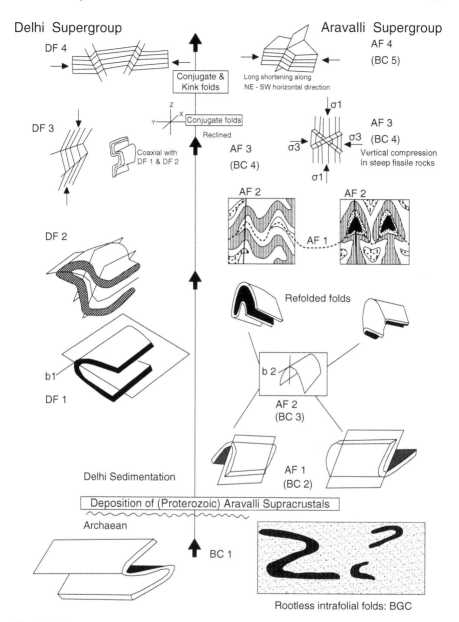

Fig. 4.2 Diagrammatic representation of the sequence of deformational phases and fold styles in the Precambrian rocks of Rajasthan, NW India (based on Naha and others, 1967, 1984; Mukhopadhyay and Dasgupta, 1978; Roy, 1988; Srivastava et al., 1995)

rock underwent nearly isobaric cooling to form the corona. Incipient corona of garnet-clinopyroxene ± quartz is also noticed in the norite dyke (Sharma et al., 1987). The granulite terrain near Gyangarh shows intrusion of granodiorite-charnockite comagmatic series in which norite inclusion is noticed at one place. This comagmatic series gave 1723 ± 14 Ma age by U–Pb zircon geochronology (Sarkar et al., 1989). In the vicinity of this felsic intrusive series, the pelitic granulites and gneisses in the Gyangarh-Thana area have developed andalusite and hornfelsic texture (Joshi et al., 1993). This suggests that the granodiorite-charnockite series was emplaced in the granulite complex at shallower depths, not exceeding 12 km (corresponding to the location of the Al-silicate triple point of Holdaway, 1971). This implies that the granulite facies metamorphism has occurred during the Delhi orogenic event. It is emphasized that the U–Pb age of 1723 Ma denotes the age of generation of the nearly anhydrous melt of granodiorite-charnockite compositions during the end-granulite metamorphism. The recent ages near 1621 Ma obtained by zircon evaporation method (Roy et al., 2005) need different interpretation if the zircon U–Pb age of 1723 Ma for the enderbite-charnockite intusion is taken to denote time of granulite facies metamorphism in the Delhi orogeny (see below).

From geothermobarometric data and textural relationships of the Al-silicate polymorphs in the pelitic granulites in which sillimanite grew between two kyanite-forming events, the P-T path for the granulite complex is found to be anticlockwise (Sharma, 1988; Joshi et al., 1993). This implies that thickening of the crust during Delhi orogeny was due to magmatic underplating/interplating and not due to continent-continent collision (England and Thompson, 1984). The structural position of the basic granulite under the pelitic granulite and gneisses of the Sandmata area supports this mechanism of crustal thickening by magma addition which, in turn, caused high temperature metamorphism in the lower crust, as revealed by pyroxene geothermometry in the basic granulites (Sharma, 1988; Joshi et al., 1993). Subsequently, the granulite complex was exhumed to shallower depths along shear zones—a proposition supported by the presence of a peripheral shear zone in the Sandmata area (Sharma, 1988) and recently confirmed by DSS studies (Tewari et al., 1995).

The proposition of Roy et al. (2005) that the granulite facies metamorphism and exhumation of the 1723 Ma old granulites are linked with the rift opening stages of the Delhi orogenic cycle, cannot be accepted for the following reasons:

(1) It is rather impossible to retain the granulite assemblage if they had acted as basement rocks in the rifted (extensional) basin for the Delhi sediments and later involved with them in the Delhi orogenic cycle.
(2) It is difficult to explain their scattered occurrence in the Delhi metasediments, and complete absence in the Aravalli Supergroup rocks, if they were not excavated during the Delhi orogenic cycle.
(3) The entire granulite complex at Sandmata cannot be a reactivated rock-suite as claimed by Roy et al. (2005). The pelitic granulite is evidently a reactivated basement due to their polymetamorphic assemblages (Sharma, 2003), but the basic granulites and the enderbite-charnockites have assemblages of only a single tectono-thermal event related to the Delhi orogenic cycle.

(4) If the extensional environment is responsible for both excavation of the granulites along shears and rifting of the Delhi basin, there must have been abundant crustal melting in the concerned earth's sector (south and central Rajasthan) to give rise to granitoid intrusion of around 1700 ± 50 Ma old. There is no documentation of such decompression melting in the entire southern part of the Delhi fold belt (SDFB).

Recently, Saha et al. (2008) suggested that there was older medium-pressure granulite facies metamorphism (preserved only within the enclaves of the BGC). This was followed by a kyanite-grade high-presure event and finally amphibolite facies overprint, mostly confined to shear zones. This hypothesis is not supported by the available regional geological and geochronological evidence. First, the metamorphic rocks of the Aravalli and Delhi fold belts show mineral assemblages of only one regional metamorphism and hence one orogeny, excluding thermal effects by late tectonic intrusions. Second, the granulites are completely absent from the terrain occupied by the Aravalli Supergroup. They occur largely in the BGC rocks and as fragments in the metasediments of the Delhi supergroup (Fareeduddin and Kroner, 1998). From this it is concluded that the granulite metamorphism occurred between Post-Aravalli regional recrystallization and 1.7 Ga ago, being the age of undeformed charnockite-enderbite intrusion in the area (Sarkar et al. 1989). Thirdly, there is no possibility for an event of increased load pressure after granulite facies metamorphism (as the last thermo-tectonic event) and the hypothesis of kyanite-grade metamorphism at increased pressure (after granulite facies metamorphism) is untenable.

4.6 Agreed Observations and Facts

Before we consider an acceptable evolutionary model for the Aravalli Mountain we must keep in mind the facts and agreed geological observations and the inferences deduced thereby about the Precambrian crystalline rocks of the Rajasthan terrain. They are enumerated as follows:

Considering these facts and the deduced inferences, we could now discuss the different evolutionary models proposed for the Aravalli Mountain ranges of Rajasthan. Plate tectonic model requires recognition of:

(a) Crustal plates that collided to produce the Mountain belt.
(b) Spreading zones of crustal creation.
(c) Shortening zones of crustal destruction.
(d) Magmatic Arc or its remnant.
(e) Occurrence of ophiolites etc.
(f) Presence of nappes/thrust sheets.

When applied to Rajasthan terrain, we fail to recognize most, if not all, of these features. The anticlockwise path deduced for the granulites is not compatible with crustal thickening by continent-continent collision.

S. No.	Observations	Inference
1.	Delhi supracrustals rest directly upon the BGC and are devoid of clastics of Aravalli Supergroup	Delhi rocks were deposited on BGC, away from Aravalli Provenance
2.	Delhi Supergroup rocks do not record early F1 folds (isoclinal/reclined) that are seen in Aravalli metasediments and underlying BGC	Aravallis are older than Delhis
3.	Both Proterozoic units, viz. the Aravallis and Delhis show mineral assemblages of only one regional metamorphism	Each of the supracrustal units underwent one orogeny only
4.	Despite being older, the Aravalli rocks from the type area (S of Udaipur) are lower grade (greenschist facies) than the juxtaposed Delhi metasediments (amphibolite facies) on the W	The two cover units (now fold belts) were temporally spaced apart during orogenesis and were accreted subsequently
5.	Delhi metasediments have abundant granite intrusions while older Aravallis are almost devoid of granite intrusions	Geothermal gradient was perhaps higher during Delhi metamorphism
6.	Delhi metasediments, like the BGC, contain granulites (as fragments) but the Aravalli rocks are without granulites	Geodynamic environment for the two Proterozoic cover rock-units was different

The Palaeoproterozoic Aravalli fold belt and the Mesoproterozoic Delhi fold belt are embedded in the Archaean crystalline complex (the BGC) which was reworked in all probability in the Proterozoic tectono-thermal events. In recognition of the disposition of the fold belts, a number of workers (Sychanthavong and Desai, 1977; Sychanthavong and Merh, 1984; Sinha-Roy, 1988; Sugden et al., 1990) tried to explain the evolution of these Proterozoic fold belts in terms of plate tectonics. The Delhi Supergroup rocks north of Ajmer (i.e. in North Delhi fold Belt, NDFB) occur in various large and small basins in Jaipur, Alwar, Bayana, Lalsot and Khetri. A similarity in patterns of litho-successions and sedimentlogical characteristics enabled some geologists to make basin-wide correlations (cf. Heron, 1953; Singh, 1988). Patterns of disposition of the folded rocks along the ductile shear zone also suggested a NE-SW trend of the basins in the Khetri Fold belt and Southern Delhi Fold Belt (SDFB) (Gupta, 2004). The platformal sediments in the Bhim Group and the presence of basement gneiss wedges within the belts indicate that the sediments were floored by continental crust (Mukhopadhyay and Bhattacharya, 2000). The Basantgarh-Barotiya belt in the SDFB is characterized by mafic rocks, implying oceanic floor of these basins or the presence of the rift fractures reaching to sub-crustal depths to facilitate outpouring of mafic melts as well as of acid volcanism (due to crustal melting) in these rifted basins. Impressed by the linearity of the mafic rocks along with ultrabasic bands, Gupta et al. (1980) erroneously interpreted the mafic rocks of the SDFB as "metamorphosed thrust wedges of oceanic crust and designated them as the Phulad Ophiolite Belt (see Sharma, 1989, p. 216).

The older Aravalli fold belt lies on the east of the Delhi fold belt and shows one deformation earlier than the earliest deformation (DF1) in the Delhi fold belt.

The Aravalli rocks from the type area show deeper facies in the west and shallower in the eastern part, although they uniformly show greenschist facies metamorphism (Paliwal, 1988). The Aravalli rocks north of Udaipur show amphibolite facies assemblages and do not extend beyond the line joining Sambhar-Jaipur-Dausa in the northeast.

4.7 Evolutionary Models and Discussion

From this and other geological information, mainly from the incorrect identification of rocks in field, especially amphibolites as ophiolites, different workers have advanced different evolutionary models for the Aravalli and Delhi fold belts.

Model 1. Applying the Wilson cycle in the evolution of the Proterozoic fold belts of Rajasthan, Sinha-Roy (1988) proposed that Rajasthan craton was subjected to extension about 2000 Ma ago, with the development of a triple junction, FRR (Fault-Ridge-Ridge) near Nathdwara. From this junction emanated three arms. One was an Aravalli Rift (Rifting Stage I) that later gave rise to two intracontinental (ensialic) Aravalli basins, namely Pur-Banera-Dariba-Bhinder (PBDB) belt and Jahazpur (J) belt (Fig. 4.3a). Of the remaining two arms, Sinha-Roy (loc. cit) postulated two major strike-slip faults originating from the triple junction. One fault with NE-SW trend opened during Middle Proterozoic to give rise to South Delhi rift while the other arm is believed to have originated from the Aravalli rift (giving no reason for not emanating from the triple junction) near Salumber and splayed into multiple strike-slip faults along which pull-apart basins are assumed to have developed in the NE Rajasthan. According to Sinha-Roy (1988) the Aravalli rift south of

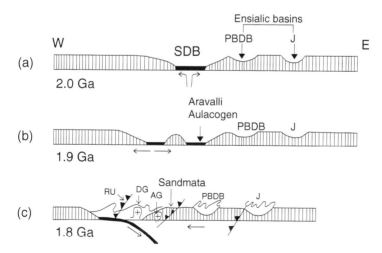

Fig. 4.3 Evolutionary model of the Aravalli fold belt and related belts shown by cartoons (redrawn after Sinha-Roy, 1988). Abbreviations: AG = Anjana Granite; DG = Derwal Granite; J = Jahazpur belt; PBDB = Pur-Banera-Dariba-Bhinder belt; RU = Rakhabdev ultramafics. SDB = South Delhi Ocean Basin. Toothed line is a major dislocation

Udaipur migrated westward and had its floor of oceanic material (Fig. 4.3a). This was followed by another rifting stage, termed Aravalli Rift-II by Sinha-Roy (1988), during which an aulocogen on the eastern side of the Aravalli ocean-floored basin was developed (Fig. 4.3b). Further spreading of this rifted basin resulted into eastward subduction under the Mewar crustal block or Mewar terrain (carrying its cover sediments of the Aravalli Supergroup; Sinha-Roy, 2004). Sinha-Roy believes that the Derwal Granite (DG), Anjana Granite (AG) and nearby coeval granites (1900 ± 80 Ma) represent magmatic arc (Fig. 4.3c). Later, obduction of the Aravalli oceanic crust is believed to have appeared as ophiolites along Rakhabdev suture zone (RU) (Fig. 4.3c). According to Sinha-Roy (1988), the Aravalli suduction is considered to have changed its character at N and S of Antalia (see Fig. 4.1); it was oceanic (ensimatic) in the south and continental (ensialic) in the north. A consequence of this differing nature of subducting material is that it resulted in the opening of Delhi-age (Mesoproterozoic) rifts. In the north (north of Antalia), the rifts formed in succession towards west from Bayana, Alwar and Khetri, giving rise to the North Delhi Fold Belt (NDFB), while in the south the rifting gave rise to the South Delhi Fold Belt (SDFB) with Basantgarh, Barotiya, Sendra, Bhim basins. Sinha-Roy assumed a transcurrent fault, named by him (Sinha-Roy, 1988) as Sambhar-Jaipur-Dausa Fault (SJDT), demarcating a boundary between the rift systems of NDFB and SDFB. The fault is believed to have played the role of transcurrent fault during the opening of the North Delhi Pull-apart basins (Fig. 4.4a) whilst it behaved as transform

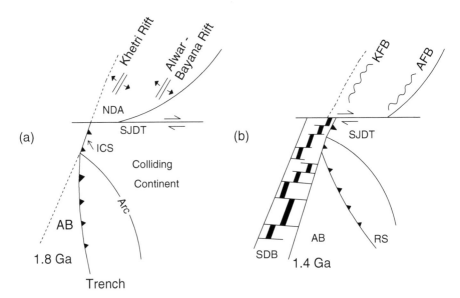

Fig. 4.4 Shows relation of the Aravalli subduction and opening of (**a**) North Delhi Rifts and (**b**) South Delhi Ocean Basin (SDB), illustrating diachronous nature of the North and South Delhi fold belts (after Sinha-Roy, 1988). Abbreviations: AB = Aravalli Fold Belt; AFB = Alwar-Bayana Fold Belts; ICS = Intracontinental Subduction; RS = Rakhabdev Suture; SDF = South Delhi Ocean Opening (Basin); SJDT = Sambhar-Jaipur-Dausa Transcurrent Fault

4.7 Evolutionary Models and Discussion

fault during the opening of the ocean basin of the SDFB (Fig. 4.4b). According to Sinha-Roy, the north Delhi rift was abortive (aulacogen) while the south Delhi rift developed gradually by ca. 1300 Ma into an ocean-floored Rift, extending all along the Main Delhi Synclinorium of Heron (1953).

Sinha-Roy (1988) further states that at ca. 1300 Ma ago, the oceanic crust started subducting toward west, during which trench sediments or accretionary prism, made of carbonate dominated turbidites of the Ajabgarh Group (Ajabgarh Series of Heron), and volcanic arc (of Sendra granite) and marginal basin to its west, came into existence. With the closing of the marginal basin, ophiolite mélange (recognized as Phulad ophiolite by Sinha-Roy, 1988) developed in the SDFB. The different stages of the evolution of the Delhi fold belt and its accretion with the Aravalli fold belt is depicted in Fig. 4.5 (stages from d to h).

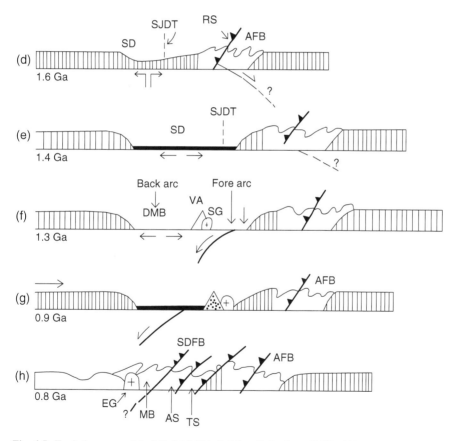

Fig. 4.5 Evolutionary model of Delhi fold belt (after Sinha-Roy, 1988). Abbreviations: AFB = Aravalli Fold Belt; AS = Arc Sequence; DMB = Delhi Marginal Basin; RS = Rakhabdev Suture; SG = Sendra Granite; SD = South Delhi Rift; SJDT = Sambhar-Jaipur-Dausa Transcurrent/Tansform Fault; SDFB = South Delhi Fold Belt; VA = Volcanic Arc. See text for details

In 2004, Sinha-Roy modified his earlier model, assuming the SDFB as a transpressional orogen and the Khetri belt as a suspect terrane, merely on the supposition that the north Aravalli sequence extended as a Great Aravalli Basin (GAS) under the SDFB. The GAS subsequently, according to Sinha-Roy (2004), moved northward along a fault to occupy the present position in the Khetri region, hence becoming a suspect terrane in the Delhi fold belt. In the revised model, Sinha-Roy (2004) considers the evolution of the NDFB earlier than that of the SDFB.

The models of Sinha-Roy have several weak points and contradictions as enumerated below:

(1) Sinha-Roy (1988) considered the triple junction as FRR but his description indicates that the three arms emanating from it are two faults and a rift (FFR). One of the fault is considered by Sinha-roy to develop from the Aravalli Rift, instead of from the triple junction. The triple junction RRF (so also the FFR) visualized by Sinha-Roy (1988) is not stable due to the relative motion of the plates along the fault; the RRF can exist only for a short instant in geological time (see Chap. 1).

(2) If there is an eastward subduction of the Aravalli block under the BGC-Bundelkhand (-Berach) block, why do we not have linearly located granites (as remnants of magmatic arc) to the east of the subduction zone (along Rakhabdev suture zone)?

The Derwal granite and Anjana granite cannot be the components of this magmatic arc believed by Sinha-Roy (1988), because the Derwal granite is a syntectonic migmatite developed during F1 folding of the Aravalli metasediments (Naha et al., 1967), if not Pre-Aravalli intrusion as argued by Heron (1953). The Anjana granite intruding the BGC rocks has been dated by single zircon ion-probe at 1641 ± 4 Ma (Roy and Jakhar, 2002) and hence much younger than the subduction time suggested by Sinha-Roy (1988, p. 100).

(3) The Aravalli rocks from the type area south of Udaipur are low-grade (greenschist facies) over a vast terrane and increase of grade is along strike from south to north (Sharma, 1988, p. 66) which is incompatible with the subduction polarity suggested by Sinha-Roy (1988).

(4) The isofacial or isogradic nature of regional metamorphism in the Delhi rocks from Ambaji-Deri area and the underlying Aravalli rocks (Type locality) on the west and east side of the contact line suggests that the two terrains were accreted nearly at same crustal depth.

(5) The near E-W trending Sambhar-Jaipur-Dausa fault assumed by Sinha-Roy as a transcurrent-transform fault is not identifiable in lineament map of Rajasthan (Bakliwal and Ramasamy, 1987). While emphasizing on this point, Roy and Jakhar (2002, p. 363) state that no important E-W trending lineament is known in Rajasthan. These observations make the model of Sinha-Roy untenable for the Proterozoic fold belts of Rajasthan. There are NE-SW faults showing strike-slip displacement, but they are all related to Neotectonism (see Sinha-Roy, 1988).

4.7 Evolutionary Models and Discussion

(6) The mafic rocks of the western basin of the SDFB are not ophiolite slices. This mafic volcanism is found to be synsedimentary and often interbedded with shallow clastics, thereby ruling out being allochthonous slices of the oceanic crust (Bose et al., 1990; Gupta et al., 1991). Moreover, these volcanic rocks are not restricted to the lower part of the stratigraphic column in the Barotiya-Sendra Groups. There is complete absence of deep-sea deposits in the SDFB and also the absence of typical ophiolitic components in the Phulad Ophiolitic Belt (Sharma, 1990; Bose et al., 1990).

(7) The rocks of the SDFB show a low to medium pressure metamorphism and not a high-pressure metamorphism (Sharma, 1988).

(8) The diachronous evolution of the Delhi fold belt is already dismissed (see earlier section), particularly when a uniform pattern of structural and metamorphic evolution is observed in the entire Delhi basin. Shear zones are the last event of Delhi orogenesis and the concept of suspect terrane in Rajasthan is not supported by stratigraphy and tectonics in the NDFB and SDFB (cf. Sinha-Roy, 1984).

Model 2. Sugden et al. (1990) proposed another model for the evolution of these Proterozoic fold belts. The model is somewhat similar to that of Sinha-Roy but the Aravalli subduction is considered eastward, not westward as conceived by Sinha-Roy. According to these authors, first there was a rifting of a rigid Archaean crust (BGC) at about 2000 Ma ago, developing a northern rift basin and a southern rift basin (Fig. 4.6a). The former did not develop and became an aulacogen, giving rise to the present Bhilwara belt. The southern rift traversed the BGC block south of Udaipur, separating this continental block into two, one on each side of the rift basin. The resulting passive margin of the fragmented western block became attenuated and had shelf and deep-sea (Jharol) sediments over it. With spreading at the site of rifting there was subduction at ca. 1500 Ma ago, along a western dipping trench, under the Delhi Arc, which they (Sugden et al., 1990) believed to have developed either by eastward subduction (Fig. 4.6b) or by westward subduction of the continental block (Fig. 4.6c). Continued collision and westward subduction gave rise to obduction of the Phulad Ophiolites (located in a back arc) along the Rakahabdev lineament and deformation of the Ajabgarh and Alwar sequences in the fore-arc basin (Fig. 4.6d).

The evolutionary model of Sugden et al. (1990) can be easily understood in terms of a RRT triple junction (formed during rifting) whose northern ridge (Rn) aborted to become the Bhilwara belt (in the eastern part of the terrain), while the southern ridge (Rs) developed into a true rifted basin. Because of E-W spreading and subduction along the trench T (part of triple junction), a continental arc called Delhi island arc first appeared due to secondary spreading (see origin of marginal basin in Chap. 1). The westward moving BGC-Aravalli block subducted under this arc, whereby the shelf and deep-sea deposits were deformed to give rise to the present Aravalli-Jharol belts (cf. Sugden et al., 1990). The final collision is believed by the authors to have resulted into crashing of the Delhi island arc and accretion of the Aravallis drifting on the west-moving block. According to Sugden et al. (1990), this collision in

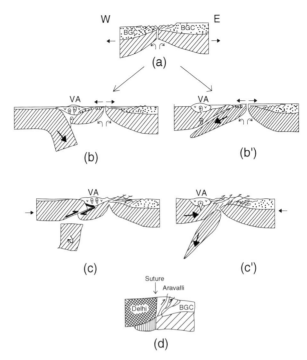

Fig. 4.6 Evolutionary model of Aravalli-Delhi fold belts (after Sugden et al., 1990). (**a**) Rifting of the Archaean lithosphere of Rajasthan, (**b**) generation of volcanic arc by eastward subduction of an oceanic crust, or (**b'**) volcanic arc generated by westward subduction of continental lithosphere, (**c**) breaking of eastward subducting slab and development of a dextral shear zone, (**c'**) breaking of westward subducting block and upthrusting of crustal rocks on the Eastern block, (**d**) crushing of the volcanic arc, upthrusting of the oceanic crust as ophiolite and accretion of Aravalli-Delhi fold belts. VA = Volcanic Arc

early stage is responsible for the obduction of the oceanic crust (now seen as the Rakhabdev ultrabasic rock, serpentinite) along the Rakhabdev lineament, oriented itself parallel to the collision zone.

The model proposed by Sugden et al. (1990) also has several weaknesses and does not explain the geological characteristics of the fold belts, stated already.

(1) The model does not take cognizance of the metamorphic assemblages in the rocks of the Aravalli fold belt (Sharma, 1988). If the Northern basin remained aulacogen, then why is the Bhliwara belt higher grade (amphibolite facies) than the belt made of the type Aravalli rocks south of Udaipur. The model fails to account for this disparity and geological setting.
(2) The model by Sugden et al. does not separate the Delhi and Aravalli fold belts in time and space. They even believe that the Delhi island arc, under which the BGC-Aravalli rocks subducted, existed in the Pre-Aravalli time. This is

4.7 Evolutionary Models and Discussion

not compatible with the stratigraphic sequence established in Rajasthan (see Heron, 1953).

(3) If there was subduction of the Aravalli-laden BGC block toward west, why are there no granitic rocks of calc-alkaline nature with ages of 1.5 Ga or older in the Southern Aravalli-Delhi fold belt. The model cannot explain this.

(4) The ophiolite in the SDFB is not a typical ophiolitic suite; nor is there any deep-sea sediment in their association. This argument is also strengthened by the occurrence of nearby granulites (Sandmata) to the east of Phulad, which implies that the geothermal gradient must have been higher, disallowing the blueschist/ophiolite formation.

(5) There are no high-pressure assemblages in the Basantgarh-Barotiya-Sendra belts, assumed to have ophiolitic mélange or ophiolitic slices.

Model 3. Roy and his coworkers (see Roy, 1990; Roy et al., 1993) also proposed an evolutionary model for the Aravalli fold belt. This model envisages three-stage evolution of the Aravalli basins, corresponding to three distinct cycles of sedimentation. These rift basins were linked at triple junction with one or more failed arms. According to Roy, the earliest basins were those of Hindoli and Khairmala belts fringing the basement rocks now represented by the Berach Granite. Roy thinks that these two basins joined at a triple junction near Chittaurgarh, without naming the nature of the third arm, whether Fault or Trench (Fig. 4.7). In the second stage of rifting two more basins, one around Udaipur and the other at Kishangarh also opened up simultaneously, again without naming the third arm of the triple junction located WNW of Nathdwara. The Bhilwara rift, according to Roy (1990) opened up

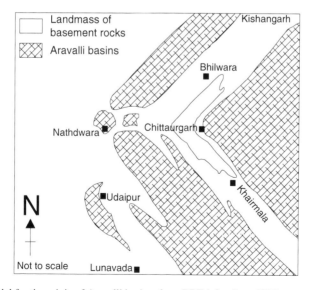

Fig. 4.7 Model for the origin of Aravalli basins along RRF (after Roy, 1990)

at a time when the Udaipur basin was deepening. This rift was aborted and remained an aulacogen (cf. Deb, 1993). The deep basins at Jharol and Lunavada were the last to form and had ultramafic intrusion.

The above model of Roy is geodynamically weak and faulty for the following reasons:

(1) The two RRFs with two closely-spaced triple junctions, at Chitttaurgarh and Nathdwara, do not seem geodynamically possible to have formed by plume responsible for lithospheric doming.
(2) The author (Roy, 1990) fails to recognise the fault (F) along which the R and R may have moved during basin evolution in the proposed model.
(3) The aulacogen of Bhilwara belt raises a fundamental question as to where have the two other elements (RT or RF) of the triple junction vanished?
(4) There seems no reason for the failed arm (the Bhilwara belt in the present case) to be higher grade than the other active belts of the Aravalli Supergroup.
(5) There is no scope of subduction, either A or B types, in the evolutionary model by Roy (1990).
(6) Finally, Roy and his coworker (see Roy and Jakhar, 2002, p. 214) believe a hiatus between the Aravalli fold belt and the Delhi fold belt. If so, it is indeed difficult to understand as to why the deformation phases in these two fold belts are correlatable; the F2 phase in the Aravalli (and BGC) is the first fold phase F1 in the Delhi rocks (see earlier section). Unlike Sinha-Roy (2004), Roy and others believe that the DFB evolved in a single stage of orogenesis, with thermal overprints and shearing post-dating the orogeny.

From the above discussion we observe that the basic assumption in the evolutionary models by Sinha-Roy (1988, 2004) and Sugden et al. (1990) are unsound and fraught with several fundamental problems. The polarity of subduction is also arbitrarily chosen; eastward subduction by Sinha-Roy, while this uncertainty led Sugden et al. (1990) to propose both westward as well as eastward subduction during the Aravalli orogeny. Both groups of workers assumed the linear outcrops of mafic/ultramafic rocks as ophiolites the existence of which are convincingly denied from detailed studies by subsequent workers (e.g. Bose et al., 1990). The model by Roy is inherently weak and confusing, as stated earlier. None of these models has taken cognizance of metamorphic grades and isograd patterns in these Proterozoic rocks. Needless to emphasize, metamorphism is the key to the understanding of tectono-thermal evolution of the fold belts.

In view of the above arguments and available geochronological data and geological information this writer proposes a more plausible model for the evolution of the Aravalli mountain belt. The evolutionary history is shown in Fig. 4.8.

Model 4. The model presented here is a modified version of the one given earlier by this writer (Sharma, 1999, 2003). In the first stage, long after the stabilization of

4.7 Evolutionary Models and Discussion

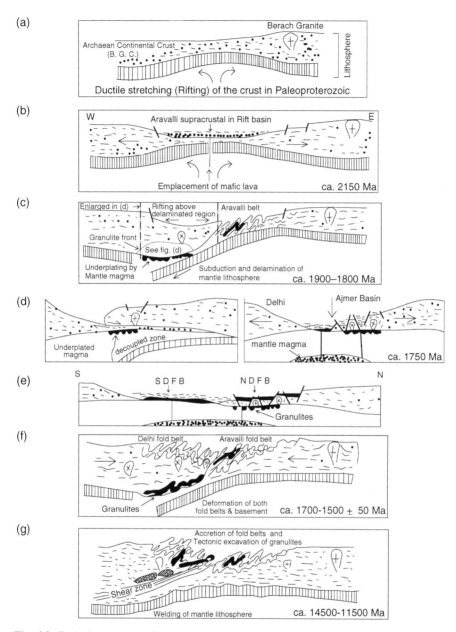

Fig. 4.8 Evolutionary stages for the Proterozoic Aravalli mountain (Aravalli and Delhi fold belts) in Rajasthan, NW India. (**a**) Plume-generated ductile stretching (rifting) of the Archaean crust of Rajasthan in Palaeoproterozoic, (**b**) deposition of Aravalli supracrustals, including volcanics of Delwara, in the rifted basin, (**c**) westward subduction and delamination of mantle lithosphere whereby mantle magma underplated the crust, forming granulites at the base and rifting of the softened crust above the delaminated region for deposition of the Delhi supracrustals, (**d**) extension of the continental crust causing a series of rift basins in the northern part of the evolving Delhi

the Archaean craton at about 2.6 Ga, the continental crust of BGC was stretched and thinned by ductile extension over a mantle plume (Fig. 4.8a). This crustal extension resulted in the formation of two major intracratonic basins meeting near Udaipur, roughly with N-S trend, both receiving shallow and deep-water sediments of the Aravalli sequence. In the southern rift, the deposition is evidently started with the emplacement of mafic and ultramafic lavas of tholeiitic to komatiitic composition (Ahmad and Tarney, 1994). The 2150 Ma Pb–Pb age of the Delwara volcanics may be taken to denote the time of this rifting (Fig. 4.8b). There is no evidence of these volcanics in the northern belt. The northern belt (here named Bhilwara belt) was possibly a failed arm or aulacogen of the triple junction whose third arm on the SW was a trench along which material of the spreading rift (S of Udaipur) subducted. The northern rift basin, like the southern basins also received both shallow and deep marine sediments, but its metamorphism became higher grade presumably due to higher geothermal gradient in this part of the rift (now Bhilwara belt). Being active, the southern rift widened and deepened during which intervening ridges of the Archaean basement (BGC) appear as unsubmerged regions, dividing the large basin into one or more smaller basins during further development process. Accumulation of the Aravalli supracrustal material resulted in further sinking of the basin. By deepening of the basin and by decay of the underlying plume with time further crustal extension ceased and the continental blocks/plates reversed their movement direction to cause shortening of the crustal segment. The compression of the basinal rocks resulted in thickening of the BGC crust and subsequent breaking/rupturing of the lithosphere with westward "subduction" of the eastern block (or plate) under its western counterpart. Alternatively, an extreme lithospheric stretching may have been responsible for rupture and subduction which, in turn, generated compressive stresses to deform the basinal rocks (Fig. 4.8c). Eastward dipping subduction of the lithospheric plate does not seem possible from the thermal structure of the exposed rocks in Rajasthan (Sharma, 1995, 1999). The westward subduction is supported by the direction of increase of metamorphic grade in the Aravalli rocks (cf. Sharma, 1988). The subduction and convergence of crustal blocks compressed the basinal rocks of the Aravalli Supergroup along with their basement and deformed them. This is the first deformation phase (AF1) of the Aravalli supracrustals (cf. Naha and Halyburton, 1974). We do not have any geological and paleomagnetic

Fig. 4.8 (continued) basin; sedimentary fill of the extensional basin by Delhi Supergroup; and decompression melting of the ascending asthenosphere and of the overlying crust whereby basic magmas emplaced in the rifted basins, including the initiated south Delhi basin, while granites intruded the crust below the rifted basins, (**e**) Enlargement of the extensional basins, including the basin initiated to the south of the rifted basins, (**f**) Convergence of the diverging blocks (due to depositional load, decay of plume responsible for lithosphere extension and change of stress direction with time), resulting into deformation of both Delhi and already deformed Aravalli fold belts, (**g**) Continued convergence of the continental blocks resulted into welding of the mantle lithosphere, accretion of the fold belts and tectonic excavation of the granulites that now occur as fragments in the BGC and Delhi metamorphic terrains. For details see text

4.7 Evolutionary Models and Discussion

evidence to support that the colliding plates/blocks moved towards each other from far off distances to compress the intracratonic Aravalli rocks. It is possible that the colliding crustal blocks were the main BGC terranes on the west (Marwar Terrain of Sinha-Roy, 2004) and the Berach/Bundelkhand granite massif (Mewar Terrain of Sinha-Roy, 2004) on the east (see Fig. 4.1b). The continent-continent collision and deformation of basinal rocks resulted in thickening of the crust and subsequently breaking/rupturing of the lithosphere with westward "subduction" of the eastern block (or plate).

During the subduction the BGC crust was delaminated or decoupled from the mantle only for a short distance, i.e. A-subduction occurred, which allowed hot mantle material to underplate the overlying crust (Fig. 4.8c,d). Above the site of underplating, high-temperature metamorphism produced pelitic granulites, high-grade gneisses and leptynites. Some of these granulites could be the residue of crustal melting; amount of melting depending on the H_2O-content of the near-Moho rocks. The crustal melting is evidenced in the form of granitic bodies and at places as H_2O deficient melt of charnockite-granodiorite which buoyed up and got emplaced in the lower/middle crustal levels, now seen associated with the gneiss-granulite rocks, for example at Gyangarh and Thana areas near Sandmata (Joshi et al., 1993). It is conceivable that the mantle-derived basic magma at the base of the BGC crust also generated anhydrous melt of charnockite-granodiorite composition. A greater melt fraction would have formed enderbite and more basic charnockite, which being denser may have remained deep in the crust here as in other high-grade terranes. The granulite complex at Sandmata and adjoining areas is thus derived from the BGC crust at depth and from asthenospheric mantle sources that gave rise to basic granulites and also the norite dykes.

Concomitant with the breaking away of the westward-subducting lithosphere and magma underplating, the overlying continental crust was softened and thinned and consequently rifted to form an elongated sedimentary basin (Fig. 4.8d,e). This long, more or less continuous basin was initially ensialic in view of the occurrence of 1750 Ma old and somewhat younger granites that are restricted in the northern part of the Delhi fold belt. These granites appear to be products of decompression melting of the rifted continental crust subjected to heating by impinging mantle material below (Fig. 4.8d). The initiation of the Delhi rift basin seems to have occurred about 1750 Ma ago, which is the age of the granites that are interpreted as intrusive or as basement-looking rocks, as stated already. With further thinning/stretching, the continental crust gave way to oceanic crust in the southern part (now SDFB) in which mafic melt of about 1700 Ma or younger age poured out by decompression melting of the upper mantle (Fig. 4.8e). Along with the mafic volcanics, the southern basin with oceanic floor also received felsic melt during sedimentation. The supracrustal sequence of the Delhi Supergroup along with the basement was subsequently deformed (DF1) on NE-trending folds which also affected the already deformed (AF1) Aravalli Supergroup and its basement BGC (Fig. 4.8f). The recrystallization of the Delhi sediments and volcanics occurred during DF1, which is evidently later than the AF1-associated metamorphism of the Aravalli sequence in the east. Because of the higher geothermal gradient towards west the Delhi metamorphic rocks experienced higher-grade metamorphism accompanied by partial melting at depth. This is seen in production of granitic melt of 1450 Ma age (Gopalan et al., 1979; Choudhary et al., 1984) and coeval mafic rocks (Volpe and Macdougall, 1990) during the Delhi orogeny. Continued compression gave rise to the uplift of the Aravalli Mountain, accretion of the Aravalli and Delhi fold belts, and development of shear zones which led to tectonic excavation (exhumation) of the deep-seated rocks, particularly the granulites (Fig. 4.8 g). It is through ductile shearing that the granulite rocks are now seen exposed within the BGC outcrops (Sharma, 1988) and also within the Delhi metasediments bordering the BGC near Ajmer and elsewhere in the Delhi fold belt (Desai et al., 1978; Fareeduddin et al., 1994), but nowhere in the metasediments of the Aravalli Supergroup. There is no geological and geochronological data that support the contention of Roy et al. (2005) that the granulites emplacement was concomitant with the opening of the Delhi basin. If it were so, one has to find an explanation as to why these granulites retained their mineralogy as well as intrusive character intact when they were involved along with the Mesoproterozoic cover rocks during the Delhi orogeny.

The closure of the Delhi orogeny is indicated by 1012 Ma (Sm–Nd) old undeformed diorites that intruded the Delhi metasediments at Ranakpur in SDFB. Subsequent granitic activities caused thermal overprinting on the Proterozoic metasediments, their basement and intrusives. As a consequence the rocks were re-equilibrated, both thermally and isotopically, to give younger ages documented in the constituent minerals of some rocks (see Volpe and Macdougall, 1990; Pandit et al., 2003; Sharma, 1999).

4.8 Geophysical Database

Recent gravity and magnetic data as well as deep seismic reflection profiling (DSS) studies are in full support of the model. The regional gravity highs over the Delhi Supergroup and the BGC indicate high-density rocks at depth (Reddy and Ramakrishna, 1988; Mishra et al., 1995). This is clearly an expression of abundant mafic rocks that occur amidst the Delhi metasediments and also as an underplated material below the crust in the region.

Deep seismic reflection studies reveal the presence of a large-scale domal structure in the lower crust, i.e. in the BGC terrain between Govindgarh and Bhinai. Its crest is located at about 16 km depth between Masuda and Bithur and with sides deepening to 27–30 km on the NW and SE, near Govindgarh and Bhinai, respectively (Tewari et al., 1995). This gravity high is interpreted as a clear manifestation of high-density material below the crust, caused by large-scale underplating envisaged in this model. The presence of high reflective zones above the Moho, although unclear in this region, can be interpreted as the seismic expression of a laminated lower crust, caused by interleaving of metamorphic and underplated magmatic rocks. Below this laminated crust a NW dipping strong reflector is reported to rise from near the Moho, extending with NE-SW orientation

(Tewari et al., 1995). This trend matches with the surface orientation of the shear zone delineated in the Sandmata region (Sharma, 1988, 1995).

The terrain of Aravalli and Delhi fold belts and of the BGC has a number of lineaments of notable extension in the NE-SW directions (Sinha-Roy et al., 1993a, 1998). These are:

1. Great Boundary Fault (GBF), running between Vindhyan Supergroup and Hindoli Group (low-grade Aravalli System of Heron) and BGC-Berach Granite.
2. Banas Lineament located between Hindoli Group and migmatite terrain (Mangalwar Complex of the Bhilwara Supergroup)
3. Delwara Lineament between Sandmata complex and Mangalwar complex.
4. Kaliguman Lineament between Delhi fold belt in the west and the Aravalli and Sandmata Complex in the East.

In view of their occurrences and identical trends, the lineaments appear to be the product of last tectonic event affecting the fold belts and the basement rocks. The Banas lineament is the only lineament that separates rocks of two different facies in central Rajasthan (Sharma, 1988).

4.9 Conclusions

The ensialic orogenesis model is a reasonable working hypothesis. It accords not only with the available structural, petrological and geochronological data on the Proterozoic rocks of the Aravalli and Delhi Supergroups (and their basement) but also explains the petrogenesis of the granulites and their deduced anticlockwise P-T-t path (Harley, 1988). The Proterozoic age of much granite, e.g. Amet, Anjana etc., in the BGC terrain nicely fits with the proposed model. Also, the geophysical data find a very satisfactory explanation in this model. In short, the proposed model elegantly explains all major geological and geophysical surface and subsurface features. By referring to this model we can answer most, if not all, of the basic questions on the Precambrian geology of Rajasthan such as: why are the supracrustals of Aravalli Supergroup lower grade than the younger metasediments of the Delhi Supergroup in their immediate contact? Why is the first deformation in the Aravalli rocks absent in the Delhi rocks? Why is the regional metamorphism of the Aravalli Supergroup older than that of the Delhi rocks despite their being involved in the Aravalli mountain orogeny? Why do the younger metamorphics of Delhi Supergroup (and their basement BGC) contain abundant granitic intrusions and granulitic rock fragments whereas the older Aravalli Supergroup is devoid of them? These and similar questions can satisfactorily be answered by this model. The proposed model does not invoke any conjectural presence of magmatic arc, trench, etc. that offer serious problems of detection in the Precambrian terrains all over the world.

This model seems an appropriate working hypothesis because it has greater explanatory power than the other models and because it is more compatible with a number of principles outlined in first chapter.

References

Abu-Hamatteh, Z.S.H., Raja, M. and Ahmad, T. (1994). Geochronology of early Proterozoic mafic and ultramafic rocks of Jharol Group, Rajasthan, northwestern India. J. Geol. Soc. India, vol. **44**, pp. 141–156.

Ahmad, T. and Tarney, J. (1994). Geochemistry and petrogenesis of Late Archaean Aravalli volcanics, basement enclaves and granitoids, Rajasthan. Precambrian Res., vol. **6**. pp. 1–23.

Bakliwal, P.C. and Ramasamy, S.M. (1987). Lineament fabric of Rajasthan and Gujarat. Rec. Geol. Surv. India, vol. **113**(7) pp. 54–64.

Biju-Shekhar, S., Yokoyama, K., Pandit, M.K., Okudaira, T., Yoshida, M. and Santosh, M. (2003). Late Proterozoic magnetism in Delhi fold belt, NW India and its implication: evidence from EPMA chemical ages of zircons. J. Asian Earth Sci. vol. **22**, pp. 189–207.

Bose, U., Fareeduddin and Reddy, M.S. (1990). Polymodal volcanism in parts of South Delhi fold belt, Rajasthan. J. Geol. Soc. India, vol. **36**, pp. 263–276.

Choudhary, A.K., Gopalan, K. and Sastry, C.A. (1984). Present status of the geochronology of the Precambrian rocks of Rajasthan. Tectonophysics, vol. **105**, pp. 131–140.

Crawford, A.R. (1970). The Precambrian geochronology of Rajasthan and Bundelkhand, northern India. Can. J. Earth Sci., vol. **7**, pp. 91–110.

Deb, M. (1993). The Bhilwara belt of Rajasthan—a probable aulacogen. In: Cassyap, S.M. (ed.), *Rifted Basins and Aulacogens*. Gyanodaya Prakashan, Nainital, pp. 91–107.

Deb, M., Thorpe, R.I., Cumming, G.L. and Wagner, P.A. (1989). Age, source and stratigraphic implications of Pb isotope data for conformable, sediment-hosted, base metal deposits, northwestern India. Precambrian Res., vol. **43**, pp. 1–22.

Deb, M., Thorpe, R.I., Krstic, F., Corfu, F. and Davis, D.W. (2001). Zircon U–Pb and galena Pb isotope evidence for an approximate 1.0 Ga terrane constituting the western margin of the Aravalli-Delhi orogenic belt, northwestern India. Precambrian Res., vol. **108**, pp. 195–213.

Desai, S.J., Patel, M.P. and Merh, S.S. (1978). Polymetamorphites of Balaram-Abu Road area, north Gujarat and southwest Rajasthan. Geol. Soc. India Mem., vol. **31**, pp. 383–394.

England, P.C. and Thompson, A.B. (1984). Pressure-temperature-time paths of regional metamorphism, I—heat transfer during the evolution of regions of thickened continental crust. J. Petrol., vol. **25**, pp. 894–928.

Fareeduddin (1995). Field setting, petrochemistry and P-T regime of the deep crustal rocks to the northwest of the Aravalli-Delhi mobile belt, north-central Rajasthan. Geol. Soc. India. Mem., vol. **31**, pp. 117–139.

Fareeduddin, Shankar, M., Basvalingu, B. and Janardhan, A.S. (1994). Metamorphic P-T conditions of pelitic granulites and associated charnockites of Chinwali area, west of Delhi fold belt, Rajasthan. J. Geol. Soc. India, vol. **43**, pp. 169–178.

Fareeduddin, Kirmani, T.R., Srivastava, B.L., Reddy, A.B. and Bhattacharya, J. (1995). Lamprophyre dykes in South Delhi fold belt near Pipela, district Sirohi, Rajasthan. J. Geol. Soc. India, vol. **46**(3), pp. 255–261.

Fareeduddin and Kroner, A. (1998). Single zircon age constraints on the evolution of the Rajasthan granulites. In: Paliwal, B.S. (ed.), *Precambrian Rocks of India*. Scientific Publishers (India), Jodhpur, pp. 547–556.

GSI (1969, 1993). *Geological Map of India, Scale 1:2000,000*. Geol. Surv. India, Hyderabad and Calcutta.

GSI (1977). Geology and mineral resources of the States of India. Part XII, Rajasthan. Misc. Publ. Geol. Surv. India, vol. **30**, 75p.

Gangopadhyaya, P.K. and Das, D. (1974). Dadikar granite: a study on a Precambrian intrusive body in relation to structural environment in north-eastern Rajasthan. J. Geol. Soc. India, vol. **15**, pp. 189–199.

Gangopadhyaya, P.K. and Sen, R. (1972). Trends of regional metamorphism: an example from "Delhi System" of rocks occurring around Balrawas, northwestern Rajasthan, India. Geol. Rundsch., vol. **61**, pp. 270–281.

References

Gopalan, K., Trivedi, J.R., Balasubramanyam, M.N., Ray, S.K. and Sastry, C.A. (1979). Rb–Sr chronology of the Khetri copper belt, Rajasthan. J. Geol. Soc. India, vol. **20**, pp. 450–456.

Gupta, B.C. (1934). The geology of central Mewar. Geol. Surv. India, Mem. **65**, Publ. GSI Calcutta, pp.107–168.

Gupta, P. (2004). Ancient orogens of Aravalli region. Geol. Surv. India Spl. Publ., vol. **84**, pp. 150–205.

Gupta, S.N., Arora, Y.K., Mathur, R.K., Iqbaluddin, Prasad, B., Sahai, T.N. and Sharma, S.B. (1980). *Lithostratigraphic Map of Aravalli Region (1:1000,000)*. Geol. Surv. India, Calcutta.

Gupta, S.N., Arora, Y.K., Mathur, R.K., Iqballuddin, Prasad, B., Sahai, T.N. and Sharma, S.B. (1997). The Precambrian geology of the Aravalli region, southern Rajasthan and northeastern Gujarat. Geol. Surv. India, Mem. **123**, Publ. GSI Calcutta, 262p.

Gupta, P. and Bose, U. (2000). An update on the geology of the Delhi Supergroup in Rajasthan. In: M.S. Krishnan Birth Centenary Seminar, Calcutta. Geol. Surv. India Spl. Publ., vol. **55**, pp. 287–306.

Gupta, P., Mukhopadhyaya, K. and Bose, U. (1988). Delhi volcanics in parts of central Rajasthan and their significance. J. Geol. Soc. India, vol. **31**, pp. 314–327.

Gupta, P., Mukhopadhyaya, K., Fareeduddin and Reddy, M.S. (1991). Tectono-stratigraphic framework and volcanic geology of the South Delhi fold belt in south-central Rajasthan. J. Geol. Soc. India, vol. **37**, pp. 431–441.

Harley, S. (1988). Granulite P-T paths: constraints and implications for granulite genesis. In: Vielzeuf, D. (ed.), *Granulites and Their Problems*. Terra Cognita, European Union of Geoscience, Wien, vol. **8/3**, pp. 267–268.

Heron, A.M. (1953). The geology of central Rajputana. Geol. Surv. India. Mem., vol. **79**, pp. 1–389.

Holdaway, M.J. (1971). Stability of andalusite and the aluminum silicate phase diagram. Am. J. Sci., vol. **271**, pp. 97–131.

Joshi, M., Thomas, H. and Sharma, R.S. (1993). Granulite facies metamorphism in the Archaean gneiss complex from northcentral Rajasthan. Proc. Natl. Acad. Sci. India, vol. **63**(A), pp. 167–187.

MacDougall, J.D., Gopalan, K., Lugmair, G.W. and Roy, A.B. (1983). The Banded Gneissic Complex of Rajasthan, India: early crust from depleted mantle at 3.5 AE. EOS Transaction Am. Geophy. Union, vol. **64** (I-26), p. 351.

Mamtani, M.A., Karanth, R.V., Merh, S.S. and Greiling, R.O. (1999). Tectonic evolution of the southern part of the Aravalli mountain belt and its environs, possible causes and time constraints. Gond. Res., vol. **3**, pp. 175–188.

Mishra, D.C., Lakshman, G., Vyagreshwar Rao, M.B.S. and Gupta, S.B. (1995). Analysis of the gravity-magnetic data around Nagaur-Jhalawar geotransect. Geol. Soc. India Mem., vol. **31**, pp. 345–351.

Mohanty, S. and Naha, K. (1986). Stratigraphic relations of the Precambrian rocks in the Salumbar area, southeastern Rajasthan. J. Geol. Soc. India, vol. **27**, pp. 479–493.

Mukhopadhyay, K. (1989). Report on stratigraphy, geochemistry and petrogenesis of Sirohi and Sindreth groups in Southern Rajasthan. GSI Report (unpublished).

Mukhopadhyay, D. and Bhattacharya, T. (2000). Tectono-stratigraphic framework of the South Delhi fold belt in the Ajmer-Beawar region, central Rajasthan, India: a critical review. In: Deb, M. (ed.), *Crustal Evolution and Metallogeny in the Northwestern Indian Shield*. Narosa Publishing House, New Delhi, pp. 126–137.

Mukhopadhyay, D., Bhattacharya, T., Chattopadhyay, N., Lopez, R. and Tobisch, O.T. (2000). Anasagar gneiss: a folded granitoid in the Proterozoic South Delhi fold belt, central Rajasthan. Proc. Indian Acad. Sci (Earth Planet. Sci.), vol. **109**, pp. 21–37.

Mukhopadhyay, D. and Dasgupta, S. (1978). Delhi-Pre-Delhi relations near Badnor, central Rajasthan. Indian J. Earth Sci., vol. **5**, pp. 183–190.

Naha, K. (1983). Structural-stratigraphic relations of the pre-Delhi rocks of southcentral Rajasthan: a summary. In: Sinha-Roy, S. (ed.), *Structure and Tectonics of the Precambrian Rocks*. Recent Res. Geol., vol. **10**. Hindustan Publishers, New Delhi, pp. 40–52.

Naha, K., Chaudhauri, A.K. and Mukherji, P. (1967). Evolution of the Banded Gneissic Complex of central Rajasthan, India. Contrib. Mineral. Petrol., vol. **15**, pp. 191–201.

Naha, K. and Halyburton, R.V. (1974). Early Precambrian stratigraphy of central and southern Rajasthan, India. Precambrian Res., vol. **4**, pp. 55–73.

Naha, K. and Halyburton, R.V. (1977). Structural pattern and strain history of a superposed fold system in the Precambrian of central Rajasthan, India. Precambrian Res., vol. **4**, pp. 85–111.

Naha, K. and Mohanty, S. (1990). Structural studies in the Pre-Vindhyan rocks of Rajasthan: a summary of work of the last three decades. In: Naha, K., Ghosh, S.K., and Mukhopadhyay, D. (eds.), *Structure and Tectonics: The Indian Scene*. Proc. Indian Acad. Sci., Bangalore, vol. **99**, pp. 279–290.

Naha, K., Mukhopadhyay, D.K., Mohanty, S., Mitra, S.K. and Biswal, T.K. (1984). Significance of contrast in the early stages of the structural history of the Delhi and pre-Delhi rocks in the Proterozoic of Rajasthan, western India. Tectonophysics, vol. **105**, pp. 193–206.

Naha, K. and Roy, A.B. (1983). The problem of the Precambrian basement in Rajasthan, western India. Precambrian Res., vol. **19**, pp. 217–233.

Paliwal, B.S. (1988). Deformation pattern in the rocks of the Aravalli Supergroup around Udaipur city. In: Roy, A.B. (ed.), *Precambrian of the Aravalli Mountain*. Mem. Geol. Soc. India, vol. **7**. Rajasthan, India, pp. 153–168, Publ. Geol. Soc. India, Bangalore.

Pandit, M.K., Carter, L.M., Ashwal, L.D, Tucker, R.D., Torsvik, T.H., Jamtveit, B., Bhusan, S.K. (2003). Age, petrogenesis and significance of 1 Ga granitoids and related rocks from the Sendra area, Aravalli craton, NW India. J. Asian Earth Sci., vol. **22**, pp. 363–381.

Pant, N.C., Kundu, A. and Joshi, S. (2008). Age of metamorphism of Delhi Supergroup rocks-electron microprobe ages from Mahendragarh district, Haryana. J. Geol. Soc. India, vol. **72**, pp. 365–372.

Poddar, B.C. and Mathur, R.K. (1965). A note on the repetitious sequence of greywacke-slate-phyllite in the Aravalli system around Udaipur, Rajasthan. Bull. Geol. Soc. India, vol. **2**, pp. 83–87.

Reddy, A.G.B. and Ramakrishna, T.S. (1988). Subsurface structure of the shield area of Rajasthan-Gujarat as inferred from gravity. In: Roy, A.B. (ed.), *Precambrian of the Aravalli Mountain*. Geol. Soc. India Mem., vol. **7**. Rajasthan, India, pp. 279–284, Publ. Geol. Soc. India, Bangalore.

Roy, A.B. (1988). Stratigraphic and tectonic framework of the Aravalli mountain range. Geol. Soc. India Mem., vol. **7**, pp. 3–31.

Roy, A.B. (1990). Evolution of the Precambrian crust of the Aravalli mountain range. In: Naqvi, S.M. (ed.), *Precambrian Continental Crust and Its Economic Resources*. Dev. Precambrian Geol., vol. **8**. Elsevier, Amsterdam, pp. 327–348.

Roy, A.B. and Jakhar, S.R. (2002). *Geology of Rajasthan (Northwest India): Precambrian to Recent*. Scientific Publishers, Jodhpur, 421p.

Roy, A.B., Kroener, A., Bhattacharya, P.K. and Rathore, S. (2005). Metamorphic evolution and zircon geochronology of early Proterozoic granulites in the Aravalli mountains of northwestern India. Geol. Mag., vol. **142**(3), pp. 287–302.

Roy, A.B., Paliwal, B.S. and Bejarnia, B.R. (1984). The Aravalli rocks: an evolutionary model and metallogenic trends. Indian J. Geol. Soc., CEISM Seminar Volume, Calcutta, pp. 73–82.

Roy, A.B. and Rathore, S. (1999). Tectonic setting and model of evolution of granulites in the central part of the Aravalli mountain belt, Rajasthan. In: Murthy, N.G.K. and Ram Mohan, V. (eds.), *Charnockite and Granulite Facies Rocks*. Geol. Assoc. TN, Spl. Publ., vol. **4**. Chennai, India, pp. 111–126, Publ. Geology Dept., Chennai.

Roy, A.B., Sharma, B.L., Paliwal, B.S., Chauhan, N.K., Nagori, D.K., Golani, P.R., Bejarnia, B.R., Bhu, H. and Ali Sabah, M. (1993). Lithostratigraphiy and tectonic evolution of the Aravalli Supergroup: a protogeosysnclinal sequence. In: Cassayap, S.M. (ed.), *Rift Basins and Auloco- gens*. Ganodiya Prakashan, Nainital.

Saha, L., Bhowmik, S.K., Fukoka, M. and Dasgupta, S. (2008). Contrasting episodes of regional granulite facies metamorphism in enclaves and host gneisses from the Aravalli-Delhi mobile belt, NW India. J. Petrol., vol. **49**, pp. 107–128.

Sarkar, G., Ray Barman, T. and Corfu, F. (1989). Timing of continental arc-type magnetism in northwest India: evidence from U–Pb zircon geochronology. J. Geol., vol. **97**, pp. 607–612.

Sengupta, S. (1984). A reinterpretation of the stratigraphy of the Delhi Supergroup of central Rajasthan between Rajosi and Chowadsiya, Ajmer district, Rajasthan, India. Indian J. Earth Sci., vol. **11**, pp. 38–49.

Sharma, R.S. (1977). Deformational and crystallization history of the Precambrian rocks in north-central Aravalli mountain, Rajasthan, India. Precambrian Res., vol. **4**, pp. 133–162.

Sharma, R.S. (1988). Patterns of metamorphism in the Precambrian rocks of the Aravalli mountain belt. Geol. Soc. India Mem., vol. **7**, pp. 33–75.

Sharma, R.S. (1989). Geotectonic evolution of the Aravalli mountain belt. In: Saklani, P.S. (ed.), *Current Trends in Geology, XII, Metamorphism and Ophiolites and Orogenic Belts*. Today and Tomorrows Press, New Delhi, pp. 211–221.

Sharma, R.S. (1990). Metamorphic evolution of rocks from the Rajasthan craton, NW Indian shield. In: Naqvi, S.M. (ed.), *Precambrian Continental Crust and Its Economic Resources*. Dev. Precambrian Geol., vol. **8**. Elsevier, Amsterdam, pp. 349–366.

Sharma, R.S. (1995). An evolutionary model for the Precambrian crust of Rajasthan: some petrological and geochronological considerations. Geol. Soc. India Mem., vol. **31**, pp. 91–115.

Sharma, R.S. (1999). Crustal development in Rajasthan craton. Indian J. Geol., vol. **71**, pp. 65–80.

Sharma, R. S. (2003). Evolution of Proterozoic fold belts of India: a case of the Aravalli mountain belt of Rajasthan, NW India. Mem. Geol. Soc. India, vol. **52**, pp. 145–162.

Sharma, R.S. and Joshi, M. (1984). Sand Mata granulites: a case of deep-level exposure of Precambrian crust in NW Indian shield. In: Saha, A.K. (ed.), *Crustal Evolution of the Indian Shield and Its Bearing on Metallogeny*. Calcutta, pp. 35–36.

Sharma, R.S., Sills, J. and Joshi, M. (1987). Mineralogy and metamorphic history of norite dykes within granulite facies gneisses from Sand Mata, Rajasthan, NW India. Min. Mag., vol. **5**, pp. 207–215.

Singh, S.P. (1988). Stratigraphy and sedimentation pattern in the Proterozoic Delhi Supergroup, northwestern India. Mem. Geol. Soc. India, vol. **7**, pp. 193–206.

Sinha-Roy, S. (1984). Precambrian crustal interactions in Rajasthan, NW India. Indian J. Earth Sci., CEISM Volume, pp. 84–91.

Sinha-Roy, S. (1988). Proterozoic Wilson cycles in Rajasthan. In: A.B. Roy (ed.), Precambrian of the Aravalli mountain range. Mem. Geol. Soc. India, vol. **7**, pp. 95–108.

Sinha-Roy, S. (2000). Precambrian terrane evolution in Rajasthan. In: M.S. Krishnan Centenary Seminar, Geol. Surv. India Spl. Publ., vol. **55**, pp. 275–286.

Sinha-Roy, S. (2004). Intersecting Proterozoic transpressional orogens, major crustal and suspect terranes in Rajasthan craton: a plate tectonic perspective. Geol. Surv. India Spl. Publ., vol. **84**, pp. 207–226.

Sinha-Roy, S. and Malhotra, G. (1989). Structural relations of the cover and its basement: an example from Jahazpur belt, Rajasthan. J. Geol. Soc. India, vol. **34**, pp. 233–244.

Sinha-Roy, S., Malhotra, G., Mohanty, M., Sharma, V.P. and Joshi, D.W. (1993b). Conglomerate horizons in south-central Rajasthan and their significance on Proterozoic stratigraphy and tectonics of the Aravalli and Delhi fold belts. J. Geol. Soc. India, vol. **44**, pp. 331–350.

Sinha-Roy, S., Mohanty, M. and Guha, D.B. (1993a). Banas dislocation zone in Nathdwara-Khamnor area, Udaipur district, Rajasthan and its significance on basement-cover relations in the Aravalli fold belt. Curr. Sci., vol. **65**(10), pp. 68–72.

Sinha-Roy, S, Malhotra, G. and Guha, d.B. (1998). Geology of Rajasthan. Publ. Geol. Soc. India, Bangalore, 262p.

Sivaraman, T.V. and Odom, A.L. (1982). Zircon geochronology of Berach granite of Chittorgarh, Rajasthan. J. Geol. Soc. India, vol. **23**, pp. 575–577.

Srivastava, D.C., Yadav, A.K., Nag, S. and Pradhan, A.K. (1995). Deformation style of the Banded Gneissic Complex in Rajasthan: a critical evaluation. Geol. Soc. India Mem., vol. **31**, pp. 199–216.

Sugden, T.J., Deb, M. and Windley, B.F. (1990). The tectonic setting of mineralization in the Proterozoic Aravalli-Delhi orogenic belts, NW India. In: Naqvi, S.M. (ed.), *Precambrian Conti-*

nental Crust and Its Economic Resources. Dev. Precambrian Geol., vol. **8**. Elsevier, Amsterdam, pp. 367–396.

Sugden, T.J. and Windley, B.F. (1984). Geotectonic framework of the early-mid Proterozoic Aravalli-Delhi orogenic belt, NW India. Geol. Mineral. Assoc. Can., vol. **94**, 68p (Abstract).

Sychanthavong, S.P.H. and Dessai, S.D. (1977). Proto-plate tectonics controlling the Precambrian deformation and metallogenic epochs in the NW peninsular India. Min. Sci., Engg., vol. **9**, pp. 218–236.

Sychanthavong, S.P.H. and Merh, S.S. (1984). Proto-plate tectonics: the energetic model for the structural, metamorphic and igneous evolution of the Precambrian rocks, Northwest peninsular India. Geol. Surv. India Spl. Publ., vol. **12**, pp. 419–458.

Tewari, H.C., Vijaya Rao, V., Dixit, M.M., Rajendra Prasad, B., Madhava Rao, N., Venkateswarlu, N., Khare, P., Keshava Rao, G., Raju, S. and Kaila, K.L. (1995). Deep crustal reflection studies across the Delhi-Aravalli fold belt, results from north-western part. Geol. Soc. India Mem., vol. **31**, pp. 383–402.

Tobisch, O.T., Collerson, K.D., Bhattacharya, T. and Mukhopadhyay, D. (1994). Structural relationship and Sm–Nd isotopic systematics of polymetamorphic granite gneisses and granitic rocks from central Rajasthan: implications for the evolution of Aravalli craton. Precambrian Res., vol. **65**, pp. 319–339.

Torsvik, T.H., Carter, L.M., Ashwal, L.D., Bhusan, S.K., Pandit, M.K. and Jamtveit, B. (2001). Rodinia refined or obscured: palaeomagnetism of the Malani igneous suite (NW India). Precambrian Res., vol. **108**, pp. 319–333.

Volpe, A.M. and Macdougall, J.D. (1990). Geochemistry and isotopic characteristics of mafic (Phulad Ophiolite) and related rocks in the Delhi Supergroup, Rajasthan, India: implications for rifting in the Proterozoic. Precambrian Res., vol. **48**, pp. 167–191.

Chapter 5
Central Indian Fold Belts

5.1 Introduction

The fold belts of the central Indian region are sandwiched between the Rajasthan-Bundelkhand craton in the north and the Bastar craton in the south. The age of 3.3 Ga of the tonalite-trondhjemite-granodiorite (TTG) rocks in these adjacent cratons of Bastar and Rajasthan-Bundelkhand (Bandyopadhyay et al., 1990) suggests that these continental nuclei once united to form a mosaic in northern Indian peninsula, as stated in Chap. 2. In the present geological set up the location of the central Indian fold belts is shown in Fig. 5.1 (overview in Fig. 5.1b). Of these four fold belts, namely the Mahakoshal, Satpura, Sakoli and Dongargarh, the former two are located in the ENE-WSW trending Central Indian Tectonic Zone (CITZ) lying between the Bundelkhand and Bastar cratons, whereas the latter two belts are located within Bastar craton and hence south of the Central Indian Shear zone (CIS) regarded as the southern limit of the CITZ.

In one aspect, the CITZ appears to divide the Indian shield into two major Archaean crustal provinces. The block to the north is the Rajasthan-Bundelkhand craton whereas the composite block to the south is an amalgamation of the Bastar, Dharwar and Singhbhum cratons (cf. Acharyya, 2003b). The CITZ is a broad belt of unequal width. Its narrow northern belt is believed to represent the Palaeoproterozoic Mahakoshal Fold Belt while the wider southern belt is represented by the Mesoproterozoic Satpura fold belt (also called Sausar fold belt) whose southern limit is demarcated by the Central Indian Shear (CIS) zone (Fig. 5.1b). The fold belts to the south of the CIS show dissimilar orographic trends; the Dongargarh belt has N-S trend while the triangular-shaped Sakoli fold belt is bounded by the N-S shear zone in the east and the near E-W trending shear zone on its northern boundary (Fig. 5.1c).

The supracrustal rocks in these fold belts are found to rest on the Archaean gneissic complex of Bastar craton and are now exposed in differently oriented fold belts. The fold belts occupy the Proterozoic terrains of Madhya Pradesh, Maharashtra, and Uttar Pradesh. The fold belts of the central Indian region received early attention of the Geological Survey of India (GSI). The geologists who contributed in a team, are Goyal and Jain (1975), Nair et al. (1995), Abhinaba. Roy and Devarajan (2000). These geologists of GSI studied the metavolcanic and metasedimentary rocks of

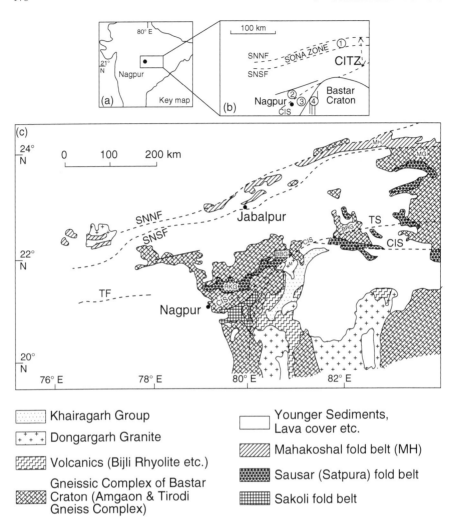

Fig. 5.1 Location map (**a**) and index map (**b**) showing the position of the different Proterozoic fold belts in central Indian region. 1 = Mahakoshal folds belt, 2 = Satpura fold belt, 3 = Sakoli fold belt, 4 = Dongargarh fold belt. Abbreviations: CITZ = Central Indian Tectonic Zone, CIS = Central Indian shear zone, SONA ZONE = Son-Narmada Lineament zone, SNNF = Son-Narmada North Fault, SNSF = Son-Narmada South Fault. (**c**) Geological map of central India (after Roy et al., 2002) showing the different Proterozoic fold belts, granulite occurrences and shear zones. M = Malanjkhand, Ts = Tan shear zone, TF = Tapti Fault. Granulites of Ramakona-Katangi (RKG), Bhandara-Balaghat (BBG), Bilaspur-Raipur (BRG) and Makrohar (MG) are all located in the basement rocks (Tirodi gneiss) of the Sausar Group making up the Satpura fold belt

upper Narmada-Son Valley (called SONA Zone) and gave the name Mahakoshal Group to the rocks earlier considered Bijawar Series (Pascoe, 1973), in view of their differences in lithology, structure, tectonics and nature of intrusives. The belt south of the SONA Zone is Satpura fold belt, made up of Sausar Group rocks which have been mapped by GSI workers under its CIRCUMSONATA Project. The geological map of this terrain is due to Narayanaswami and his team (1963), Bandyopadhyay and others (1995). Pioneering work of the fold belts within the Bastar craton is due to Sarkar (1958) who mapped and established a sequence of Kotri-Dongargarh fold belt, whereas Sakoli fold belt occurring to the western Bastar is described by Bandyopadhyay, Ghosh and Roy in 1990s.

The above-stated fold belts are described below individually with a comment on the geological/geodynamical status of the CIS.

5.2 Mahakoshal Fold Belt

5.2.1 Introduction: Geological Setting

In the northern structural belt of the Central Indian Tectonic Zone (CITZ), the Mahakoshal fold belt forms an E-W to ENE-WSW trending terrane along the Son-Narmada Lineament (SONA Zone) which is nearly parallel to the Sausar (Satpura) fold belt (bordering the Bastar craton in the south). Being deformed, the rocks making up the Mahakoshal fold belt have been designated Mahakoshal Group, different from the Bijawar Series that are undeformed and unconformably overlie the Bundelkhand Granite Gneiss (Goyal and Jain, 1975). The Mahakoshal Group comprises volcano-sedimentary rocks. This rock group, constrained within the SONA Zone, is located between Son-Narmada North Fault (SNNF) and the Son-Narmada South Fault (SNSF) (Fig. 5.1b). This belt of supracrustals extends ~350 km from Hoshangabad in Madhya Pradesh in the west to Palamau in the Jharkhand State in the east. The Mahakoshal Group rocks show an unconformable relationship with the gneissic complex (Archaean) exposed in the north as well as south. The Mahakoshal rocks within SONA Zone are undated but have been intruded by syn- to post-tectonic granites (Tamkhan and Barambara) of 1800–1865 Ma age (Sarkar, 1958; Sarkar et al., 1998). This age data constrain the upper age limit of the Mahakoshal Group, indicating Palaeoproterozoic age for the Mahakoshal orogeny. Geophysical studies along the SONA Zone revealed high density, high velocity subhorizontal layers at depth of 8–12 km down to Moho (Reddy and Rao, 2003). This is attributed to magma underplating or to intrusion of mantle material at shallower depth. Geological postulation by Auden (1949) and geophysical studies indicate that the SONA Zone is upwarped (Kaila and Sain, 1997), which led some workers to believe that there is a Trap under the SONA Zone. This subcrustal Trap might find its source from Deccan Trap or from the Re-Union Hot Spot located some 400 km westward in the Cambay region (West, 1962).

The relative ages of the basement rocks and the supracrustals of the Mahakoshal Group are unclear. However, GSI has given the following sequence, based on the geological map and relationships of the cover rocks.

Intrusives

Parsoi Formation

Agori Formation

Chitrangi Formation

······Fault/local unconformity······

Migmatite and granite gneiss

The volcano-sedimentary assemblage is briefly described below.

As stated already, the MFB is a volcano-sedimentary assemblage. The basal part of the Mahakoshal Group (MG) is dominated by subaqueous mafic to ultramafic lava (*Chitrangi Formation*) overlain by clastic to non-clastic sediments, viz. quartzites with minor basalts, carbonates and BIF (*Agori Formation*) which are overlain by a thick succession of argillites and turbidites (*Parsoi Formation*). The close of sedimentation of the Mahakoshal Group is marked by syn- to post-tectonic granite intrusions along faulted margins. The different intrusions have given Palaeo- to Meso-proterozoic ages: 1.8–1.865 Ga for Tamkhan and Barambara granites; 2.405 Ga for associated Barambara granodiorite; 1.7–1.8 Ga for Jhirgadandi monzodiorite/quartz syenite, Bori syenite and Rihand-Renusagar calc-alkaline granitoids (Sarkar et al., 1998; Roy and Devarajan, 2000).

5.2.2 Agreed Observations and Facts

With the given background of stratigraphic and geologic information it is useful to list the agreed observations and facts about the fold belt and its constituent rocks. These are:

1. The Mahakoshal fold belt (MFB) lies within the SONA Zone, demarcated by southern and northern faults of the Narmada-Son Lineament.
2. The MFB trends E-W to ENE-WSW and is nearly parallel with the Satpura fold belt on its south.
3. The bounding faults of the MFB are pre-Mahakoshal and controlled the sedimentation and volcanism of the Mahakoshal supracrustals (MG).
4. Ultramafic rocks together with gabbroic and basaltic rocks occur in the lower part of the MG, supported by gravity high along the MFB.
5. The supracrustals of volcano-sedimentary litho units show a gradual change from volcanism to sedimentary environment and are intruded by gabbroic dykes, indicating lithospheric rifting.
6. There is gravity high along the MFB, indicating high-density material below the fold belt.
7. Of three phases of folding, F1 and nearly coaxial F2 determine the orogenic trend of the MFB.
8. The major structures in the MFB suggest N-S compression of the MG supracrustals.
9. Metamorphism of the MFB is low grade (greenschist facies), with rare occurrence of andalusite and staurolite + chlorite and cordierite (Bandyopadhyay et al., 1995).
10. The MG supracrustals are older than 1865 Ma, which is the age of Tamkhan granite intruding the MG rocks.

11. The MG supracrustals show unconformable relationship with the underlying gneisses on northern as well as southern sides.
12. Linear zones of secondary silicification and ferruginization are abundant near the basin margins, suggesting basin evolution through fault controlled processes.
13. Syntectonic granite intrusion at Tamkhan within the SONA zone is dated at 1800 Ma by Rb/Sr whole-rock isochron (Sarkar et al., 1998).

5.2.3 Evolution of the Mahakoshal Fold Belt (MFB)

The evolution of the MFB started with intracratonic rifting of the continental crust (Archaean). This intracratonic rift was obviously mantle activated since mafic/ultramafic lavas extruded in the rifted basin, due to decompression melting and perhaps upwarping of the Moho between the faults. These lavas are now seen as greenstone belts (Shanker, 1991). The occurrence of a narrow linear gravity high (bounded by faults) as a consequence of high-density material below the belt clearly favours a rifted basin and rules out island arc proposition for these igneous lavas in the basal part of the Mahakoshal Group (cf. Rao and Sri Rama, 1993). The extrusion of lava was followed by deposition of clastic to non-clastic sediments. During the end phases of sedimentation the conditions seem very stable as is indicated by chemical precipitation and development of carbonate facies with organic structures. Tectonically, active sedimentation occurred in the southern part of the rift and it is documented by turbidites in the younger Parsoi formation, which is absent in the west and northwest part of the Mahakoshal (MG) supracrustals.

The geometry of F1 structures in the MG supracrustals suggests that there was N-S compression of the rifted basin and the converging plates were obviously the Bastar craton in the south and the Bundelkhand Granite massif in the north.

The deformed MG rocks were metamorphosed in greenschist facies conditions, revealed by the rare occurrence of andalusite and staurolite coexisting with chlorite. It is possible that andalusite may have developed by intrusion of granite/granodiorite plutons. The granodiorite of Barambara locality within the MG metasediments gave Rb–Sr whole rock age of 2405 ± 17 Ma (in Nair et al., 1995), while the Tamkhan granite intrusion in the MG rocks yielded 1865 Ma by Rb–Sr systematics. Available geochronlgical data indicate that the Mahakoshal supracrustals are Palaeoproterozoic, lay upon scarcely exposed Archaean or Late Archaean basement as at Sushi. In this situation, the 2405 Ma old granodiorite from the Mahakoshal fold belt (MFB), instead of being intrusive, could possible be a component of basement gneiss complex and hence a possible parent for the 1800 Ma old granite melts that intruded this terrain.

Considering the parallelism of the two fold belts, namely the Mahakoshal and Satpura (see next section), and the near coincidence of the trend of the axial planes of their respective first-generation folds, one can argue that the Satpura fold belt and the Mahakoshal fold belt had a synchronous development. But this does not appear so. Whereas the age range of 1500–1200 Ma has been set for the Satpura intracontinental tectono-thermal event, the Mahakoshal fold belt (MFB) seems to have been metamorphosed earlier at about 1800 Ma, being the age of syn- to post-tectonic Tamkhan and Barambara granites intruding the MFB. That the metamorphism of the MFB had been earlier than the Satpura belt is also supported by the mineral isochron age of 1760–1610 Ma from a lamprophyre that intruded the MFB. These rocks and the mineral data from the MFB conclusively suggest that metamorphism of the MFB is prior to that of the Satpura belt occurring to its south, despite the parallelism of their orographic trend.

The MFB is without any thrust, melange or ophiolite together and is characterized by low-grade (greenschist facies) metamorphic rocks, supporting the origin of the fold belt in the intracratonic basin setting, precluding the possibility of any ancient suture or accretionary prism in the area. If the development of Son-Narmada lineament zone (SONA Zone) is prior to the Satpura orogeny, an opinion generally accepted by most geologists including this writer (see next section), the two fold belts—Mahakoshal and Satpura—seem to be distinct in time and space. It is, therefore, proposed that the folding of the MFB in the north by converging plates of Bastar and Bundelkhand created extension in the south and finally developed the Satpura rift basin to the south of the SONA zone. It is presumed that the configuration of the SONA zone had also controlled the orientation of the Satpura rift since the time it came into existence and filled-in by sediments that were later deformed and metamorphosed along with the mafic/ultramafic lavas. The SONA zone or Son-Narmada Lineament delineates the southern boundary of the Vindhyan basin (Meso to Neo-Proterozoic) and the northern boundary of the Gondwana basin (Late Paleozoic to Mesozoic). This together with the 1813 ± 65 Ma old syn- to post-tectonic granite (denoting the age of the Mahakoshal orogeny) definitely suggest that the SONA zone is a Palaeoproterozoic rift basin for deposition of the Mahakoshal supracrustals. The SONA Zone in the north is bordered by Son-Narmada North Fault (SNNF) and in the south by the Son-Narmada South Fault (SNSF). The reactivation of Son-Narmada lineament during Satpura orogeny helped retain the existing parallelism of the Satpura fold belt with the Mahakoshal fold belt located in the SONA zone in Central India. Recent structural work on the SNSF, the southern boundary of the MFB, shows that the deformation associated with N-S compression was superposed by a strong sinistral shearing movement that may have facilitated exhumation of lower crustal rocks south of the SNSF (Roy et al., 2002).

5.3 Satpura Fold Belt

5.3.1 Introduction: Geological Setting

The supracrustal rocks occupying the terrain to the south of the SONA Zone belong to Sausar Group rocks which constitute the Proterozoic Satpura fold Belt (SFB) The fold belt strikes E-W to ENE-WSW and fringes the northern boundary of Bastar craton, (although much of it is covered by Deccan Trap and Quaternary rocks) and the southern fault (SNSF) of the Narmada-Son lineament (SONA Zone). The Satpura mobile belt is considered the southern structural domain of the *Central Indian Tectonic Zone* (CITZ). This belt is nearly parallel to the SONA Zone located in its north (Fig. 5.1b). The other structural domain in the north is the *SONA Zone* (Jain et al., 1995) bounded by Narmada-Son-North Fault (NSNF) and the Narmada-Son-South Fault (NSSF). These two structural domains are nearly parallel (Fig. 5.1b).

The Satpura fold belt is 215 km long and 35 km wide and the Sausar Group rocks constituting this belt are metamorphosed cross-bedded quartzites, pelites, carbonates and Mn-ores. Lithostratigraphy of the Sausar Group given by Narayanaswami et al. (1963) is given below:

Bichua Fm:	marble & calc-silicate
Chorbaoli Fm:	quartzite & mica schist
Mansar Fm:	mica schist, Mn-ores (Gondite) etc.
Lohangi Fm:	marble, biotite schist
Sitasaongi Fm:	quartzite, conglomerate

········unconformity········

Tirodi Gneiss Complex

From the rock types of each formation (Fm) in the above given sequence, the Satpura belt appears to be *devoid* of volcanic rocks. Consequently, a belt of bimodal volclanics and mafic-ultramafic intrusions, located between MFB and SFB in the western part, has been distinguished from the Sausar Group and is called *Betul belt*, although at one place the Betul belt connects the Sausar belt (see Ramakrishnan and Vaidyanadhan, 2008). Like the Sausar Group, the Betul belt also consists of unclassified gneisses and granites and thus may be a part of Sausar Group rocks, although dominated by mafic-ultramafic complex. The Tan shear zone is the dividing line between Betul and Sausar belts. The Betul belt, like the Sausar belt, has ENE-WSW strike and extends for about 135 km from Betul in the west to Chhindwara in the east. In the geodynamic evolution, the Betul belt may not be considered a separate belt but a part of the Sausar belt.

Basement to the Sausar Group is the *Tirodi gneissic suite,* which contains biotite gneiss, hornblende-biotite schist/gneiss, migmatites, and also granulites that occur as rafts and boudins or lensoid bodies at a few places within the gneissic suite. The conglomeratic horizon between the Tirodi gneiss suite and the Sausar Group, recently discovered by GSI (Bandyopadhyay et al., 1995, p. 455), rejects the earlier belief that the Tirodi gneisses were products of high-grade metamorphism/anatexis of the Sausar Group. The granulite occurrences are known from Ramakona-Katangi domain (RKG), the Bhandara-Balaghat domain (BBG) along the Central Indian shear zone. There are also two more occurrences at Bilaspur-Raipur domain (BRG) and Makrohar domain (MG) in the Archaean gneissic terrain of the CITZ (Fig. 5.1c). This high-grade gneissic-granulite terrain shows a positive Bouguer anomaly over the Satpura fold belt (Das et al., 2003).

As a re-worked basement, the Tirodi gneisses yielded Rb—Sr whole rock isochron age of 1500 Ma; the granulites gave an age of 1400 Ma by Rb—Sr systematics while Sm—Nd age of granulites yielded 2600 Ma as the protolith age (Ramachandra and Roy, 2001). The intrusion of Dudhi adamellite-granite within the basement rocks (Tirodi gneiss suite and granulites) gave 1500 Ma whole rock isochron age, possibly constraining the Satpura orogeny. The peak granulite facies metamorphism from Khawasa-Ramakona area is estimated at 8–10.5 kbar and 850°C, obtained from geothermobarometry and P-T grid in the KFMASH system (Bhowmik et al., 1999, 2000). The peak conditions succeeded by an isothermal decompression, 6–4 kbar at 750–800°C, indicating erosion of about 15 km crustal material from the previously thickened crust. Another implication of the isothermal decompression is that there was no loading since the granulite metamorphism occurred in the basement (Tirodi gneiss suite). That is, both cover and basement were involved during the Satpura orogeny (1500 Ma ago; see above) that gave rise to amphibolite facies assemblages in the Sausar metasediments and granulites (in the basement of Tirodi gneiss-suite) by dehydration melting reactions involving biotite and/hornblende in suitable layers. Considering the isolated occurrences of the granulites within the Tirodi gneissic complex and the consideration of nearly similar Rb—Sr ages (1450 ± 50 Ma) of the Tirodi gneiss and associated granite (Ramachandra and Roy, 2001), it is highly probable that the granulites have formed during a tectono-thermal event of Mesoproterozoic age, although Sm—Nd systematic gave protolith age of granulites at $\sim$2800 Ma (cf. Ramachandra and Roy, 2001). The possible dehydration melting reactions that may have occurred in the Tirodi gneissic suite are: (1) biotite + quartz = hypersthene + Kfeldspar + water ± garnet and (2) hornblende + quartz = hypersthene + diopside + plagioclase + water. The released water and the feldspars as product phases of the reaction in gneisses were responsible for generating granitic melts that separated as granitic intrusions in the Sausar metasediments, leaving granulites as restites within the Tirodi gneissic complex.

Three deformation phases have been established in Satpura fold belt (Roy et al., 2001). According to Bhowmik et al. (2000) the D1 developed a ductile shear zone at and near the contact of the Sausar supracrustals and the basement gneisses, and produced mylonite and layered tectonites. The D1 phase is well recognized from Deolapar with the development of nappe and klippe structure. Considering the association of regional metamorphism with D1 deformation it seems more probable that the shearing was post D1 or even later. Bhowmik et al. (ibid) also claim that the metamorphism associated with D1 developed a progressive sequence of greenschist to amphibolite facies from S to N. The authors claim that the D1 truncated the Pre-Sausar fabric in the northern granulite belt. They further state that the peak P-T conditions of Sausar metamorphism is at $\sim$7 kbar and 675°C, based on the mineralogy of calc-silicate gneisses of Deolapar (Bhowmik et al., 1999). D2 deformation is believed to have produced shallow plunging upright folds with E-W to ENE-WSW trending axial planes. D3 produced folds with NNW-SSE to NW-SE axial planes. According to Bhowmik and Pal (2000), mafic dykes of different phases were emplaced within Sausar mobile belt, synchronous with reactivation of granulites within the Tirodi biotite gneiss. These basic dykes show subsolidus cooling documented in the development of coronal garnet at the interface of orthopyroxene and plagioclase. The coronal garnets in the mafic dykes yielded a Sm/Nd isochron age of 1416 (Roy et al., 2006), indicating a Mesoproterozoic cooling event and ruling out imprint of any tectono-thermal event (i.e. an orogeny) thereafter. The youngest thermal event in the RKG belt (and also in the BBG belt) and in Tirodi gneiss occurred at about 800–900 Ma as evidenced by (i) Rb–Sr isochron age of 840 Ma in the metadolerites of the RKG belt, (ii) 820 Ma age of amphibolites in the Tirodi gneisses, (iii) 880 Ma age for the monazite within the garnet of RKG (Ramakona-Katamgi) granulite (Bhowmik and Spiering, 2004), and (iv) 860 Ma K–Ar and Rb/Sr Mineral and whole-rock ages from the basement of Tirodi gneiss (Sarkar et al., 1986), and (v) 800 ± 171 Ma age by Rb–Sr for the BBG charnockites whole-rock and its mineral fractions (Roy et al., 2006, p. 66). This coincidence of different isotopic ages in varieties of rock is significant. The Rb–Sr isotope clock is believed to have been re-set all over the Bastar craton and Sausar fold belt, leaving the older Sm/Nd isotopic signatures unaffected (cf. Roy et al., 2006, p. 72).

Granites in the Satpura fold belt are stated to be of two phases, according to Bhowmik et al. (1999). First phase granites, occurring as sheet-like bodies, are foliated and broadly coeval with D1. The second phase granite is intrusive and massive and was emplaced at the terminal stage of D3. The Rb/Sr whole rock isochron age of ~1120 Ma for the tourmaline-bearing granite from RKG belt also suggests that the region remained unaffected by any orogeny (tectono-thermal event), although thermal imprints of 1100 Ma or even younger (800–900 Ma) age are documented in the terrain. Ramachandra and Roy (2001) and Acharyya (2003) suggest that Rb–Sr mineral and whole rock dates (1.5–1.4 Ga) of mafic granulites at the northern sheared end of the Bhandara-Balagaht granulite belt (Ramchandra, 1999) and their retrogression to amphibolite facies assemblages also correspond to the Satpura orogeny. On the basis of $^{40}Ar/^{39}Ar$ ages of 950 Ma recorded in cryptomelane from Sitapar Mines, Ramakona area (Lippolt and Hautman, 1994) and K–Ar mineral isochron ages of 860 Ma from the Tirodi biotite gneiss (Sarkar et al., 1986), Bhowmik et al. (2000) believe that the tectono-metamorphic event of the Satpura mobile belt closed during the terminal Mesoproterozoic (1.0 Ga). Nevertheless, the preceding discussion clearly suggests that Sausar orogeny occurred ca. 1500 Ma ago with intermittent cooling and heating events, the last being at 800–900 Ma.

5.4 Agreed Observations and Facts

Before we discuss the evolution of the Satpura fold belt, we summarize below the agreed observations and facts about this fold belt:

1. The Satpura fold belt trends E-W to ENE-WSW.
2. It is made up of metamorphosed Sausar Group metasediments of stable shelf environment.
3. The fold belt is devoid of volcanic rocks, barring the nearby Betul belt.

4. The Sausar Group rocks are metamorphosed in the amphibolite facies conditions.
5. Granulites occur as rafts and boudins within the Tirodi gneiss complex, near and away from the shear zones.
6. The basement of Tirodi gneissic suite has yielded a Rb–Sr whole rock isochron age of 1525 ± 70 Ma for biotite gneiss and a mineral isochron age of 860 Ma (Sarkar et al., 1986).
7. The 1525 Ma age, by most workers, is interpreted as the main phase of regional metamorphism during which the basement reactivated to produce the migmatites and granulites.
8. The age of 1.5 Ga for the adamellite-granite intrusion within the basement rocks (Tirodi gneissic suite and granulites) constrain the age of Sausar orogeny (Sarkar et al., 1998)
9. The Tirodi gneissic suite has a conglomerate band at the contact with the overlying Sausar group and contains intricately folded fragments of other schist belt and Mn-ores, perhaps due to infolding (i.e. synclinal keels).
10. The structure of the fold belt is characterized by F1 isoclinal folds that are overturned toward N, while the northern region contains south-directed recumbent folds/thrusts.
11. Bimodal mafic and volcanic-plutonic rocks (? Age) from the southern as well as northern granulite belts suggests continental rift setting (Roy et al., 1997).

5.4.1 Evolution of the Satpura Fold Belt (SFB)

An evolutionary model needs a time frame. The Satpura belt is most likely to have formed in the Mesoproterozoic time, most probably between 1525 Ma, (being the metamorphic age of the Tirodi gneiss) and 1400–1600 Ma which is the age of the Semri Series of the undeformed Vindhyan Supergroup now delimited by Satpura belt. The N-S trending basement gneisses (pre-Tirodi gneiss complex) were rifted and the intracratonic basin thus formed, received the cratonic sedimentary assemblage of mature cross-bedded sandstone, pelite and carbonates and Mn-ores. The basinal sediments were subsequently compressed in the Satpura orogeny, documented by the presence of recumbent folds (overthrusts?) along the northern and southern margins of the belt and by the increase of metamorphic grade toward the N or NNW, with granulite facies rocks near the northern margin of the fold belt (BBG belt excluded due to its location within Amgaon gneiss of Bastar craton and due to its anticlockwise P-T path, just opposite to that of the RKG belt granulites) (cf. Bhowmik et al., 1999; Bhowmik and Roy, 2003).

The orogenic trend and the overturning of the folds from both sides towards a crystalline core suggest that stable crustal blocks to the N and S of the orogen (in its present orientation) collided to evolve the Satpura fold belt. The structural pattern in the Sausar belt is, however, different from that found in other mobile belts where the vergence and thrusting are away from the centre of the orogen; in the Satpura belt the fold vergence is toward the centre of the orogen. This may be explained by overriding of crustal areas during compression, but needs another explanation, especially when there is no evidence of any subduction or magma underplating. The author thinks that partial melting of basement rocks in a narrow zone supplied the required heat to recrystallize the folded and the overturned basinal rocks of the Sausar Group. The reactivated Tirodi gneiss and its enclosed granulite pods and boudins are a clear evidence of partial melting of the basement. Tectono-metamorphic events of the Satpura fold belt evidently closed during the end of Grenville period (Mesoproterozoic, 1.0 Ga), as revealed by of $^{40}Ar/^{39}Ar$ cooling ages of 950 Ma recorded in the cryptomelane from Sitapar Mines, Ramakona granulite belt (Lippolt and Hautman, 1994). A dominant thermal event during 800–900 Ma is documented over the entire terrain of the Satpura belt and its basement Tirodi gneiss complex during which the Rb/Sr ages have been re-set in all rocks of the fold belt and re-worked basement of the Tirodi gneiss, as already stated. A K–Ar mineral age of 860 Ma from the Tirodi biotite gneiss (Sarkar et al., 1986) also supports this proposition of the author (see also Roy et al., 2006).

The preceding geochronological and petrological interpretation does not support the hypothesis of Roy et al. (2006) that the Satpura orogeny occurred during Grenvillian period ($\sim$1000 Ma), assigning this age to the peak metamorphism (10.5 kbar/$\sim$775°C) of Ramakona-Katangi granulites studied by Bhowmik et al. (1999). Following the peak, the granulites were found to show nearly isothermal cooling and then isobaric cooling. Unlike the RKG granulites of the northern belt, the Bhandara-Balaghat granulites (BBG) occurring within the Amgaon gneiss complex of Bastar craton in the south show anticlockwise P-T path (Bhowmik et al., 1999, 2000). Considering Sm/Nd isochron age of 2672 Ma for the charnockite of BBG, Roy et al. (2006) assigned Pre-Sausar age to the peak metamorphism of the granulite. These two granulite belts are thus not coeval and need not be compared while attempting evolution of the Satpura fold belt. The Bastar craton south of CIS seems to have remained unaffected by the Satpura orogeny but certainly documents the 800–900 Ma old thermal (not tectono-thermal) event that evidently affected the entire region of the Sausar and Pre-Sausar rocks in central India.

An important matter of inquiry is about recognizing the crustal plates in the north that had converged with the BC craton to south of the Satpura fold belt. On a broad scale, the stable block on the north is presumably the Bundelkhand massif of Rajasthan craton. On a regional scale, one can also consider the Satpura fold belt to have originated during the convergence of the Aravalli-Bundelkhand Protocontinent against the Bastar (-Bhandara) craton (Fig. 5.2) For each of the propositions a serious problem arises due to the presence of Narmada rift between the Satpura belt and the Bundelkhand massif. If the rift is older than the Satpura belt, the Satpura orogeny need to have occurred during the closure of the rift basin, by compression at right angles to the rift margin. If so, the present topographic expression of the rift is the result of reactivation of a former rift area. Alternatively, if the rift is younger than the Satpura orogeny, the Satpura belt owes its origin to intracratonic deformation of the siliceous and platformal deposits. And the rift probably developed along

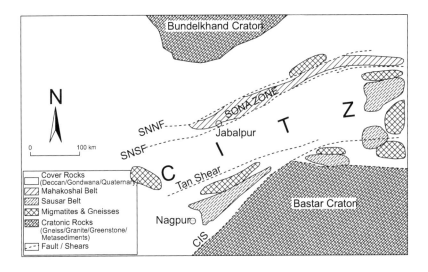
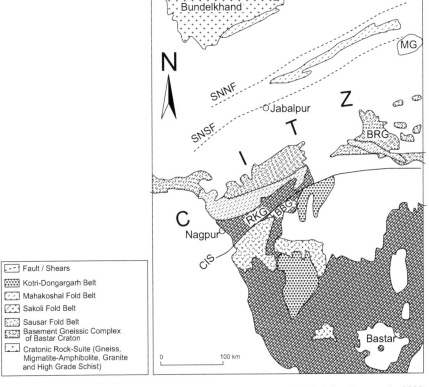

Fig. 5.2 Geological map of the Central Indian Tectonic Zone (CITZ) (after Roy et al., 2002) sandwiched between Bundelkhand craton in the north and Bastar craton in the south (*top figure*). The location of the four fold belts and shear zones of central India as well as the four granulite localities (Fig. 5.1) are also shown in *bottom figure*. All abbreviations as in Fig. 5.1

the orogenic belt because of the tendency of new continental rifts to develop in areas that have been tectonically thickened (cf. Naqvi and Rogers, 1987).

It is interesting to note that the Satpura orogenic belt/CITZ is shown colinear with the Albany Fraser mobile belts of Australia in the Gondwana reconstructions (Harris, 1993; Yoshida, 1995).

5.5 Geological Significance of the Central Indian Tectonic Zone

The Central Indian Tectonic Zone (CITZ), first proposed by Radhakrishna (1989), is regarded by Acharyya and Roy (2000) as the terrain between the Son-Narmada North Fault (SNNF) and the Central Indian Shear (CIS) zone (Fig. 5.2, see also Fig. 5.1c). It means that the CITZ includes (i) the SONA zone as a rift zone of Palaeo- to Mesoproterozoic age occupied by low-grade (greenschist facies) Mahakoshal Group supracrustals and the syntectonic granite of 1.8 Ga, and (ii) the polymetamorphic rock-suite of high-grade Tirodi gneiss-migmatite containing scattered granulites in form of rafts and boudins at four areas, namely the Ramakona-Katangi (RKG), Bhandara-Balaghat (BBG), Bilaspur-Raipur (BRG) and Makrohar (MG). Amidst this rock-suite occurs the Sausar supracrustals that show amphibolite facies metamorphism.

We do not have precise age-data on the granulites but the 1.7 Ga old granite intrusions in the Tirodi gneiss complex (gneiss-migmatite-amphibolite-granulite) adjacent to the Makrohar granulite domain indicates that granulite-forming event was near 1700 Ma (Roy and Hanuman Prasad, 2003; Solanki et al., 2003). This needs a critical evaluation in light of metamorphism and relevant geochronological data. As already stated, the occurrence of granulites as boudins or lensoid bodies within the Tirodi gneiss complex, indicates that these isolated granulite rafts are the product of partial melting of the gneisses during which biotite and hornblende became unstable to give rise to hypersthene + feldspar, with or without garnet and clinopyroxene. In other words, the granulite pods have been derived directly from the rocks hosting them. It is interesting to note that the granulites of Ramakona-Katangi domain were earlier mapped only as Tirodi gneiss, perhaps because the granulites occur only as small boudins or rafts within the gneiss complex that shows sheared contacts with the Sausar group rocks (cf. Bhowmik et al., 1999). If it be so, then the age of granite (as products of anatexis of Tirodi gneisses) may also be taken as the age of high-grade metamorphism and granulite formation. This event of melting/recrystallization is undoubtedly associated with the Satpura orogeny that deformed and metamorphosed both the cover (Sausar group) and the basement Tirodi gneiss complex. According to this interpretation the Satpura orogeny is Mesoproterozoic (1.7 Ga old). The age range of 1.7–1.5 Ga for the granite intrusions in the Satpura fold belt could be the manifestation of variable cooling rates due to emplacement at different depths of these anatectic granites that may have also formed in more than one phase of partial melting (Subba Rao et al., 1999).

The above interpretation does not support the proposition of Acharyya (2001, 2003b) that the Pre-Sausar orogeny occurred at 1.6–1.5 Ga and that the Satpura orogeny at 1.0 Ga. It also contradicts the proposition of Bhowmik and others (Bhowmik et al., 1999) that the Satpura orogeny is Grenvillian—the proposition based on the interpretation of $^{40}Ar/^{39}Ar$ age data of 950 Ma (Lippolt and Hautman, 1994) for the cryptomelane from Sitapar Mines in the Ramakona area of Ramakona-Katangi Granulite (RKG) domain. Lippolt and Hautman (1994) interpreted this age as a thermal overprinting event (not an orogeny). This thermal event is possibly related to the 1147 ± 16 Ma old tourmaline granite intrusion (in the Sausar) from Nainpur-Lalbara area in the RKG domain (cf. Pandey et al., 1998). The youngest age clusters of Rb–Sr data at 973–800 Ma for the crystalline rocks of the CITZ (Ramachandra and Roy, 2001) can also be attributed to thermal imprints related to the tourmaline granite and related intrusions of 1150 Ma. It seems, therefore, that the post-tectonic intrusion of the 1100 Ma old tourmaline granite (undeformed) is responsible for the thermal imprint in both cover and basement rocks of the CITZ. Furthermore, if the granulites of the RKG domain are co-relatable with the granulites of Bilaspur-Raipur granulite (BRG) domain, as claimed by Bhattacharya and Bhattacharya (2003), then there should be only one widespread granulite metamorphic event (Mesoproterozoic) in the CITZ, most probably related to the Satpura orogeny. Until granulites from the CITZ are dated, it is misleading to assign granulite events to Late Archaean or to Grenvillian age, or to correlate it with the Archaean Limpopo belt (Mazumder et al., 2000), merely on the assumed continuity of structure. On the basis of the Mesoproterozoic ages of the CITZ rocks and the mobile belts within it, Harris (1993) proposed that the Albany mobile belt of Australia is an extension of the CITZ, while Zhao et al. (2003) consider the Trans-North China orogen as the westward extension of the CITZ at 1.8 Ga, based on the possible juxtaposition of Eastern Block of North China with the Southern Indian craton.

The *SONA zone* is considered as a major tectonic divide, separating the southern Indian peninsular block from a northern block, both of which are considered to have distinct evolution (West, 1962). The zone is occupied by the Mahakoshal group rocks (uncertain age) having Archaean basement as at Sushi. These rift sediments are intruded by 1813 ± 65 Ma old granite (Sarkar et al., 1998), indicating that the age of Mahakoshal orogeny is Palaeoproterozoic. This is also supported by the fact that the SONA Zone delineates the southern boundary of the Vindhyan basin (~1400 Ma). The Son-Narmada North Fault of the SONA Zone is considered to define the northern limit of the CITZ (Acharyya, 2003). This zone, however, extends more than 1600 km with the E-W trend and is obviously a rift zone, as discussed earlier (Das and Patel, 1984).

The southern limit of the CITZ is defined by the brittle-ductile Central Indian Shear (CIS) zone (Jain et al., 1991; Radhakrishna, 1989; Acharyya and Roy, 2000; Acharyya, 2001, 2003a, b). Across the CIS there is a marked contrast in lithologies and metamorphic grade. High-grade metasedimentary rocks and granulites (Satpura fold belt) occur on the northern side and the low-grade volcanic sequences are in the S in the Sakoli and Kotri-Dongargarh belts. Considering this contrast, Yedekar et al. (1990) proposed that the CIS is a suture zone or collision zone along which southward subduction had occurred at 2.5 Ga ago. These authors further assumed that the southward subduction gave rise to the island arc represented by Nandgaon volcanics of (Kotri-) Dongargarh belt. Bhowmik et al. (2003) rejected this proposition of subduction on the consideration that granite plutons of Dongargarh and Malanjkhand in the overriding plate (i.e. Bastar cratonic block) show N-S trend, and not ENE trend required by the arc system. Moreover, there is a conspicuous cataclastic/mylonitic zone on the E and W ends of the CIS, implying that it is merely a brittle-ductile shear zone than a subduction zone. Furthermore, the CIS cuts the NW margin of the 2.5 Ga old Malanjkhand batholith and also overprints the dominant N-S structural grain of the Kotri-Dongargarh belt (Bandyopadhyay et al., 1995). This again supports the brittle-ductile nature of the CIS shear zone. In conformity with this conclusion, the CIS is found to change its initial NE-SW trend on its eastward extension (as a 2.5 km wide and 500 km long brittle-ductile shear zone) and then bifurcates and swing abruptly with steep northerly dips (see Fig. 5.1b).

The above discussion thus rules out the proposition of some workers (Yedekar et al., 1990; Jain et al., 1991; Misra et al., 2000) that the CIS brought together the two cratons—the north Indian Bundelkhand protocontinent and the south Indian protocontinent comprising Dharwar, Bastar, and Singhbhum cratons.

A very vital point now arises in regard to the age of the CIS zone. Since the CIS clips the 2.5 Ga old Malanjkhand batholith (Stein et al., 2004) and overprints the N-S structural grain of the Dongargarh belt (Bandyopadhyay et al., 1995), the shearing must post-date the emplacement of the 2.5 Ga old granitoids. The location of the CIS also shows that this shear zone is a single zone as well as a bifurcated zone transecting the Sausar group rocks that were deformed and metamorphosed in Mesoproterozoic Satpura orogeny. This means that the CIS must be synchronous or younger than the Satpura orogeny, dated at 1.7 Ga (see earlier section).

The CIS is a shear zone affecting the metamorphosed Sausar (Satpura) belt of Mesoproterozoic age, but its shear sense is not known. In light of the brittle-ductile nature and Mesoproterozoic age of the CIS, this writer considers it an unreasonable attempt to join the Aravalli-Delhi fold belts in the west with the Singhbhum belt in the east (cf. Jain et al., 1991; Stein et al., 2004).

In conclusion, the CIS in all certainty is a shear zone, unlike the SONA zone. Hence, it raises a vital question whether the two zones, namely the SONA zone and the Zone south of it up to CIS, having different tectonic evolution and ages, could be combined together under a single CITZ. Surely, detailed mapping and isotopic ages in the CITZ are required to unravel the evolution of this mega structure relative to cratonic regions to the north and south.

5.6 Dongargarh Fold Belt

5.6.1 Introduction: Geological Setting

The Dongargarh fold belt, also called Kotri-Dongargarh fold belt, is located to the south of the Central Indian Shear (CIS) zone (Fig. 5.2b). It has N-S orographic trend which is nearly perpendicular to the nearly E-W trending fold belts of the CITZ (Mahakoshal and Satpura fold belts). The Dongargarh fold belt (DFB) is made up of igneous rocks, rhyolites and granites that unconformably overlie the

Amgaon gneiss complex—an Archaean basement complex composed of augen gneiss, banded gneiss, migmatites, and amphibolites (Sarkar et al., 1981). The supracrustal rocks of the DFB are dominated by greenschist facies *Nandgaon Group* that includes the rhyolite-basalt bimodal volcanic sequences, intruded by igneous sequences of Dongargarh and Malanjkhand granitoids. In the Malanjkhand area, the Nandgaon volcanic sequence consists of basalts, andesites and dacite (Bhargava and Pal, 2000). The Nandgaon volcanics to the south of Malanjkhand include rhyolite, rhyolite tuff of *Bijli Formation* and overlying mafic volcanic rocks of the *Pitepani Formation*. Based on the geochemical data, the Nandgaon sequences have been described as a bimodal suite with intrusion of A-type granitoids of Dongargarh—all attributed to a rift environment (Deshpande et al., 1990; Ramachandra and Roy, 1998), like the Malani Igneous suite of western Rajasthan. The bimodal suite is overlain by shale and sandstone and some volcanics of *Khairagarh Group* (Fig. 5.3), all of which are seen to overlie directly upon the Amgaon gneissic complex, as shown below in the generalized stratigraphic succession (after Sarkar et al., 1981; Sarkar, 1994).

	Kotima Volcanics
	Ghogra Formation
	Mangikhuta volcanics
Khairagarh Group	Karutola Formation
	(Intertrappean shale)
	Basal shale
	Bortalao Formation

··························Unconformity··························

	Dongargargh Granite
	Pitepani Volcanics
Nandgaon Group	Bijli Rhyolites

··························Amgaon Orogeny··························

Amgaon Group	Amgaon gneiss complex (gneiss, banded gneiss,
(Archaean)	migmatites, amphibolites & quartzites)

Initially, the Amgaon Group, Nandgaon Group, Dongargarh Group and Khairagarh Group have been placed together under the Dongargarh Supergroup by Sarkar et al. (1981) who recognized three orogenies in the region, merely on the basis of unconformity between the Groups. It is now accepted that, excluding the Archaean age Amgaon orogeny, the supracrustals of the DFB are clearly the outcome of a single orogeny called the Dongargarh orogeny.

The supracrustal rocks of the Khairagarh Group in the Dongargarh and Malanjkhand area are somewhat problematic in terms of their age, stratigraphic continuity in N-S direction and their contact relationships with the underlying volcanics and intrusive rocks of the Dongargarh belt.

5.6 Dongargarh Fold Belt

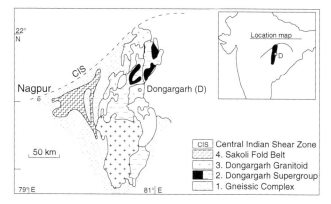

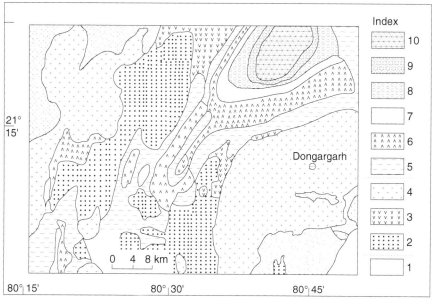

Fig. 5.3 Geological map of Dongargarh fold belt, central India (after Sarkar, 1957; Deshpande et al., 1990). 1 = Amgaon gneiss complex, 2 = Bijli rhyolite, 3 = Pitepani volcanics, 4 = Dongargarh granitoid, 5 = Bortalao formation, 6 = Sitagota volcanics, 7 = Karutola formation, 8 = Manjikhuta formation, 9 = Ghogra formation, 10 = Katima volcanics (5–10 lithounits belong to Khairagarh Group)

Geochronological work involving Rb/Sr systematics is known to have some problems, perhaps due to disturbed systematics (Sensarma and Mukhopadhyay, 2003). The Bijli rhyolites gave Rb–Sr whole rock isochron ages of 2503 ± 35 Ma and 2180 ± 25 Ma, while the Dongargarh granite (as intrusive into the volcanics) gave Rb–Sr whole-rock isochron ages at 2465 ± 22 Ma and 2270 ± 90 Ma (Sarkar et al., 1981; Krishnamurthy et al., 1988, 1990). The Malanjkhand batholith, described as intrusive into the Nandgaon equivalents gave Rb/Sr whole-rock isochron ages, ranging from about 2470 to 2100 Ma, with large uncertainties (cf. Ghosh et al., 1986; Panigrahi et al., 2002). Recently, U–Pb zircon ages on grey and pink samples of Malanjkhand granite (equivalent to Dongargarh granite) gave 2478 ± 9 and 2477 ± 10 Ma (Panigrahi et al., 2002). A similar age of 2500 Ma is obtained by Re–Os geochronology in molybdenum from the Malanjkhand Cu–Mo–Au deposits (Stein et al., 2004).

5.6.2 Agreed Observations and Facts

Before we discuss the evolution of the Dongargarh fold belt (DFB), it is useful to list the agreed observations and facts about the rocks of the DFB. These are:

1. The fold belt is 150 km long and 90 km wide.
2. The DFB has N-S trend similar to that of the basement Amgaon gneissic suite
3. A bimodal suite with 4500 m thick rhyolite (2180 ± 25 Ma) at the base overlies the basement gneisses.
4. The rhyolite, called Bijli Rhyolite, has been intruded by Dongargarh Granite (2270 ± 90 to 2466 Ma old).
5. The bimodal volcanic suite of rhyolite-basalt is overlain by shale, sandstone and volcanics in alternating manner, which are now seen as quartz-sericite schist, quartzite, and amphibolite (Sarkar et al., 1990).
6. Metamorphism of the DFB is in greenschist facies (low to medium grade).
7. The DFB is traversed by lineaments (ductile to brittle-ductile) on its east and west margins along which gabbroic rocks were emplaced.
8. The DFB is affected at its northern end by the ENE-WSW trending Central Indian Shear Zone (CIS).
9. The earliest ENE-trending structures in the Sausar fold belt have truncated and possibly rotated the N-S oriented structures in the Dongargarh belt (Bhowmik et al., 2000).

5.6.3 Evolution of the Dongargarh Fold Belt (DFB)

The evolution of the Dongargarh fold belt (DFB) is typical of intracontinental rift because of the occurrence of bimodal volcanics and arkose conglomerate association. Initially, the rifting of the Amgaon gneissic basement started in an extensional regime with rift margins trending N-S. In this rifted basin bimodal volcanics were emplaced the felsic component of which is the Bijli rhyolite dated at 2180–2503 Ma. The rhyolite has been intruded by comagmatic Dongargarh granite (2270–2466 Ma). Both rhyolite and the granite have nearly the same initial strontium ratio of 0.7092 and clearly appear to be partial melts of continental crust at depth (Sensarma et al., 2004). Recently, Sensarma (2007) reports bimodal igneous province with equal volumes of coeval felsic and mafic volcanic rocks in Dongargarh fold belt. This he attributed to plume tectonic setting. However, the longer time of magmatic activity between 30 and 70 Ma is too long to be explained by plume model. Therefore, this author suggests extensional tectonic regime in which mantle and crustal sources were mobilized more or less simultaneously to give rise to this large igneous province of central Indian region. Following the bimodal volcanism, the rifted basin received intercalations of shale and sandstone and volcanics (Khairagarh Group). Later, the rocks of the basin were deformed by E-W compressive forces and metamorphosed in the greenschist to lower amphibolite facies during which process the isotopic systematics in the igneous rocks remained (?) undisturbed (Sarkar et al., 1994). The deformation of the cover rocks has been clearly controlled by the basement structures in the terrain. Because ENE-WSW trending Central Indian Tectonic Zone (CITZ) affects the DFB, the Khairagarh orogeny was much earlier than the Mahakoshal and Satpura orogenies. The present shear zones on the east and west sides of the DFB could represent the reactivated margins of the N-S rift that had formed during Early Proterozoic time in this part of the Bastar craton.

From above it thus becomes clear that the stable continental crust in central Indian region was rifted only intracratonically and the rifts were presumably aborted before the oceanic crust could develop in response to tensile stresses.

5.7 Sakoli Fold Belt

5.7.1 Introduction: Geological Setting

It is a triangular shape fold belt and constitutes a part of the northern Bastar craton in central India. The Sakoli fold belt (SKFB) occupies a key position at the convergence of the Dongargarh fold belt (DFB) and the Satpura fold belt (SFB). As a result, the SKFB bears the impress of both these events. The triangular outcrop of the SKFB is because of the two prominent ductile shear zones, one along the eastern margin with N-S strike (which also defines the trend of the DFB) and the other along the NW margin with ENE-WSW strike (which is also the trend of the SFB) (Fig. 5.3a).

The basement to the Sakoli Group is the Amgaon gneiss complex which consists of gneisses, migmatites, Cr-bearing metamorphosed ultramafic rocks and Pre-Sakoli supracrustal assemblage of quartzite, garnet-staurolite-biotite schist, kyanite-sillimanite bearing metapelites, calc-silicate rocks and marble, and cordierite-gedrite-anthophyllite rocks (Bandyopadhyay et al., 1995). The Amgaon basement complex is unconformably overlain by the Sakoli Group rocks. The Sakoli Group contains one of the most significant volcano-sedimentary sequences in the Central India. It has four formations that from base to the top are: (1) conglomerate and arkose with minor carbonaceous phyllite, ferruginous quartzite and BIF, (2) meta basalts with minor metapelite, chert bands and meta-ultramafic rocks, (3) the Bhiwapur Formation consisting of metapelites with interbanded bimodal volcanics dominated by rhyolite, rhyodacite, tuff, breccia, epiclastic rocks and minor basalts, and

(4) slate, phyllite, debris flow deposits, and meta-arkose and quartzite (Bandyopadhyay et al., 1995). The Sakoli Group has several quartz veins, quartz reefs, silicified shear zones, granite-pegmatites, tourmaline granite, and gabbro-dolerite. The Sakoli Group rocks are overlain unconformably by the Gondwana Supergroup and the Deccan basalt. Isotopic ages are almost unavailable on the Sakoli rocks. Poorly constrained Rb–Sr data for meta-volcanics and tuff intercalated with metapelite and granitoid give ages of 1295 ± 40 Ma and 922 ± 33 Ma but are interpreted as reset ages of earlier event related with Sakoli orogeny (Bandyopadhyay et al., 1990).

The above description shows that metapelitic rocks dominate the Sakoli Group rocks but lower part contains the bimodal volcanic-arkose-conglomerate-quartzite association, indicating intracontinental rift setting. The existence of basement gneissic rocks of Amgaon Group surrounding the SKFB suggests that the supracrustals of SKFB had an ensialic rift setting. There is gravity high all over the SKFB to suggest further that an anomalous mantle material underlies the rift zone. Precise age of the Sakoli fold belt rocks is not known.

The supracrustals of SKFB have been subjected first to E-W compression to produce F1 folds with N-S striking axial planes and variable plunges of the fold axis. The shear zone appears to represent the terminal event of this E-W compression. The ductile shear zone along the eastern margin of the SKFB and the F1 fold in the Dongargarh fold belt seem to be time equivalent and belong to the same family of intraplate N-S trending crustal zones of weakness. Although F2 folds are coaxial to F1, the F3 folds are upright and asymmetric with ENE-WSW striking axial planes. A possible time relationship is observed by Bandyopadhyay et al. (1995) between the ENE-WSW trending shear zone in the NW margin of the Sakoli fold belt and the F3 phase of Sakoli orogeny, and is thus correlatable with the structures belonging to the Satpura orogeny—a major intracontinental tectono-thermal event that produced the Satpura fold belt in central India. As a consequence, the NW Sakoli rocks and their basement near the contact bear the imprints of Mesoproterozoic (1.5–1.2 Ga) tectono-thermal event. This event was also attended by a widespread crustal softening, deformation, and reactivation of the basement. One cannot ensure a time equivalence between the similarly oriented folds in the Sakoli and Sausar rocks, although some geologists propose that the Mesoproterozoic F3 folds of SKFB are coeval with the F1 fold in the Sausar belt (SFB) (cf. Bandyopadhyay et al., 1995). If this is so, the time equivalence of F1 in SKFB and DFB lead to suggest that the Sakoli supracrustals are older than the Sausar Group of rocks. But this may not be conclusive.

5.7.2 Agreed Observations and Facts

Before we summarize the evolutionary history of the Sakoli fold belt, it becomes useful to get a synoptic view of the geological characteristics and the agreed observations or facts on the Sakoli fold belt. These are:

5.7 Sakoli Fold Belt   197

1. Sakoli fold belt (SKFB) is at the convergence of N-S trending Dongargarh fold belt (DFB) and E-W trending Satpura fold belt (SFB) in central India.
2. The Sakoli belt, like DFB, is surrounded by the Archaean Amgaon Gneissic complex. But the contact between the cover and basement is unrecognizable.
3. The SKFB is a volcano-sedimentary belt.
4. Precise age of the Sakoli rocks is not known.
5. The supracrustal sequence of Sakoli is intruded by pre-tectonic dolerite dykes and synkinematic granitoids.
6. Although poorly constrained, Rb/Sr data for meta-volcanics and tuff intercalated with metapelites and granitoids provide ages of 1295 ± 40 Ma and 922 ± 33 Ma, interpreted as reset ages.
7. Four deformations are recorded. F1 produced folds with N-S axial planes and the N-S trending ductile shear zone in end phase of F1.
8. F3 folds are asymmetric upright with ENE-WSW striking axial planes.
9. The NW margin of the Sakoli belt has ENE-WSW shear zone developed during F3. The shear zone along the E-margin of the belt strikes N-S.
10. The ENE shear zone and the associated F3 structures in the SKF are correlatable with the structures belonging to the Satpura orogeny such that $F3_{(Sakoli)} = F1_{(Satpura)}$. This means that Sakoli rocks are older than Sausar (Satpura) fold belt.
11. Metamorphism in the Sakoli belt is post F1 but overlaps the F2 and F3 fold phases.

5.7.3 Evolution of the Sakoli Fold Belt (SKFB)

To propose an evolutionary model for the Sakoli fold belt (SKFB), a time frame is required but geochronological data are scarcely available on the rocks in question. Therefore, one can rely only on the lithological relationship and structural patterns of the Sakoli rocks to attempt a geotectonic evolution of the fold belt. There is also no serious work on relating the four deformation phases with the metamorphic episode(s). The following account is drawn from the published work and needs modification by future research on the Sakoli rocks.

The existence of Archaean gneissic rocks of Amgaon Group all around the triangular shaped Sakoli fold belt indicates that the supracrustal rocks of SKFB had an ensialic rifting. The rifting was initiated by crustal extension and the rift was filled with coarse clastics followed by basic volcanics that are obviously the result of decompression melting of the crustal lithosphere. This is expressed in the gravity high all over the SKFB. These were overlain by thick pelites and bimodal volcanics that are indicative of deposition in a rapidly subsiding trough that was forerunner of the Sakoli orogeny. The end of sedimentation is marked by deposition of thin clastics over the pelites, indicating shallow water sedimentation.

The Sakoli supracrustals have been deformed by 4 episodes of folding (Bandyopadhyay et al., 1995). The volcano-sedimentary rocks of the Sakoli Group were first subjected to E-W compressive stresses that produced F1 folds with N-S striking axial planes. With advanced stage of F1 (or coaxial F2) developed the N-S trending ductile shear zone that borders the Sakoli fold belt on its east. This structural pattern seems to have been controlled by the N-S trending rift zone, similar to that of the Dongargarh fold belt. The F3 later superposed the F1 and F2, but the map pattern of the Sakoli fold belt is the result of superposition of westerly plunging F3 folds with ENE-WSW trending axial foliation on F1. The F3 folds are with sinistral sense of vergence. During F3 folding, there also developed ENE-WSW trending shear zone along the western margin of the Sakoli fold belt. The NE to ENE extension of the Sakoli rocks along their NW margin is attributed to the presence of this shear zone developed during F3. All the Sakoli rocks and the basement gneissic complex have been rotated into parallelism with the shear zone boundary. It may then be suggested that the ENE shear zone and the associated structures in the Sakoli rocks are correlatable with the structures characterizing the Sakoli orogeny that is a remarkable ensialic fold belt orogeny in the central Indian metamorphic province.

5.8 Discussion of Evolutionary Models of the Central Indian Fold Belts

The multiple linear rift systems generated in the Archaean crust of central India controlled the development of Early and Middle Proterozoic fold belts, all of which are ensialic. Of the four Precambrian fold belts, the Mahakoshal and Satpura (Sausar) fold belts are E-W to ENE-WSW trending, whereas the Dongargarh and Sakoli fold belts have N-S trend. The triangular-shaped Sakoli fold belt, however, has the structural patterns of both fold systems as it occurs at the convergence of Dongargarh and Satpura fold belts.

In the absence of geochronological data on the constituent rocks of the fold belts, it is difficult to establish their chronological evolution. However, based on the structural patterns, fold orientations and the strike of main axial foliation as well as the bounding shear zones, different models have been advanced.

Model 1 (Yedekar et al., 1990)

Accepting that the Central Indian Suture (CIS) was a subduction zone, Yedekar et al. (1990) proposed that the Bundelkhand craton subducted ca. 2.7 Ga ago beneath the Bastar craton along the CIS. As a result, the Mahakoshal belt originated in a rift setting and the Sausar belt as a margin sediments of the Bundelkhand craton. The Dongargarh volcanics and also the Sakoli Group were considered by the authors as arc-related volcanism. The basic granulites of CIS, according to Yedekar et al. (loc. cit.), are MORB basalt which was metamorphosed at depth and obducted during the convergence of the stated cratons.

This model has a basic weakness for assuming the CIS as Subduction zone. It has been clearly shown that the CIS is a brittle-ductile shear zone. Furthermore, the southward subduction along CIS cannot be accepted because the 2.3 Ga old granite plutons (Dongargarh and Malanjkhand) in the overriding plate (Bastar craton) should have subduction-parallel orientation of E-W or ENE-WSW and not the N-S

5.8 Discussion of Evolutionary Models of the Central Indian Fold Belts   199

trend shown by the granites (if formed as magmatic arc) Moreover, the CIS bifurcates with steep northerly dip, and hence do not favour the model of Yedekar et al. (1990).

Model 2 (Roy and Hanuman Prasad, 2003)

This is also a plate tectonic model but the polarity of subduction is considered reverse to that in Model-1. During subduction of Bastar plate under the Bundelkhand craton, there occurred delamination of the crust and granulites were formed first due to magma underplating. The Mahakoshal is assumed to have originated as back arc ocean at 2.2 Ga and was followed by cal-alkaline magmatism (1800 Ma) which continued up to 1500 Ma during which Betul belt formed as an intra-arc rift filled with bimodal volcanism. The authors propose that Betul basin closed at 1500 Ma as evidenced by syntectonic granite rocks accompanied by mafic-ultramafic magmatism (Pichai Muthu, 1990). The subduction is believed to have culminated at 1500 Ma with continent-continent collision along the Ramakona-Katangi granulite belt along which both Bastar and Bundelkhand cratons accreted with exhumation of granulites. The authors also state that the amalgamated craton formed the basement for the Sausar Group sediment (shelf). These sediments were later deformed at 1100 Ma during which younger granites were generated.

The model by Roy and Hanuman Prasad (2003) considers pre-Grenvillian collision at 1500 Ma and extensive re-working during the Grenvill orogeny at 1.1–1.0 Ga. It implies in the model that the Mahakoshal belt, Betul belt, and granulites were formed in the same event of subduction (2.2 Ga). But the Balaghat-Bhandara Granulites (BBG) are considered older by the authors. This cannot be substantiated by the available isotopic data. The exhumation of different granulite belts of small extension (mostly as pods and lenses) is assumed to have occurred along shear zones amongst which the CIS has southward movement. Earlier discussion has clearly shown that the granulites formed from Tirodi gneisses during anatexis in the Satpura orogeny. The CIS in all probability is a shear zone and not a zone of subduction, whether occurred northward or southward.

Model 3 (Bandyopadhyay et al., 1995)

The Satpura fold belt, according to Bandyopadhyay et al. (1995), is a major intracontinental tectono-thermal mobile belt of central India. And the triangular shaped Sakoli belt is the outcome of two prominent ductile shear zones, one along the eastern margin with N-S strike and the other along the NW margin with ENE-WSW strike. The N-S trending shear zone is considered as a result of termination of F1 fold phase. Consequently, this shear zone on the eastern margin of the Sakoli fold belt, according to these authors, is time equivalent to F1 fold in the Dongargarh fold belt that strikes N-S. The F1 fold in the Sausar (Satpura fold belt) becomes time equivalent to the F3 fold in the Sakoli because the shear zone in the NW margin of the Sakoli fold belt strikes ENE-WSW, similar to the F1 structures in the Satpura belt. It means that the Sakoli rocks are older than the Sausar Group and that the F1 fold of the Dongargarh Group and Sakoli Group are nearly coeval.

Because of the parallelism of the Sausar (Satpura) belt and the Mahakoshal belt, the structures in these two Proterozoic fold belts are considered synchronous. But it

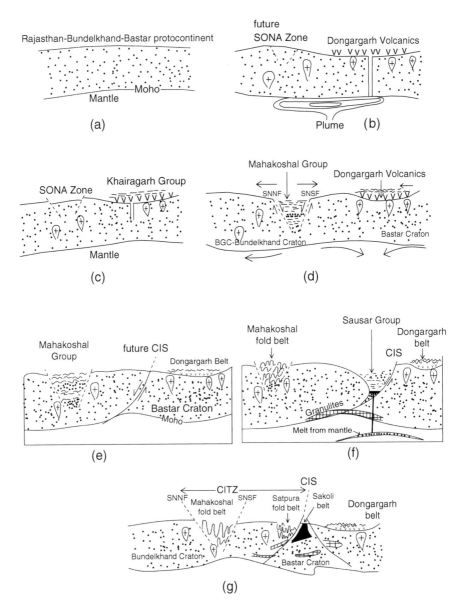

Fig. 5.4 A plausible model for the evolution of the fold belts of the central Indian region. (**a**) Existence of a large Indian protocontinent (united cratons of Rajasthan-Bundelkhand-Bastar, RCBKCBC). (**b**) Plume-generated early Proterozoic igneous activity in the Indian protocontinent, initiating rifting of the protocontinent in the western region (Bundelkhand-BGC cratonic region) and also resulting into igneous activity of the Dongargarh bimodal volcanics (2.5 Ga old) with minor ultramafics and granite intrusions in eastern part (Bastar cratonic region) of the protocontinent. (**c**) Development of the rift basin that became the SONA Zone bounded by north and south faults (called Son-Narmada North Fault, SNNF, and Son-Narmada South Fault, SNSF, respectively). The SONA Zone became the site for deposition of the Mahakoshal supracrustal rocks,

5.8 Discussion of Evolutionary Models of the Central Indian Fold Belts

is also possible that this parallelism is the result of rotation of structures of one belt into parallelism with the other during a later deformation.

In this model there is no clear relationship between the Mahakoshal belt and the Satpura belt and their relationship with the N-S trending Dongargarh belt. Therefore, this author proposes another model which is consistent with the available age of volcano-sedimentary rocks of the fold belts and which also meets the requirement of the necessary geodynamics of the region including basinal development and its deformation by converging crustal plates.

A Plausible Model: The model proposed by this writer (Fig. 5.4) is based on the documentation that there was a widespread thermal event in the Central Indian shield in the age range of 2.5–2.4 Ga, straddling the Archaean-Proterozoic boundary. This is revealed by the 2.5 Ga old granite intrusions in the Bundelkhand massif (Mondal et al., 1998, 2002) and of the same age intrusions (Untala, Gingla, Ahar River and Berach) in the Archaean Banded gneissic complex (BGC) in Rajasthan (Wiedenbeck et al., 1996). In the BGC of Rajasthan, Wiedenbeck and Goswami (1994) also report a $^{207}Pb/^{206}Pb$ age of 2.54 Ga from overgrowths on zircons. These ages conclusively suggest that the Rajasthan craton (BGC-Berach Block) and Bundelkhand craton evolved as a single crustal unit beginning at ~2.5 Ga. The age range of 2.5–2.45 indicate stabilization of craton and crust-forming event by addition of granitic material. Besides geochronology, geodynamic setting in both Rajasthan craton and Bundelkhand craton is also the same (cf. Mondal, 2003, Table 2). That is, the two cratons evolved as a single large protocontinent named here as RCBBC. It is interesting to note that this thermal event at Archaean-Proterozoic boundary is also recorded in the Dharwar craton in southern India (Peucat et al., 1993). In the Dharwar craton 2.5 Ga old granulite metamorphism accompanied by the same age anatectic granite and intrusive granitoid of Closepet are shown by robust geochronology (Friend and Nutman, 1992; Jayananda et al., 1995). Furthermore, this protracted thermal event of 2.5–2.4 Ga is also found in other cratonic areas when Proterozoic reconstruction of supercontinent is attempted (Moores, 1981; Stein et al., 2004).

Evidence indicates that the northern Indian protocontinent of the Indian shield was a large cratonic assembly of Rajasthan craton (RC, comprising Archaean BGC-

Fig. 5.4 while sinking of the softened crust of the Bastar craton developed a basin for the Khairagarh supracrustals. (**d**) Deformation of Dongargarh-Khairagarh Group rocks due to basinal subsidence and also by compression due to tensional forces associated with the development of SONA Zone. (**e**) Formation of shear, the Central Indian Shear Zone (CIS), due to ductile stretching of the region of Bundelkhand-Bastar cratons, caused by sub-lithospheric mantle activity and also by deformation (F1) of the Mahakoshal supracrustals in the SONA Zone. (**f**) Development of an ensialic basin as a result of thinning of continental crust due to shearing (CIS) and deposition of the Sausar Group supracrustals therein. Magma underplating, due to crustal delamination, is believed to have developed granulites in the lower crust of the Bundelkhand-Bastar cratonic region. (**g**) Convergence of the once rifted Bundelkhand and Bastar cratons, resulting in deformation (F2) of the Mahakoshal rocks and the Sausar Group rocks along with their basement gneisses

Berach Granite, and Bundelkhand craton, BKC) and Bastar craton, BC (Fig. 5.4a). It is believed that at about 2.5–2.4 Ga ago the united cratonic mass was subjected to plume-generated igneous activity, manifested by Dongargarh bimodal volcanics with minor ultramafics and 2.5 Ga old granite intrusions (Fig. 5.4b). In Bastar craton we have 2.5 Ga old igneous activity to the south of the Central Indian shear zone, CIS, manifested by extrusions of bimodal volcanics, Bijli rhyolite and Pitepani basalt, that are intruded by 2.5 Ga old granites of Dongargarh and Malanjkhand. The increased density of the volcanics, mainly basalt (with minor ultramafic) and bimodal volcanics, resulted into sinking of the sialic crust and consequently in the development of a basin for deposition of rock sequences of Khairagarh Group (Fig. 5.4c). Geological criteria further suggest that western part of the united craton (protocontinent) was also subjected to this large-scale thermal event, initiating rifting between the Rajasthan-Bundelkhand cratonic region (RC) and Bastar craton (Fig. 5.2b). This juvenile rift later developed into what is familiarly known the Son-Narmada Lineament Zone or SONA Zone, bounded by north and south faults, known as Son-Narmada North Fault (SNNF) and Son-Narmada South Fault (NSSF), respectively (Fig. 5.4c). The *SONA zone* became the site for deposition of the Mahakoshal Group rocks. The transtensional forces associated with the development of the SONA rift basin deformed the Khairagarh Group rocks and the underlying igneous suite of Early Palaeoproterozoic age, giving rise to the evolution of Dongargarh fold belt. In view of similar trends of the Amgaon gneiss complex and the overlying supracrustals (comprising igneous suite and the unconformably overlying Khairagarh Group), it is inferred that the Amgaon basement gneiss complex controlled the tectonic orientation of the cover rocks when both basement-cover were deformed by transtensional forces. It is for this reason that the deformed and weakly metamorphosed Dongargarh-Khairagarh Group rocks, making up the Dongargarh fold belt attained N-S trend similar to the underlying Amgaon gneissic complex (Fig. 5.4d). A shear zone seems to have developed at the interface of the cover and basement since the Dongargarh fold belt is bordered on its west by a nearly N-S striking shear zone. It should be recognized here that the development of SONA Zone generated two separate cratonic regions, Bastar craton in the south or southeast and the Rajasthan-Bundelkhand craton in the north or northwest. It is presumed that by sub-lithospheric convection currents these two cratons collided at some stage to deform and recrystallize the Mahakoshal Group rocks, giving rise to the nearly E-W trending Mahakoshal fold belt (MFB) (Fig. 5.2d). Since the MFB is without any thrust, mélange or ophiolite, it is believed that the fold belt originated in the intracratonic setting and precludes the possibility of any ancient suture or accretion prism (see Sharma, 2007).

At some stage there occurred a major shear event due to ductile stretching of the Bundelkhand-Bastar cratonic region, as a consequence of sub-lithospheric mantle convection. The ductile stretching associated with an important shear, now recognized as Central Indian Shear Zone (CIS), developed an ensialic basin. The basin formed as a result of ductile stretching and crustal thinning was floored with Archaean crust (now Tirodi gneisses) and became the depositional site for the Sausar Group of rocks (Fig. 5.4e). The sedimentary basin also received lavas whereby the volcano-sedimentary rocks of the Sausar Group were heated by the proximity of this hot mantle material, resulting into granulite formation, presumably by anatexis of the basement gneisses. The possible magma underplating caused by crustal delamination may also have developed granulites in the lower crust of the Bundelkhand-Bastar craton (Fig. 5.4f). Perhaps because of sediment load, decay of mantle "plume" below, as well as change of vertical stress components, the diverging crustal blocks/plates reversed their motion from extension to nearly N-S compression whereby the Sausar sequence (with their basement) was deformed into folds with dominantly ENE-WSW axial-plane foliation. Nearly N-S convergence of the once rifted Bundelkhand and Bastar cratons deformed the Sausar Group rocks along with their basement gneisses, giving rise to the Satpura fold belt during Mesoproterozoic time.

During collision of the Bundelkhand and Bastar cratonic blocks the crust was thickened around 1450 ± 50 Ma ago when the Sausar Group rocks were deformed and metamorphosed along with the basement gneisses, both showing high T-P metamorphism documented by the northern granulite belt, especially the RKG (cf. Bhowmik et al., 1999). Bhowmik and Spiering (2004) deduced changing P-T conditions from compositional variations in zoned garnets from the metapelites of the Sausar region. These authors deciphered four stages (first garnet growth, followed by peak metamorphism at $\sim$9.5 kbar/850°C when biotite dehydration melting occurred which succeeded by M3 stage of isothermal decompression at 6 kbar/825°C and finally having post-decompression cooling) in the protracted metamorphic history. The different stages from M1 to M4 deciphered by Bhowmik and Spiering (2004) are part of a single tectono-thermal event since these events occur in a very short span of time (see also Bhowmik and Spiering, ibid.). These authors also showed clockwise P-T path consistent with a model of crustal thickening due to continent collision, followed by rapid vertical thinning. It is presumed that gravitational instability caused the thickened crust to collapse around 1100 Ma, resulting in the emplacement of basic dykes (Srivastava et al., 1986) and granite bodies, especially the tourmaline granite (cf. Pandey et al., 1998). With the elevation of the deformed-recrystallized basinal rocks into the Satpura fold belt there seems to have occurred upthrusting of the lower crustal rocks along a NE-dipping shear, marking the termination of Satpura orogeny in the Central Indian region. Although we do not have any age data on the CIS, but its bifurcation in the deformed-recrystallized Sausar Group clearly suggests that the CIS is later than the Satpura orogeny (Mesoproterozoic) and hence younger than the SONA zone whose E-W trend is similar to the trend of the Mahakoshal rocks in the SONA zone (see below). The interesting region between the NE dipping shear and the CIS resulted in the formation of the triangular Sakoli belt in the Bastar craton (Fig. 5.4 g).

The Sakoli fold belt occupies a key position at the convergence of the DFB and SFB as it bears the impresses of both orogenic events. According to this consideration, the evolution of the triangular fold belt of Sakoli, is attributable to the genesis of the N-S shear zone and E-W shear zone, lying to the south of the CIS. According to this argument the Sakoli fold belt is as old as or younger than the Satpura fold belt. If the time equivalence of fold phases is considered wherein the F3 fold phase in the Sakoli belt is equated with F1 fold phase in the Sausar rocks (Bandyopadhyay et al., 1995) the Sakoli becomes older than the Satpura orogeny. A remarkable gravity high all over the Sakoli belt suggests that the Sakoli rift basin is underlain by mantle material. It would be an interesting study if it could be proved that the Sakoli fold belt developed in a basin which was filled-in by sediments and igneous lavas and later deformed by E-W stresses as well as N-S stresses, having the impresses of both the orogenic events, namely the Dongargarh and Satpura orogeny. It is proposed that the stresses and igneous activities reactivated the Son-Narmada lineament zone as a result of which ENE-WSW or the E-W trending shear zones developed in the western margin of the Sakoli fold belt. The E-W shear zone perhaps also affected the northern end of the Dongargarh belt.

This model proposed by the author clearly explains that the two sets of fold belts are not related temporally and each set has its origin in a particular time interval, yet to be discerned. The chronology of the four fold belts can now be placed in the order: Dongargarh fold belt, Mahakoshal fold belt, and Satpura fold belt. The Sakoli fold belt to the south of the CIS lies at the intersection of the two shear zones, the N-S shear zone abounding the Dongargarh fold belt, and the E-W shear zone that aligns with the shear zones traversing the Satpura and Mahakoshal fold belts in the north. And it is matter of enquiry if the Sakoli fold belt is a real fold belt formed by convergence of two cratonic blocks responsible for evolution of the two fold belts of the CITZ (namely the Mahakoshal and Satpura fold belts), or is simply a result of superimposition of the two stated shear zones (see above) and hence not a true fold belt.

References

Acharyya, S.K. (2001). Geodynamic setting of the Central Indian Tectonic Zone in central, eastern, and northeastern India. Geol. Surv. India Spl. Publ., vol. **64**, pp. 17–35.

Acharyya, S.K. (2003a). The nature of Mesoproterozoic Central Indian Tectonic Zone with exhumed and reworked granulites. Gondwana Res., vol. **6**, pp. 197–214.

Acharyya, S.K. (2003b). A plate tectonic model for Proterozoic crustal evolution of Central Indian Tectonic Zone. Gond. Geol. Mag. Spl., vol. **7**, pp. 9–31.

Acharyya, S.K. and Roy, A. (2000). Tectonothermal history of the Central Indian Tectonic Zone and reactivation of major faults/shear zones. J. Geol. Soc. India, vol. **55**(3), pp. 239–256.

Auden, J.B. (1949). Geological discussion of Satpura hypothesis. Nat. Inst. Sci. India, vol. **15**, pp. 315–340.

Bandyopadhyay, B.K., Bhoskar, K.G., Ramachandra, H.M., Roy, A., Khadse, V.K., Mohan, M., Barman, T., Bishui, P.K. and Gupta, S.N. (1990). Recent geochronological studies in parts of the Precambrian of central India. Geol. Surv. India Spl. Publ., vol. **28**, pp. 199–211.

Bandyopadhyay, B.K., Roy, A. and Huin, A.K. (1995). Structure and tectonics of a part of the central Indian shield. In: Sinha-Roy, S. and Gupta, K.R. (eds.), Continental Crust of Northwestern and Central India. Geol. Soc. India Mem., Bangalore, vol. **31**, pp. 433–467.

Bhargava, M. and Pal, A.B. (2000). Cu–Mo–Au metallogeny associated with Proterozoic tectonomagmatism in Malanjkhand porphyry copper district, Madhya Pradesh. J. Geol. Soc. India, vol. **56**, pp. 395–413.

Bhattacharya, A. and Bhattacharya, G. (2003). Petrotectonic study and evaluation of Bilaspur-Raigarh belt, Chhatisgarh. Gond. Geol. Mag. Spl. Publ., vol. **7**, pp. 89–100.

References

Bhowmik, S. and Pal, T. (2000). Petrotectonic implication of the granulite suite north of the Sausar mobile belt in the overall tectonothermal evolution of the Central Indian mobile belt. *GSI Progress Report* (unpublished).

Bhowmik, S.K., Pal, T., Pant, N.C. and Shome, S. (2000). Implications of Ramakona cordierite gneiss in the crustal evolution of Sausar mobile belt in central India. Proceedings Volume of the International Seminar on *Precambrian Crust in Eastern and Central India*, IGCP-368, Bhubaneswar, 1998, Geol. Surv. India Spl. Publ., vol. **57**, pp. 131–150.

Bhowmik, S.K., Pal, T., Roy, A. and Pant, N.C. (1999). Evidence for Pre-Grenville high-pressure granulite metamorphism from the northern margin of the Sausar mobile belt in central India. J. Geol. Soc. India, vol. **53**, pp. 385–399.

Bhowmik, S.K. and Roy, A. (2003). Garnetiferous metabasites from the Sausar Mobile Belt: Petrology, P-T path and implications for the tectonothermal evolution of the Central Indian Tectonic Zone. J. Petrol., vol. **44**, pp. 387–420.

Bhowmik, S.K., Sarbadhikari, A.B., Spiering, B. and Raith, M.M. (2003). Mesoproterozoic reworking of Palaeoproterozoic ultrahigh temperature granulites in the Central Indian Tectonic Zone and its implications. J. Petrol., vol. **46**, pp. 1085–1119.

Bhowmik, S.K. and Spiering, B. (2004). Constraining the prograde and retrograde P-T paths of granulites using decomposition of initially zoned garnets: an example from the Central Indian Tectonic Zone. Contrib. Mineral Petrol., vol. **147**, pp. 581–603.

Das, L.K., Agarwal, B.N.P., Das, A.K., Naskar, D.C., Roy, K.K., Chowdhury, K. and Mazumder, R.K. (2003). Evolutionary history of the central Indian shield, re-examined by an integrated geophysical study. In: Roy, A. and Mohabe, D.M. (eds.), *Advances in Precambrian of Central India*. Gond. Geol. Mag. Spl., Nagpur, vol. **7**, pp. 33–50.

Das, B. and Patel, N.P. (1984). Nature of the Narmada-Son lineament. J. Geol. Soc. India, vol. **25**, pp. 267–276.

Deshpande, G.G., Mohabey, N.K. and Deshpande, M.S. (1990). Petrography and tectonic setting of Dongargarh volcanics. Geol. Surv. India Spl. Publ., vol. **28**, pp. 260–288.

Friend, C. and Nutman, A.P. (1992). Response of U–Pb isotopes and whole rock geochemistry to CO_2 induced granulite facies metamorphism, Kabbaldurga, Karnataka, south India. Contrib. Mineral Petrol., vol. **111**, pp. 299–310.

Ghosh, P.K., Chandy, K.C., Bishui, P.K. and Prasad, R. (1986). Rb–Sr age of granitic gneiss in Malanjkhand area, Balaghat district, Madhya Pradesh. Indian Minerals, vol. **40**, pp. 1–8.

Goyal, R.S. and Jain, S.C. (1975). Geology of parts of Jungel-Bariapan area, district Mirzapur, Uttar Pradesh. Report Geol. Surv. India (unpublished).

Harris, L.B. (1993). Correlations of tectono-thermal events between the Central Indian Tectonic Zone and the Albany mobile belt of Western Australia. In: Findley, R.H., Unrug, R., Banks, M.R. and Veevers, J.J. (eds.), *Gondwana Eight: Assembly Evolution and Dispersal*. Balkema, Rotterdam, pp. 165–180.

Jain, S.C., Nair, K.K.K. and Yedekar, D.B. (1995). Geology of the Son-Narmada-Tapti lineament zone in Central India. In: *Geoscientific Studies of the Son-Narmada-Tapti Lineament Zone*. Geol. Surv. India Spl. Publ., Nagpur, vol. **10**, pp. 1–154.

Jain, S.C., Yedekar, D.B. and Nair, K.K.K. (1991). Central Indian shear zone: a major Pre-Cambrian crustal boundary. J. Geol. Soc. India, vol. **37**, pp. 521–531.

Jayananda, M., Janardhan, A.S., Sivasubramanian, P. and Peucat, J.J. (1995). Geochronological and isotopic constraints on granulite formation in the Kodaikanal area, southern India. Mem. Geol. Soc. India, vol. **34**, pp. 373–390.

Kaila, K.L. and Sain, K. (1997). Variation of crustal velocity structure in India as determined from DSS studies and their implications on regional tectonics. J. Geol. Soc. India, vol. **49**, pp. 395–407.

Krishnamurthy, P., Chaki, A., Pandey, B.K., Chimote, J.S. and Singh, S.N. (1988). Geochronology of granite-rhyolite suites of the Dongargarh Supergroup, central India. In: *Proceedings of Fourth National Symposium on Mass Spectrometry*, EPS. 2/1-EPS-2/3, Hyderabad.

Krishnamurthy, P., Sinha, D.K., Rai, A.K., Seth, D.K. and Singh, S.N. (1990). Magmatic Rocks of the Dongargarh Supergroup, central India—their petrological evolution and implications on metallogeny. Geol. Surv. India Spl. Publ., vol. **28**, pp. 309–319.

Lippolt, H.J. and Hautmann, S. (1994). Ar^{40}/Ar^{39} ages of Precambrian manganese ore mineral from Sweden, India and Morocco. Mineral. Deposita, vol. **30**, pp. 246–256.

Mazumder, R., Bose, P.K. and Sarkar, S. (2000). A commentary on the tectonosedimentary record of the pre-2.0 Ga continental growth of India vis-a-vis a possible pre-Gondwana Afro-Indian subcontinent. J. African Earth Sci., vol. **30**, pp. 201–217.

Misra, D.C., Singh, B., Tiwari, V.M., Gupta, S.B. and Rao, M.B.S.V. (2000). Two cases of continental collisions and related tectonics during the Proterozoic period in India—insight from gravity modeling constrained by seismic and magnetotelluric studies. Precambrian Res., vol. **99**, pp. 140–169.

Mondal, M.E.A. (2003). Are the Bundelkhand craton and the Banded gneissic Complex of Rajasthan parts of one large Archaean Protocontinent? Evidence from the ion microprobe $^{207}Pb/^{206}Pb$ zircon ages. D.C.S. News Letter, DST New Delhi, vol. **13**(2), pp. 6–10.

Mondal, M.E.A., Goswami, J.N., Deomurari, M.P. and Sharma, K.K. (2002). Ion microprobe $^{207}Pb/^{206}Pb$ ages of zircons from the Bundelkhand massif, northern India: implications for crustal evolution of the Bundelkhand-Aravalli protocontinent. Precambrian Res., vol. **117**, pp. 85–100.

Mondal, M.E.A., Sharma, K.K., Rahman, A. and Goswami, J.N (1998). Ion microprobe $^{207}Pb/^{206}Pb$ ages for gneiss-granitoid rocks from Bundelkhand massif: evidence for Archaean components. Curr. Sci., vol. **74**, pp. 70–75.

Moores, E.M. (1981). Ancient suture zones within continents. Science, vol. **213**, pp. 41–46.

Nair, K.K.K., Jain, S.C. and Yedekar, D.B. (1995). Stratigraphy, structure and geochemistry of the Mahakoshal greenstone belt. Geol. Soc. India Mem., vol. **31**, pp. 403–432.

Naqvi, S.M. and Rogers, J.J.W. (1987). *Precambrian Geology of India*. Oxford University Press, Oxford, 233p.

Narayanaswami, S., Chakraborty, S.C., Vemban, N.A., Shukla, K.D., Subramanyam, M.R., Venkatesh, V., Rao, G.V., Anandalwar, M.A. and Nagrajaiah, R.A. (1963). The geology and manganese ore deposits of the manganese belts in Madhya Pradesh and adjoining parts of Maharashtra. Bull. Geol. Surv. India Ser. A, vol. **22**(1), 69p.

Pandey, B.K., Krishna, V. and Chabria, T. (1998). An overview of the geochronological data on the rocks of Chhotanagpur gneiss-granulite complex and adjoining sedimentation sequences, eastern and central India. Proceedings of International Seminar on *Crust in Eastern and Central India* (UNESCO-IUGS-IGCP 368). Geol. Surv. India, Bhubaneswar, India. Abstract, pp. 131–135.

Panigrahi, M.K., Misra, K.C., Brteam, B., and Naik, R.K. (2002). Genesis of the granitoid affiliated copper-molybdenum mineralization at Malanjkhand, central India: facts and problems. Extended Abstract. In: *Proceedings of 11th Quadrennial IAGOD symposium and Geocongress*, Windhoek, Namibia.

Pascoe, E.H. (1973). A manual of geology of India and Burma. Geol. Surv. India, Calcutta, vol. **1**, 485p.

Peucat, J.J., Mahabaleswar, R. and Jayananda, M. (1993). Age of younger tonalitic magmatism and granulite metamorphism in the southern Indian Transition Zone (Krishnagiri area): comparison with older peninsular gneisses from Gorur-Hasan area. J. Metam. Geol., vol. **11**, pp. 879–888.

Pichai Muthu, R. (1990). The occurrence of gabbroic anorthosites in Makrohar area, Sidhi district, Madhya Pradesh. In: *Precambrian of Central India*. Geol. Surv. India Spl. Publ., Nagpur, vol. **28**, pp. 320–331.

Radhakrishna, B.P. (1989). Suspect tectono-stratigraphic terrane elements in the India subcontinent. J. Geol. Soc. India, vol. **34**, pp. 1–24.

Ramachandra, H.M. (1999). Petrology of Bhandara and Balaghat granulite belt in parts of Maharashtra and Madhya Pradesh. Report Geol. Surv. India, vol. **56**, pp. 1–55.

Ramachandra, H.M. and Roy, A. (1998). Geology of intrusive granitoids with special reference to Dongargarh granite and their impact on tectonic evolution of the Precambrian in central India. Indian Minerals., vol. **52**, pp. 15–32.

Ramachandra, H.M. and Roy, A. (2001). Evolution of the Bhandara-Balaghat granulite belt along the southern margin of the Sausar mobile belt of Central India. Proc. Indian Acad. Sci., (Earth Planet. Sci.), vol. **110**, pp. 351–368.

Ramakrishnan, M. and Vaidyanadhan, R. (2008). *Geology of India*, vol. **1**. Geol. Soc. India, Bangalore, 556p.

Rao, M.S.V. and Sri Rama, B.V. (1993). Final report on studies in phase-II of special project CRUMANSONATA. *Geol. Surv. India* (unpublished).

Reddy, P.R. and Rao, I.B.P. (2003). Deep seismic studies in central Indian shield—a review. Geol. Soc. India, Mem., vol. **53**, pp. 79–98.

Roy, A., Bandyopadhyay, B.K. and Huin, A.K. (1997). Geology and geochemistry of basic volcanics from the Sakoli schist belt of central India. J. Geol. Soc. India, vol. **50**, pp. 209–221.

Roy, A. and Devarajan, M.K. (2000). A reappraisal of the stratigraphy and tectonics of the Palaeo-Proterozoic Mahakoshal supracrustal belt, central India. Geol. Surv. India Spl. Publ., vol. **57**, pp. 79–97.

Roy, A., Devarajan, M.K. and Hanuman Prashad, M. (2002). Ductile shearing and syntectonic granite emplacement along the southern margin of the Palaeoproterozoic Mahakoshal supracrustal belt: evidence from Singrauli area, Madhya Pradesh. J. Geol. Soc. India, vol. **59**, pp. 9–21.

Roy, A. and Hanuman Prasad, M. (2003). Tectonothermal events in Central Indian Tectonic Zone (CITZ) and its implications in Rodinia crustal assembly. J. Asian Earth Sci., vol. **22**, pp. 115–129.

Roy, A., Hanuman Prashad, M. and Bhowmik, S.K. (2001). Recognition of pre-Grenville and Grenville tectonothermal events in the Central Indian tectonic zone: implications on Rodinia crustal assembly. Gondwana Res., vol. **4**, pp. 755–757.

Roy, A., Kagami, H., Yoshida, M., Roy, A., Bandyopadhyay, A., Chattopadhyay, A., Khan, A.S., Huin, A.K. and Pal, T. (2006). Rb–Sr and Sm–Nd dating of different metamorphic events from the Sausar mobile belt, central India: implications for Proterozoic crustal evolution. J. Asian Earth Sci., vol. **26**, pp. 61–76.

Sarkar, S.N. (1957–1958). Stratigraphy and tectonics of the Dongargarh system, a new system in the Precambrians of Bhandara-Durg-Balaghat area, Bombay and Madhya Pradesh. J. Sci. Engg. Res., IIT Kharagpur, India, vol. **1**, pp. 237–268, and vol. **2**, pp. 145–160.

Sarkar, S.N. (1994). Chronostratigraphy and tectonics of the Dongargarh Supergroup Precambrian rocks in Bhandara-Durg region, central India. Indian J. Earth Sci., vol. **21**, pp. 19–31.

Sarkar, A., Bodas, M.S., Kindu, H.K., Mamgain, V.V. and Ravishankar. (1998). Geochronology and geochemistry of Mesoproterozoic intrusive plutonites from the eastern segment of the Mahakoshal greenstone belt, central India. Abstract in IGCP-368 Seminar on *Precambrian Crust in Eastern and Central India*, Bhubaneswar, pp. 82–86.

Sarkar, S.N., Gopalan, K. and Trivedi, J.R. (1981). New data on the geochronology of the Precambrians of Bastar-Drug, central India. Indian J. Earth Sci., vol. **8**(2), pp. 131–151.

Sarkar, A., Sarkar, G., Pal, D.K. and Mitra, N.D. (1990). Precambrian geochronology of the central Indian shield—a review. Geol. Surv. India Spl. Publ., vol. **28**, pp. 453–482.

Sarkar, S.N., Sarkar, S.S. and Ray, S.L. (1994). Geochemistry and genesis of the Dongargarh Supergroup Precambrians of Bhandara-Durg, central India. Indian J. Earth Sci., vol. **21**, pp. 117–126.

Sarkar, S.N., Trivedi, J.R. and Gopalan, K. (1986). Rb–Sr whole rock and mineral isochron age of the Tirodi gneiss, Sausar Group, Bhandara district, central India. J. Geol. Soc. India, vol. **27**, pp. 30–37.

Sensarma, S. (2007). A bimodal large igneous province and the plume debate. In: Foulagar, G.R. and Jurdy, D.M. (eds.), *Plates, Plumes, and Planetary Processes*. Geol. Soc. Am. Spl. Publ., vol. **430**, pp. 831–840.

Sensarma, S., Hoernes, S. and Mukhopadhyay, D. (2004). Relative contributions of crust and mantle to the origin of the Bijli rhyolite in a Palaeoproterozoic bimodal volcanic sequence (Dongargarh Group), central India. Proc. Indian Acad. Sci. (Earth Planet. Sci.), vol. **113**(4), pp. 619–648.

Sensarma, S. and Mukhopadhyay, D. (2003). New insights on stratigraphy and volcanic history of Dongargarh belt, central India. Gond. Geol. Mag., vol. **7**, pp. 129–136.

Shanker, R. (1991). Thermal and crustal structure of "SONATA", a zone of mild continental rifting in Indian shield. J. Geol. Soc. India, vol. **37**, pp. 211–220.

Sharma, R.S. (2007). Evolution of the central Indian fold belts: A geodynamic model. In: C. Leelanandam, I.B. Rama Prasad Rao, Ch. Sivaji and M. Santosh (Eds.), The Indian Continental Crust and upper Mantle, Publ. International Association of Gondwana Res. Mem. **10**, pp. 41–45.

Solanki, J.N., Sen, B., Soni, M.K., Tomar, N.S. and Pant, N.C. (2003). Granulites from southeast of Waidhan, Sidhi district, Madhya Pradesh in NW extension of Chhotanagpur Gneissic Complex: Petrography and geothermobarometric Estimation. Gond. Geol. Mag., vol. **7**, pp. 297–311.

Srivastava, R.K., Hall, R.P., Verma, R. and Singh, R.K. (1996). Contrasting Precambrian mafic dykes of the Bastar craton, central India: Petrological and geochemical characteristics. J. Geol. Soc. India, vol. **48**, pp. 537–546.

Stein, H.J., Hannah, J.L., Zimmerman, A., Markey, R.J., Sarkar, S.C. and Pal, A.B. (2004). A 2.5 Ga porphyry Cu–Mo–Au deposit at Malanjkhand, central India: implications for Late Archaean continental assembly. Precambrian Res., vol. **134**, pp. 189–226.

Subba Rao, M.V., Narayana, B.L., Divakar Rao, V. and Reddy, G.L.N. (1999). Petrogenesis of the protolith for the Tirodi gneiss by A-type granite magmatism: geochemical evidence. Curr. Sci., vol. **76**, pp. 1258–1264.

West, W.D. (1962). The line of Narmada-Son valley. Curr. Sci., vol. **31**, pp. 143–144.

Wiedenbeck, M. and Goswami, J.N. (1994). An ion-probe single zircon ^{207}Pb/^{206}Pb age from the Mewar gneiss at Jhamakotra, Rajasthan. Geochim. Cosmochim. Acta, vol. **58**, pp. 2135–2141.

Wiedenbeck, M. and Goswami, J.N., Roy, A.B. (1996). Stabilization of the Aravalli Craton of northwestern India at 2.5 Ga: an ion microprobe zircon study. Chem. Geol., vol. **129**, pp. 325–340.

Yedekar, D.B., Jain, S.C., Nair, K.K.K. and Dutta, K.K. (1990). The central Indian collision suture. In: *Precambrian of Central India*. Geol. Surv. India Spl. Publ., Nagpur, vol. **28**, pp. 1–43.

Yoshida, M. (1995). Assembly of East Gondwanaland during the Mesoproterozoic and its rejuvenation during the Pan-African period. In: Yoshida, M. and Santosh, M. (eds.), *India and Antarctica During the Precambrian*. Geol. Soc. India, Mem., Bangalore, vol. **34**, pp. 25–45.

Zhao, G., Sun, M. and Wilde, S.A. (2003). Correlation between the eastern block of the North China craton and the South Indian block of the Indian shield: an Archaean to Palaeoproterozoic link. Precambrian Res., vol. **122**, pp. 201–233.

Chapter 6
Singhbhum Fold Belt

6.1 Introduction: Geological Setting

The Singhbhum fold belt (SFB) or Singhbhum Mobile belt, also called North Singhbhum belt (Saha, 1994) in eastern India is sandwiched between the Singhbhum craton in the south and Chhotanagpur Granite-Gneiss terrain (CGGC) in the north (Fig. 6.1). Although slightly curvilinear, the fold belt has nearly E-W strike like that of the Satpura fold belt (Chap. 5) and is believed to have evolved in the Proterozoic Satpura orogeny (Sarkar and Saha, 1962). The Singhbhum fold belt is made up of metasedimentary and metavolcanic rocks categorized in three domains. Gupta and Basu (2000) of Geological Survey of India gave the following lithostratigraphy of the Singhbhum fold belt.

    Dalma Group

    Singhbhum   {Dhalbhum Formation
      Group   {Chaibasa Formation

    Dhanjori Group

- - - - - - - - - - - - -unconformity- - - - - - - - - - - -

    Singhbhum Granite basement

The *Singhbhum Group* which consists of pelitic and semipelitic schists, quartzites, amphibolites, felsic and intermediate tuffs recognized near the craton as *sericite schists and phyllite in upper part* (Dhalbhum Formation, DH) *and mica schists, quartzites, amphibolites in lower part* (Chaibasa Formation, CB), occurring north of the Dhalbhum Formation (Fig. 6.1). The Singhbhum Group unconformably overlies the Singhbhum Granite phase III (~3.12 Ga old) and its equivalent Bonai Granite and Chakraddharpur Granite Gneiss (CH) (Bandyopadhyay and Sengupta, 1984; Saha, 1994; Sengupta et al., 1991). Second is the *Dalma Group or Dalma volcanics* comprising ultramafic lavas and high Mg-basalts overlain by low-K pillowed tholeiites in the middle part. Third group is a low-grade metamorphosed volcano-sedimentary assemblage further north, contacting the CGGC along a shear zone, called the Northern Shear zone. Another shear zone occurs in the southern boundary

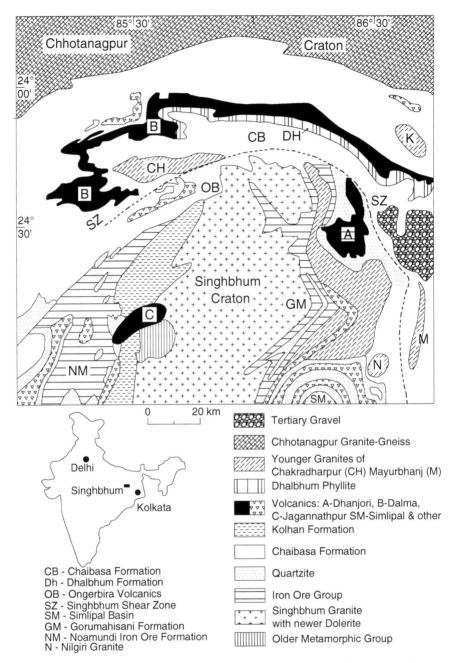

Fig. 6.1 Location of Singhbhum fold belt between Singhbhum craton and Chhotanagpur Granite Gneiss Complex (Chhotanagpur craton) (after A.K. Saha, 1994). (A) = Dhanjori volcanics, (B) = Dalma volcanics, (C) = Jaganathpur volcanics, CB = Chaibasa formation, CH = Chakradharpur granite, DH = Dalma formation, GM = Gorumahisani formation, K = Kuilapal Granite, N = Nilgiri Granite, NM = Noamandi Iron-Ore formation, M = Mayurbhanj Granite, OB = Ongerbira volcanics, SM = Simlipal basin, SZ = Singhbhum shear zone

6.1 Introduction: Geological Setting

of the Singhbhum fold belt against its contact with the Singhbum Granite of the Singhbhum craton in the south and is familiarly known as the Singhbhum shear zone, SZ (see Fig. 6.1). The fold belt, like the Singhbhum craton, is most intensely studied region of the Precambrian terrains of India (Saha, 1994).

The Chaibasa Fm of domain (1) is reported to occur in a geoanticline (Dunn, 1929; Dunn and Dey, 1942), later termed Singhbhum anticlinorium by Sarkar and Saha (1962). This anticlinorium is bound on the north by Dalma syncline of domain (2) and by another anticlinorium further north, corresponding to domain (3) (Fig. 6.1). The Chaibasa Formation of domain (1), encompassing the high-grade metamorphic rocks, has a confusing history in regard to its geologic position. Dunn and Dey (1942) placed the Chaibasa (stage) in the lower part of the Iron-Ore Series (Chap. 2), while Sarkar and Saha (1962) placed them in the Singhbhum Group. Later, Iyengar and Murthy (1982) re-named the Chaibasa stage as Ghatsila Formation and placed it at the base of the Dhanjori sequence (equivalent to Dalma volcanics). Recently, Srivastava and Pradhan (1995) found a thrust relationship between them due to which the Chaibasa was thrust upon the Dhanjori. It may be noted that the southern limit of the Chaibasa Formation is the Copper Belt Thrust or the Singhbhum shear zone (SZ). The Chaibasa Stage/Formation was also traced by Dunn and Dey (1942) westwards in the Rajkharswan-Chakradharpur area where the Iron Ore Group (IOG), owing to its westward plunge, successively overlies them. Northward, the high-grade rocks of Chaibasa gradually merge into an overlying sequence of low-grade rocks (phyllite, chlorite schist, carbon phyllite and orthoquartzite) that occur in a belt south of the Dalma volcanics. The low-grade rock-suite was grouped into the Iron-Ore Stage by Dunn and Dey (1942) and into Dhalbhum Formation by Sarkar and Saha (1962) who placed this formation in the Singhbhum Group.

The domain (2) is made of the *Dalma volcanics* which have been called Proterozoic greenstone assemblage by Gupta et al. (1982) who recognized them as low-K, high Fe-tholeiites with pillowed structures; and also minor high-Mg volcanics as flows and komatiitic intrusives. Associated with these volcanics are high-Mg volcaniclastics, tuff and carbon phyllites. The Dalma volcano-sediments have been metamorphosed under the greenschist facies.

To the north of the Dalma volcanics the domain (3) is an assemblage of low—grade metamorphic association of volcanics (acid and basic/ultrabasic), volcaniclastics and metasediments, collectively named *Chandil Formation* by Ray et al. (1996). This assemblage, according to Ray et al. (loc. cit), is distinct from the dominantly metasedimentary assemblage exposed to the south of the Dalma volcanic belt (see Acharyya, 2003). The contact between the Chandil Formation with the CGGC on the north is marked by a shear zone and by carbonatite intrusions (Gupta and Basu, 2000). The acid tuff of the Chandil Formation from Ankro yielded Rb–Sr whole-rock isochron age of 1487 ± 34 Ma (Sengupta and Mukhopadhyay, 2000), indicating a thermal event in the Singhbhum orogeny. The Chandil supracrustals are intruded by a number of oval granite bodies. Amongst them, the Kuilapal Granite (see K in Fig. 6.1) yielded Rb–Sr whole-rock isochron age of 1638 ± 38 Ma (Sengupta et al., 1994 in Acharyya, 2003) that fixes the upper age limit of the Chandil Formation.

The age of the Dalma volcanics is not satisfactorily constrained. A gabbro-pyroxenite intrusion into the Dalma volcanics from Kuchia, located in the eastern part of the Dalma belt, has yielded 1619 ± 38 Ma age by Rb–Sr whole-rock isochron method (Ray, 1990; Roy et al., 2002b). Trace element, REE and Sr and Nd isotopic studies on these intrusives indicate highly depleted nature of the parent mantle (Acharyya, 2003). Bose (2000) also inferred depleted nature of the mantle beneath the Dalma volcanics. In view of the very similar character of trace elements and REE, both gabbro-pyroxenite intrusive and the Dalma volcanics are inferred to have been derived from the same depleted mantle source (see Acharyya, 2003). Thus, we can conclude that both the Chandil Formation and the Dalma volcanics were metamorphosed together and were intruded by granite and coeval gabbro-pyroxenite at about 1.6 Ga ago. The age of the extrusion is, however, not yet definitely known but may not be much different (cf. Acharyya, 2003).

It may not be out of place to state here that south of the Singhbhum shear zone there are a number of volcano-sedimentary basins, especially the Dhanjori (A), Simlipal (SM), Jagannathpur (C) and Malangtoli basins, which are circum-cratonic (Fig. 6.1). These basins with prominent tholeiitic magmatism are believed to have generated along circum-cratonic peripheral rifts around the Singhbhum craton (or ACCR of Mahadevan, 2002) in the Early Proterozoic or Late Archaean. The Simlipal volcanics are intruded by Simlipal granite of 2.2 Ga old (cf. Mahadevan, 2002). This means that the Simlipal basin and Dhanjori basin (with 2.2 Ga old Mayurbhanj Granite intrusion in the Dhanjori sequence; Gupta et al., 1982) are older than 2.2 Ga while the Dalma basin may be older than 1.6 Ga. According to Sarkar et al. (1992) the Dalma volcanism was the culmination of the mafic volcanism that manifested in Dhanjori, Jagnnathpur, Simlipal basins, resulting due to crustal thinning and mantle upwelling (Iyengar and Anand Alwar 1965).

To the south or southwest of Chakradharpur granite there is another volcanic sequence, called *Ongerbira volcanics* (OB), which resembles the assemblage of the Dalma rocks and the Chandil volcanics. The Ongerbira volcanics is a sequence of ash beds, tuffaceous metasediments and low-grade metapelites resting on mafic/ultramafic rocks. Like the Dalma syncline, the Ongerbira volcanics occur in an E-W trending syncline. Earlier the Ongerbira volcanics were correlated with Newer Dolerite (Neoproterozoic) by Dunn (1929) who later suggested them to be a continuation of the Dalma volcanics. Sarkar and Saha (1962, 1977), on the other hand, considered the Ongerbira volcanics as part of the Iron Ore Group (IOG) and hence within the Singhbhum craton, separated from the mobile belt by the Singhbhum shear zone (see also Gupta et al., 1981; Gupta and Basu, 2000). Subsequently, many other workers, chiefly Sarkar (1982), Sarkar and Chakraborti (1982) and Mukhopadhyay et al. (1990) considered these volcanics as the extension of the disrupted and folded sequence of the Dalma. This appears to be quite plausible if one observes the synclinal nature and similarity of rocks of the Ongerbira with the Dalma syncline. The Ongerbira volcanics could be a detached portion of the refolded Dalma rocks since their continuity with the Ongerbira volcanics is disrupted by the Chakradharpur Granite Gneiss and the shear zone, that affected these rocks in its vicinity. Although the Ongerbira syncline is disrupted from the Dalma volcanics, there is no evidence favouring an extension of the shear zone to separate these two volcanic domains. The Ongerbira magmatic suite with E-W trending folds belong to the Singhbhum mobile belt and appears tectonically juxtaposed against the NNE-striking IOG rocks (Noamandi-gua) belonging to the Singhbhum craton. Perhaps due to the presence of the Singhbhum shear zone (Copper Belt Thrust), Ray (1990) inferred the Ongerbira volcanics to occur as klippe of the overthrusted Dalma volcanics overriding the rocks of southern Singhbhum belt and the IOG in the Singhbhum craton. Recently, the GSI disclosed that both northern and southern boundaries of the Ongerbira volcanics are faulted (see Mazumdar, 1996). That the Ongerbira volcanics are detached portion of the Dalma volcanics is supported by the comparable major and trace element data from Ongerbira volcanics and Dalma basalt (Blackburn and Srivastava, 1994). The chemical data on the Ongerbira basalts are interpreted for continental rift tholeiites with oceanic affinity or back-arc rift setting (Blackburn and Srivastava, 1994), which is not much different from the interpretation of Sarkar (1982) for ensimatic mid-oceanic rift. The geochemical data of the Dalma volcanics also indicate back-arc setting. The Ongerbira volcanics are distinct from the primitive volcanics from the IOG (Sengupta et al., 1997; Bose, 2000). The asymmetric distribution of the Dalma volcanics and the development of acid-dominant Chandil belt exposed to the north of the Dalma belt, and the absence of 1.6–1.5 Ga old charnockites in the southern part of the CGGC belt flanking the Dalma belt, do not support the postulation by A. Roy et al. (2002a) that the Dalma volcanics are generated by Mid-Proterozoic plume-related thermal event.

6.2 Deformation

The rocks of the Singhbhum fold belt show three major fold phases each characterized by their linear and planar structures. There is a progressive change in the geometry and morphology of these structural elements across the fold belt, from the fold belt to the thrust or shear zone. This means that that the deformation changed in style in different domains of the fold belt. The first generation planar structure is the metamorphic banding formed by F1 folding in the recrystallized rocks of the Singhbhum Group. In the lower formation (Chaibasa Formation) of the Singhbhum Group, the F1 folds are few and small which are characterized by reclined geometry, found at places as rootless hinges with mineral lineation L1 due to intersection of S_0/S_1 (see Acharyya, 2003). The second fold phase (F2), generally coaxial with F1, gave rise to E-W regional folds with a strong axial plane foliation (S2) that is recognized as regional foliation in the terrain. In the eastern sector (e.g. in Ghatsila) of the SFB, the F2 folds are represented by the antiforms and synforms in pelitic schists (Mukhopadhyay, 1988; Gupta and Basu, 2000). These structures were earlier identified as F1 folds by Sarkar and Saha (1962) and also by Naha (1965). According to Mukhopadhyay et al. (1990) a secondary foliation has developed at low angles to the bedding, defining the blunt-hinged synformal closure (as at Ghatsila) and puckered nature of the S_0. The large-scale folds in the bedding schistosity are considered the outcome of F2 in the Galudih area near Ghatsila (Mukhopadhyay et al., 1990). These authors also described steeply plunging U-shaped F2 folds with closures facing in opposite directions to form steeply plunging sheath folds with acute hairpin curvature of fold axis.

The F2 folds in the northern belt are asymmetric and indicate that rocks in the N have moved upwards relative to the rocks on the south (Sengupta and Chattopadhyay, 2004). In the southern part the F2 folds are upright in nature with regional foliation maintaining a vertical attitude. A little to the south, the folds maintain the same sense of asymmetry but the axial plane dips in a northerly direction. According to Sengupta and Chattopadhyay (2004), the variation in the plunge of these F2 folds in the southern sector is pronounced in contrast to the consistent gentle plunge of folds in the north. Interestingly, Dunn and Dey (1942) recognized the Singhbhum anticlinorium as a refolded fold whose asymmetric digitations shown by the outcrops of the Dalma volcanics at its closure region are the second-generation folds (F2). The F2 folds are upright, asymmetric, and in the middle part of the Singhbhum fold belt they show northerly-dipping axial planes (S2). The F2 folds become progressively overturned to the south as the Singhbhum shear zone is approached. Here the schistosity is more intensely developed, its dips moderate towards N and the fold axes are conspicuously parallel to the down-dip stretching lineation. All shear sense indicators in the zone suggest upward transport of the northern block relative to that occurring towards south.

Thus it can be inferred, following Sengupta and Chattopadhyay (2004), that the second deformation (F2) across both fold belt and the shear zone has been contemporaneous and non-co-axial. Consequently, the folds changed from being upright in the north to overturn in the south. Plunges of the fold also show change from subhorizontal to steep in the same direction, seen from N to S. Sengupta and Chattopadhyay (2004) interpret this variation in the fold geometry to be the result of progressive non-axial deformation in the ductile shear zone.

Third generation folds (F3) occur mostly within the Singhbhum shear zone. They are superposed on F1 and F2 fold structures (Bhattacharya and Sanyal, 1988). The F3 axes have variable plunges towards NNE and the F3 folds gradually die out northward and southward.

The F4 folds appear as macroscopic folds in the SE part of the Singhbhum fold belt, e.g. near Hathimara. These folds are open and upright, developed on S3 and earlier S-planes. Axial planes are often marked by a fracture cleavage (S4) that has NNE strike and subvertical dips towards E or W. The fold axes plunge at low angles towards NNE. F4 folds diminish toward N and W into minute crinkles on S4 (Bhattacharya and Sanyal, 1988).

6.3 Metamorphism

Compared to structural studies, metamorphic investigation of the rocks of the Singhbhum belt is limited. The highest grade rocks are the Chaibasa Fm that show amphibolite facies assemblages and are exposed all along the middle part of the anticlinorium. The association of the highest grade with the central part of the anticlinorium is most striking but not correlatable with the granitic intrusions in the region. From this high-grade central part of the anticlinorium the metamorphic grade decreases northward to greenschist facies (see review in Gupta and Basu, 2000). High-grade rocks of amphibolite facies are also found in the Kuilapal migmatite complex of domain (2). Here as well as in Sini area the metamorphism is Barrovian type (Lal et al., 1987). The progressive regional metamorphic sequence from chlorite, biotite, garnet, staurolite, and kyanite zones are clearly seen in eastern Singhbhum (Ghatsila) (Naha, 1965). A similar sequence is also found in central Singhbhum (Gamaria) (Roy, 1966). In all these areas, the higher-grade rocks occur against the direction of axial plunge and the regional metamorphism is found to be syn- to post-kinematic with respect to F1 and pre-kinematic with respect to F3 (Bhattacharya and Sanyal, 1988). Migmatization is slightly later and the associated rocks are found to contain sillimanite in suitable compositions. Although evidence of partial melting of lower continental crust is seen in the occurrence of the Arkasani granophyre and associated Soda granites, the Singhbhum fold belt, like the Singhbhum craton, is devoid of granulitic rocks. The granulites are conspicuously present as boudins in the CGGC craton occurring in the north of the Singhbhum fold belt.

6.4 Geochronolgy

The Singhbhum (-Orissa) craton in the eastern Indian shield is found to attain stability at around 3.1 Ga, which is the age of the Singhbhum Granite phase III. No such dates are, however, recorded in the Chhotanagpur Gneissic Complex (CGGC), although > 2.3 Ga old gneissic rocks of the complex contain metasedimentary enclaves, presumably Archaean in age (Saha et al., 1988). Like other Archaean gneissic complexes, the CGGC is also a polymetamorphic gneissic complex, but without record of Archaean history, perhaps due to resetting of ages in the constituent rocks. It, therefore, seems that the CGGC was once united with the Singhbhum (-Orissa) craton and this united cratonic block (here called SC-CGGC) subsequently rifted (Sarkar and Saha, 1977, 1983; Bose, 1990) to give rise to a basin for the deposition of the volcano-sedimentary rocks of the Singhbhum fold belt. No definite date is available for the time of rifting and the lower age of these supracrustal rocks cannot be constrained. As stated earlier, the supracrustal volcano-sedimentary succession, collectively called the Singhbhum Group (Saha, 1994; Gupta and Basu, 2000) unconformably overlies the 3.12 Ga old Singhbhum Granite phase III and its equivalent Bonai Granite and Chakradharpur Granite gneiss (see Saha, 1994). This means the rifting could be Mesoarchaean or much later after the cratonization.

The supracrustal sequences of the Singhbhum Group are seen intruded by granites. Amongst these is the undeformed Mayurbhanj Granite (Sarkar and Saha, 1977), having U—Pb zircon age of 3.09 Ga (Misra, 2006). Since this Pb—Pb zircon age of the Mayurbahnj Granite (see locality M in Fig. 6.1) is very close to the age of the Singhbhum Granite Phase III, it is highly probable that the analyzed zircons in the sample (BG-64; Misra, 2006, p. 367) have ^{206}Pb proportion comparable to that of the Singhbhum Granite. This becomes significant because the Mayurbhanj granite is considered as an anatectic product of the SBG-III (Misra, 2006, p. 367). Furthermore, this granite is found to have three phases of intrusions (Naha, 1965; Sarkar et al., 1979), out of which only the third phase postdated the Singhbhum Shear Zone (SSZ). It is in this context that the whole-rock Rb–Sr age between 2.37 and 2.08 Ga for the MBG (Vohra et al., 1991; Iyengar et al., 1981) is relevant to constrain the upper age limit of the Singhbhum Group rocks.

The Singhbhum Group was followed by a major mafic volcanism known as the Dalma Volcanics (Bose, 1994; Gupta and Basu, 2000; Saha, 1994). Field study reveals that the sequence starts with shale-phyllite, carbon phyllite-tuff with interlayered volcanics followed upward by ultramafic volcaniclastics and komatiitic intrusives. The ultramafic horizon in the lower part is followed by tholeiitic lava flows, separated from each other by pyroclastic rocks. The Dalma volcanics which overlie the Singhbhum Group yielded whole-rock Rb–Sr age of $\sim$ 2.5–2.4 Ga (Misra and Johnson, 2005). This means that the temporal equivalency of the Dalma Volcanics with the 2.8 Ga old Dhanjori volcanics should be seen in terms of tectono-thermal history of the Singhbhum fold belt. Dalma volcanics are deformed and weakly metamorphosed.

Deformational study shows that both supracrustal sequences of Singhbhum Group and Dalma Volcanics were subjected to superposed folding attended with regional metamorphism (syn- to post F1). The Singhbhum fold belt records tectonomagmatic activity at 2.2, 1.6, and 1.0 Ga. The 2.2 Ga event within the fold belt is documented by the emplacement of 2.2 Ga old Soda granite while 1.6 Ga old event is recorded by: (i) intrusion of 1.6 Ga old Kuilapal Granite within the fold belt (Sengupta et al., 1994, whole-rock Rb–Sr age), and (ii) copper mineralization at 1.65–1.7 Ga (Johnson et al., 1993, whole-rock Pb–Pb age). Furthermore, an isolated gabbro-pyroxenite body intruding the Singhbhum Group rocks just north of the Dalma Volcanics yielded whole-rock Rb–Sr age of 1.6 Ga (Roy et al., 2002b).

These afore-said geochronological data lead us to conclude that the Singhbhum orogeny began with rifting somewhat earlier in Mesoproterozoic, perhaps in Palaeoproterozoic, and the sedimentation-volcanism in the rifted basin terminated at about 2.2 Ga. The main orogenic event involving deformation, metamorphism attended by granite intrusion evidently occurred in the time span of 2.2–1.6 Ga, as documented by granite intrusions (Soda Granite, Kuilapal Granite) and 1.54 Ga K–Ar ages for hornblende from amphibolite dykes (Sarkar et al., 1979). The K–Ar ages between 1.18 and 0.84 Ga for muscovites and biotites from schists perhaps denote end stage of regional metamorphism, with or without shearing. The age of 1.0 Ga is also recorded by the shearing of Arkasani Granite along Singhbhum shear zone (Sengupta et al., 1994, whole-rock Rb–Sr age).

When the Singhbhum fold belt experienced Proterozoic deformation and metamorphism, the Singhbhum craton recorded mafic magmatism by way of emplacement of Newer Dolerites, which have yielded K–Ar dates at 2.0, 1.6, and 1.12–1.0 Ga (Mallik and Sarkar, 1994; Sarkar et al., 1979).

6.5 Agreed Observations and Facts

Before we discuss these evolutionary models we take an inventory of the relevant facts and agreed observations about the constituent rocks of the Singhbhum fold belt. They are stated as follows:

1. The Singhbhum fold belt with E-W trend is made up of metasedimentary and metavolcanic rocks.
2. The constituent rocks are dominantly arenaceous-argillaceous (Singhbhum Group) in the south, the Dalma volcanics in the middle and a suite of volcano-sedimentary rocks in the north.
3. The Singhbhum fold belt is located between the Singhbhum craton (SC) in the south and the Chhotanagpur Granite Gneiss Complex (CGGC) in the north.
4. The evolution of the SMB started with rifting of the once united SC-CGGC Archaean block before the 1.7 Ga, the age of the sulphide mineralization. The rocks suffered the last thermal overprinting or cooling at about 900 Ma ago which is the closing time of the Satpura orogeny (Bhattacharya and Sanyal, 1988).
5. The volcano-sedimentary rocks of the rift basin have been deformed with earliest episode, producing E-W trending folds which have been superposed by folds with NE-SW to NNE-SSW striking axial planes.
6. The age of metamorphism coeval with F2 is dated at 1700 Ma (Sarkar et al., 1979); metamorphic isograds generally show an accordant relation with structural trends and stratigraphic boundaries.
7. Post-dating these folds is a 200 km long and up to 25 km wide arcuate Singhbhum Shear Zone (SSZ) or Copper Belt Thrust, between the Singhbhum belt in the north and the Singhbhum-Iron Ore cratonic Province in the south. The SSZ also shows localized folding.
8. The uranium ores in SSZ are dated at about 1600 Ma (Rao et al., 1979; Pb–Pb uraninite age).
9. The Singhbhum shear zone is developed near the southern boundary of the North Singhbhum fold belt.
10. The Dalma volcanics "spine" subdivides the Singhbhum fold belt into two segments.
11. The granite rocks in the fold belt have been dated between 2080 Ma (age of Mayurbhanj Granite) and 1700 Ma (age of Soda Granite) (Iyengar et al., 1981; Sarkar et al., 1985).

With these relevant informations we can now discuss the different evolutionary models for the Singhbhum fold belt (SMB).

6.6 Evolution of the Singhbhum Fold Belt (SMB)

Several tectonic models have been proposed for the evolution of the Singhbhum mobile belt. But preference of one over the other seems difficult because sedimentation, volcanism, or tectonism in the fold belt are not well constrained, geochronologically. However, nearly all models believe that the Singhbhum fold belt formed in a rift basin generated during Palaeoproterozoic time. In this basin we have a thick sequence of arenaceous and argillaceous rocks, now seen as deformed and recrystallized metasediments at the periphery of the Singhbhum craton (or the Archaean Core Craton Region, ACCR, of Mahadevan, 2002) at the southern contact. The dominant arenaceous rocks (Chaibasa Formation) often alternate with argillaceous components. The Chaibasa Formation is described as tidal flat, shallow marine deposits (Bose et al., 1997). The oldest member of the formation is a lithicwacke with lenses of matrix-supported conglomerate, which occurs extensively as a basal unit immediately overlying the Archaean basement to the south. The remaining younger units in the Chaibasa Formation are pelites and psammi-pelites, now seen as quartz-mica schist, garnetiferous mica schist, quartz-kyanite schist, and quartzites with well-preserved sedimentary structures. The presence of sedimentary structures, coarse grain size, and detrital plagioclase suggest that these rocks were deposited in a nearshore basin, proximal to the Singhbhum craton (Bhattacharya, 1991). The Chaibasa Formation is overlain in the north by the Dhalbhum Formation (younger unit of the Singhbhum Group), which is dominantly argillaceous (now metapelitic schists) with subordinate quartzite. The quartzites are fine-grained, resembling metachert. Besides this, Dunn and Dey (1942) reported tuffaceous rock, implying that the Dhalbhum Formation possibly contains volcanogenic material. These features indicate that the Dhalbhum Formation in the north represent a deep-water facies rock association (eugeosynclinal deposits). The clastic arenaceous facies is taken over by the dominantly argillaceous facies up to the linear belt of the Dalma volcanics. Beyond Dalma volcanics in the north there is a broad belt of volcano-sedimentary rocks, similar in lithology to the rocks in the south of the Dalma belt, that include tuff, acid and basic volcanics, all continuing up to the high-grade CGGC in the north. It seems that the Singhbhum Group and the Dalma volcanics overlap in time; the former developed in the rift basins that initiated the latter due to thinning of the continental lithosphere (cf. Mahadevan, 2002).

All the different lithologies occur as linear units with E-W orientation parallel to the trend of the fold belt. This regional trend is the consequence of compression of the volcano-sedimentary belt by the N-S collision of the Singhbhum craton in the south against the Chhotanagpur Granite Gneiss Complex in the north. During the collision of the crustal blocks both volcano-sedimentary rocks and Dalma volcanics were metamorphosed and intruded by granite and coeval gabbro-pyroxenite at about 1.6 Ga ago (Ray, 1990; Roy et al., 2002a, b). The Proterozoic granites of Kharswan, Arkasani, Mosabani etc., occurring as linear bodies parallel to the extent of the mobile belt, are possibly the expression of partial melting of the underlying crust of the fold belt. The Chandil supracrustal in the volcano-sedimentary belt north of the Dalma volcanics is also intruded by granites, amongst them the

Kuilipal Granite is dated by Rb—Sr whole-rock isochron at 1638±38 Ma (Sengupta et al., 1994). Thus, the collision of the Precambrian blocks is expressed in the E-W trending folds, regional foliation and E-W running metamorphic isograds (see previous section). Regional metamorphism of the Singhbhum Group rocks was during 1.6 Ga is also supported by the Rb—Sr whole-rock isochron age of the Soda granites (Sarkar et al., 1985), syntectonically intruding the Chaibasa Formation. Following Sengupta and Chattopadhyay (2004), this writer also thinks that the Singhbhum fold belt developed without closure of any large ocean. The presence of marked asymmetry of the fold belt and the reactivated basement fully accord with the origin of the fold belt by continental collision. The nature of the sediments and deformation style in the Singhbhum fold belt suggest them to be similar to a fold-and-thrust belt developed due to collision tectonics. It has involved shallow-water sediments and their continental basement and is therefore of intra-continental nature (cf. Sengupta and Chattopadhyay, 2004).

The southerly vergence of the structures in the fold belt suggests that the domain of collision should lie beyond the Dalma volcanic "spine", but certainly not to include the E-W trending Chhotanagpur Granite Gneiss Complex (CGGC) as postulated by Acharyya (2003). He (Acharyya, loc. cit.) regarded the CGGC and the Singhbhum fold belt (SFB) as the southern fold belt of the Central Indian Tectonic Zone in its eastern sector and recognized that they (CGGC and the SFB) are formed by collision of the Bundelkhand craton in the north and the Singhbhum craton in the south (Acharyya, 2003, p. 11). This proposition is not supported by the progressive underthrusting of the Singhbhum craton deep under the Northern region as a result of which several bodies of ultramafic rocks and mafic schists (Gupta and Basu, 2000) occur all along the shear zone (Banerji, 1975; Saha, 1994).

The southern boundary of the Singhbhum fold belt (SFB) against the eastern and northern margins of the Singhbhum craton is demarcated by a shear zone, called the Singhbhum Shear Zone (SSZ) (Mukhopadhyay et al., 1975). Discontinuous sheets of smaller bodies of granites, namely Soda Granite (2.2 Ga), Arkasani Granophyre (1052 Ma), Mayurbhanj Granite (some of which may even represent wedges of basement granitoids, see Fig. 6.1 for location), occur close to the shear zone and in variable state of deformation. The SSZ with its narrow belt of mylonites, according to Dunn, 1929) and Mukhopadhyay (1984), tapers out westward. However, some workers (Sarkar and Saha, 1962; Gupta et al., 1981; Gupta and Basu, 2000) think that the SSZ also extends along the NW margin of the Singhbhum Granite and along the southern margin of the Chakradharpur Granite (CH), which represents the largest tectonic wedge of the basement granitoid exposed to the north. These workers also infer that the SSZ separates the domain hosting the Ongerbira volcanics and the craton. This proposition, however, is not accepted by Sarkar and Chakraborti (1982) who recorded lithological similarity of rocks across the supposed western extension of the SSZ. A continuity of structures and absence of mylonite belt were also established across the supposed extension of the SSZ by Mukhopadhyay et al. (1990). Recent mapping by the GSI shows the absence of any major dislocation along the southern margin (Mazumdar, 1996). Beyond Porhat (location not given in Fig. 6.1), the SSZ possibly grades into a high-angle gravity fault and extends SW along the western boundary of the low-grade Iron-Ore Group (IOG) and then along the western margin of the Bonai Granite (Saha, 1994, p. 177).

The SSZ is an arcuate belt of high strain and is characterized by ductile shearing, soda-granite magmatism and polymetallic mineralizations. The major movement along the shear zone is mainly vertical, as evidenced by steeply plunging a-lineation. Perhaps impressed by the curvilinear nature of the shear zone, presence of crushed rocks and mylonites and retrogression of the different rocks along the SSZ, Dunn and Dey (1942) regarded the SSZ as a late orogenic feature of the deformation affecting the SFB. Other workers (e.g. Mukhopadhyay et al., 1975; Mukhopadhyay, 1984; Gupta and Basu, 1985) consider the SSZ as an early feature, the earliest recognizable phase of deformation. The early features are believed to have been obliterated by phases of deformation and protracted ductile shearing (cf. Gangopadhyay and Samantha, 1998). Recently, Pradhan and Srivastava (1996), on the basis of their study in Chakradharpur area, showed that the ductile shearing occurred between F1A and F2 group of folds. This was followed by brittle deformation, claim the authors (Pradhan and Srivastava, 1996). The authors also recognized four phases of ductile folding separated by two phases of brittle deformation in the shear zone. Since the SSZ traverses rocks of different ages and separates contrasted metamorphic facies on its either side, it is highly likely that the SSZ is coeval with or post-dates regional metamorphism of the Singhbhum Group rocks and hence younger than 1.6 Ga.

The evolution of the Singhbhum shear zone is considered multiepisodic (Misra, 2006) at 2200, 1800, 1600, and 1000 Ma. The soda granite is emplaced at 2.2 Ga; Copper mineralization occurred at 1.8 Ga; Kuilapal Granite and Uranium mineralization occurred at 1.6 Ga and Arkasani Granite intrusion at 1.05 Ga. The granitic rocks have been sheared and occur as detached bodies. It is also possible that the Singhbhum shear zone was the last event to have deformed all the granitoids emplaced prior to shearing, and the mobilization during shearing is responsible for the U and Cu mineralization.

6.7 Evolutionary Models and Discussion

Keeping in view the above discussion, we now critically evaluate below the different evolutionary models for the Singhbhum fold belt.

Model 1: Intraplate Subduction Model (Sarkar and Saha, 1977, 1983)

On the assumption that the Singhbhum Shear Zone (SSZ) is a deep suture between the Proterozoic Singhbhum fold belt and the Archaean Singhbhum craton, Sarkar and Saha (1977, 1983) proposed that the rocks north of the SSZ were developed in a geosynclinal basin and were later involved in the Satpura orogeny (1600–900 Ma). The evolutionary model can be summarized in the following stages:

- Long after cratonization ($\sim$3.1 Ga) a geosyncline (ocean) developed around 2.2 Ga in which Proterozoic sediments of the North Singhbhum Mobile belt were deposited.

6.7 Evolutionary Models and Discussion

- The geosynclinal sediments of the Singhbhum Group were subsequently deformed and metamorphosed due to northward subduction of the Singhbhum craton (along the SSZ) under the lithospheric plate to the north.
- There occurred a regional tension phase during which tholeiitic lavas of Dhanjori-Dalma were erupted and gabbro-anorthosites were emplaced.
- Subsequently there was a renewed subduction of the Singhbhum plate on the south, whereby partial melts in the upper part of the subducting plate were developed and emplaced as Soda Granite, Arkasani Granite, and the Mayurbhanj and Kuilapal granite suites.
- Renewed compression continued deformation of the Singhbhum Group and the Dalma-Dhanjori volcano-sedimentary formations.
- Final stage is marked by transcurrent faulting and thrusting (North shear zone) as well as the renewed shearing along the Singhbhum Shear zone (SSZ).

The above model was challenged on several accounts. First relates to controversies of constituent rocks in respect of their structural and stratigraphic status as outlined by Sarkar and Saha (1977, 1983). Second is the absence of typical subduction-related assemblages along the SSZ. Third is the uncertain reason for generating a tensional phase between two compressive regimes acting in nearly the same direction. Fourth is lack of explanation for the tholeiitic ocean floor (that characterize the Dalma lavas) on the overriding plate and lack of any large-scale movement across the SSZ (Mukhopadhyay et al., 1990; Sarkar et al., 1992).

Model 2: Microplate Collision Model (Sarkar, 1982)

A.N. Sarkar (1982) proposed a model of converging microplates to interpret the tectonic evolution of the Singhbhum and Chhotanagpur Granite Gneiss Complex (CGGC) regions. In this model the CGGC block represents an overriding plate and the Singhbhum microplate as the subducting plate. The collision of these continental microplates took place around 1600 Ma ago. The model considers convergence and collision of the Singhbhum microplate against a stationary Chhotanagpur microplate (CGGC) in three cycles. The first cycle (2000–1550 Ma) relates with the northward movement of the Singhbhum microplate and its collision with the CGGC microplate. In this event it is believed that Dalma volcanics was emplaced as ophiolite in a flysch environment. In the second cycle (1550–1000 Ma) the Singhbhum plate is assumed to have rotated clockwise towards NE and generated F2 folds, including the NW-SE trending fold of the Dhanjori rocks. The third cycle (1000–850 Ma) relates to the overriding of the Singhbhum plate onto the CGGC plate in a NNW-SSE direction, obduction of the continental lithosphere in the southern part of the Singhbhum fold belt, and also F3 deformation and M3 metamorphism. At the close of the orogenic cycle the Singhbhum fold belt was uplifted and subjected to erosion.

The collision model of A.N. Sarkar implies a very long period of subduction history, spanning over 300 million years. The model is based on poor database and necessitates extremely slow motion or very long distance journey of the Singhbhum plate.

Model 3: Marginal Basin Model (Bose and Chakraborti, 1981; Bose, 1990, 1994)

In this model, it is proposed by Bose and his co-authors (see Bose, 1990, 1994) that secondary spreading (rifting) of the Singhbhum craton occurred due to heating of the craton above a subducting slab. This resulted into separation of a continental mass making the continental arc of the Chhotanagpur Gneissic Complex (CGGC), what Bose (1990) called the "fossil island Arc" lying on a supra-subduction zone. The rifted basin between the Singhbhum craton and the separated continental arc of CGGC became the marginal basin which had a spreading ridge, called Dalma volcanic ridge (Fig. 6.2), located somewhere in the middle. This ridge separated the marginal basin into two sub-basins (see Fig. 6.2) that received supracrustals of the Singhbhum Group and the Dalma volcanics.

Deformation and metamorphism of the Proterozoic supracrustals in the marginal basin is stated to have occurred due to southward subduction of a lithospheric plate. The subduction zone in this model, according to Bose and his coauthor (see Bose, 1994), was to the north of the Chhotanagpur Gneissic Complex in what is today the Ganga Basin (Fig. 6.2). The N-S convergence of the south-directed plate against the Singhbhum craton situated on the south gave rise to the E-W trending Singhbhum fold belt.

Although the marginal basin model satisfactorily explains the high-temperature mineralogy of the CGGC (see Chap. 2), as a continental arc, and also other metamorphic deformation characteristics of the Singhbhum fold belt, it fails to explain some important geodynamic questions. The model envisages subduction zone somewhere to the north of the CGGC but the North shear zone (NSZ), located at the contact of the CGGC-Singhbhum fold belt, cannot be considered as a subduction zone. This is because there are no ultramafics associated with it and the NSZ is too impersistent to be considered as a subduction zone. Again, there is no evidence of arc-type volcanism or plutonism within the CGGC terrain. Finally, it is a matter of debate whether the Indo-Gangetic plain is also a part of the Singhbhum-Orissa microplate because northerly extension of the Singhbhum (-Orissa) carton is questionable. Lastly, the

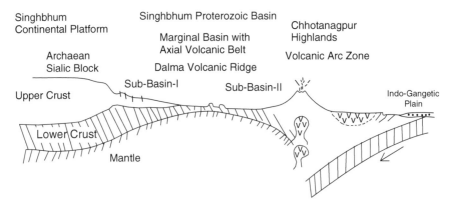

Fig. 6.2 Diagram showing tectonic setting of Singhbhum marginal basin and associated morphostructural unit (redrawn from Bose, 1994)

6.7 Evolutionary Models and Discussion

model starts with the second phase of the Wilson cycle without any reference to the first stage (Gupta and Basu, 2000).

Model 4: Ensialic Orogenesis Model (Gupta et al., 1980; Mukhopadhyay, 1984; Sarkar et al., 1992)

A least controversial model is the ensialic orogenesis model in which the Archaean crust of the region is assumed to have attenuated and rifted in response to mantle heat cell during which the Dalma volcanics erupted while sedimentation was initiated. Later, extension is believed to have been replaced by plate convergence during which the subducting lithospheric plate was delaminated, similar to that proposed for the Aravalli fold belt. This event of A-subduction not only caused deformation but also regional metamorphism of the Proterozoic rift sediments along with the Dalma lavas (Gupta et al., 1980; Mukhopadhyay, 1984). This model of ensialic orogenesis was first applied by Gupta et al. (1980) and later by Mukhopadhyay (1984) and Sarkar et al., (1992) and further refined with additional data and discussed by S.C. Sarkar et al. (1992) and Gupta and Basu (1985). The ensialic orogenesis model for the Singhbhum fold belt is summarized by Gupta and Basu (2000) and is briefly reviewed here.

Mantle plume below the continental lithosphere in this part of the Indian shield rifted a united craton of Singhbhum-Chhotangpur gneiss Complex. This gave rise to a rift basin between the Singhbhum craton in the south and the CGGC in the north. The basin became the site of deposition of the Singhbhum Group supracrustals; shallow deposits near the Singhbhum craton and distal facies sediments with contemporary volcanic-plutonic rocks farther. Dalma volcanics were in the central part of the basin while volcanism near the cratonic margin appeared in Dhanjori and equivalent basins. This volcanism occurred with intermittent effusion of acid/basic tuffs, alkali basalt and co-magmatic mafic-ultramafic intrusions. The carbonatite intrusion occurred along the northern boundary of the North Singhbhum Mobile belt and locally along the Dalma belt. It must be stated here that geochemistry of the volcanics in the mobile belt is mainly MORB type (cf. Bose, 1990) and is therefore indicative of rift environment. At some stage of the volcanic activity the stable continent also developed cracks along which mafic melt intruded as what is now called the Newer Dolerite dykes in the Singhbhum craton. The model envisages Singhbhum as a foreland block for the south-directed stresses generated by plate convergence and overriding of the north plate, the CGGC. Partial melting of the sialic upper crust of the southern plate seems to have generated granites of Arakasani, Mayurbhanj etc.

The ensialic orogenesis model is supported by the geophysical evidence, which indicates continuation of the continental crust below the entire width of the fold belt (Verma et al., 1984). The model explains many of the geological observations, but it assumes that both Singhbhum Group and Dalma volcanics were deposited at the same time. The geochronological data suggest the Dalma Volcanics are younger than the Singhbum Group. The model also fails to explain as to why granulites are absent in the fold-and-thrust terrain of the Singhbhum when so much quantity of mafic magma appeared during rifting and lithospheric delamination.

The metamorphic history in this ensialic orogenesis model by Mukhopadhya and Mukhopadhyay (2008) suggest that the early stage of heating was by asthenospheric upwelling and extension in the Singhbhum craton. This event (M1) formed andalusite-bearing paragenesis. The next stage was a collision related compressional deformation followed by M2 metamorphism (Barrovian) which was followed by cooling and decompression. The last event (M3) is retrogression and post D3.

The formation of andalusite prior to sillimanite and kyanite is not compatible with the stability of the Al-silicate minerals. If magmatic underplating was there, why did not granulites form and why did the temperature remain near $500°C$ to from andalusite. It is a serious question as to how did the early andalusite remain stable in D2 and D3 deformation and P-T conditions that were beyond the stability of this low-pressure mineral.

Model 5: Slab-Breakoff Model

Slab-Breakoff Model by the present writer takes into account the age considerations of the Dalma Volcanics and the Singhbhum Group rocks and also the presence/absence of the granulites in the colliding plates on the north and south of the Singhbhum fold belt. The proposed model is similar to the Model 4 described above, but it explains the features of fold belt and its constituent rocks more elegantly. The model, shown in Fig. 6.3, envisages that ductile stretching and rifting of the once united craton of Chhotanagpur Granite Gneiss Complex (CGGC) and Singhbhum craton (SC) gave rise to sedimentary basin (s) in the Palaeoproterozoic time, which became the site of sedimentation of the Singhbhum Group rocks (Fig. 6.3a). Further stretching developed large ensialic rift, the deeper parts of which in the north seems to have received minor tuff (Dunn and Dey, 1942) (Fig. 6.3b). This was followed by collision of CGGC and SC at about 1.6 Ga ago, resulting in the folding of the supracrustal and the underlying basement (Fig. 6.3c). This convergence (perhaps oblique in nature) of the N and S crustal blocks with subduction of southern block (Singhbhum block) was followed by a narrow rifting and slab weakening (location by small double arrows, Fig. 6.3d). Because of slab rupture and underplating, extrusion of mafic magma (Dalma Volcanics) occurred amidst sediments of the Singhbhum Group, and more volcano-sedimentary deposits took place with high-level intrusion of mafic (gabbro-pyroxenite) dykes at 1.6 Ga (Fig. 6.3e). As a consequence of slab rupture and underplating there occurred extrusion and intrusion of mafic magmas as well as 1.6 Ga old granites (Kuilapal etc.) and Granophyre (at Arkasani), formed as a result of partial melting within the crust due to heat from underplated magma (Fig. 6.3f). Further slab break off and its downgoing, i.e. sinking of the slab, generated melt of calc-alkaline composition that appeared in various forms as volcanics (Chandil) and acid tuff Ankro area of Chandil (1487 ± 34 Ma; Sengupta and Mukhopadhyay, 2000) (Fig. 6.3 g). Finally, the slab sinks away and melts at depth, producing the last phases of granitoid liquid while dykes (Newer Dolerite; 900–935 Ma; Saha, 1994) were injected from the underplated magma into the overlying crust (Fig. 6.3 h).

In this model proposed by the author it is claimed that the SFB developed without closure of any ocean basin. The basement for the volcano-sedimentary sequence of

6.7 Evolutionary Models and Discussion

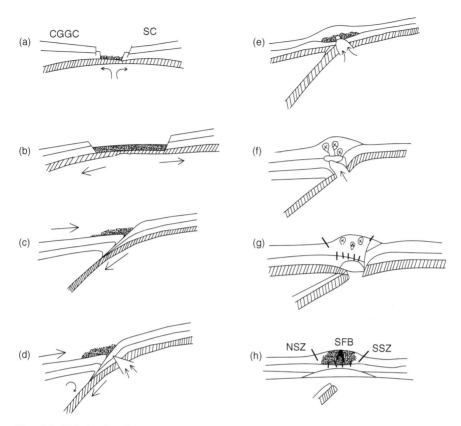

Fig. 6.3 Slab break off evolutionary model for the Singhbhum fold belt, based on available geological-geochronological data. (**a**) Ductile stretching and rifting of Archaean craton (SC+CGGC) in Palaeoproterozoic (ca. 2.1 Ga) and sedimentation of Singhbhum Group. (**b**) Development of ensialic crust and deposition of more sediments with minor tuffs (cf. Dunn and Dey, 1942). (**c**) Start of collision at about 1.6 Ga ago and folding of supracrustal and basement rocks. (**d**) Convergence (perhaps oblique) and subduction of southern block (SC) followed by narrow rifting and slab weakening. (**e**) Slab rupture, underplating and extrusion of mafic magma (Dalma volcanics) amidst sediments of Singhbhum Group. Additional volcano-sediment deposition and high level intrusion of mafic (gabbro-pyroxenite dykes at 1.6 Ga). (**f**) With underplating following slab rupture, mafic magma extruded and intruded. Heat from the underplated magma also induced partial melting within the crust to produce 1.6 Ga old granites (Kuilapal etc.) and granophyre (Arkasani). (**g**) Further slab break off and slab downgoing/sinking generated melt, giving rise to calc-alkaline magma that appeared as volcanics (Chandil) and acid tuff (Ankro area of Chandil) at 1487 ± 34 Ma (Sengupta and Mukhopadhyay, 2000). (**h**) Slab sinks away and melting at depth produced the last phases of granitoid liquid while dykes (Newer Dolerite; Saha, 1994) were injected from the underplate into the overlying crust. Abbreviations: CGGC = Chhotanagpur Granite Gneiss Complex, SC = Singhbhum Craton, SFB = Singhbhum Fold Belt, SSZ = Singhbhum Shear Zone, NSZ = Northern Shear Zone

the Singhbhum belt was evidently continental rock(s) of Archaean age. The Dalma volcanics are the outcome of slab breaking that gave access to the melt produced by decompression. The model rules out the proposition of some authors that the Dalma volcanics represent island arc magma or an ophiolite belt, because the metamorphic rocks are no higher than amphibolite facies in the entire fold belt. The deformed and metamorphosed rocks of the Singhbhum fold belt are bordered on both sides by shear zones against the cratonic blocks, which were rigid mass to squeeze the rocks and elevate them into what we call the Singhbhum fold belt. Several detached granitic bodies, namely the Chakradharpur granite, Arkasani granophyre and soda granite near Kharswan, occur along the shear zone, specially the Singhbhum shear zone. The Soda granite is retrograded into feldspathic schists when it underwent intense deformation.

In the Singhbhum fold belt the metasediments occurring north of Darjin Group and to the west of structural closure of the Dalma rocks are not included, because their metamorphic-deformational history is not yet established. These rocks are known as the *Gangpur Series* or *Gangpur Group* (Mahalik, 1987). The rock-association consists of calcareous, psammopelitic and Mn-metasediments that have been named the Gangpur Series by Krishnan (1937) who, mainly because of easterly plunging antiformal structures, inferred these to be older than the Iron-Ore Series on the south and east. Contrary to this, Banerjee (1968) finds that the eastern plunging folds are inverted towards the core and constitutes a reclined fold, later re-folded in an antiform during Satpura orogeny. Banerji considered the Raghunathpalli conglomerate as the base of the Gangpur Group. The rocks of the Gangpur belt also shows three fold phases (F1–F3), F1 and F2 coaxial while F3 are trending N-S upright folds. The Ekma granite pluton intruding the Gangpur metasediments yielded Rb—Sr whole rock isochron age of 1024 ± 4 Ma (Pandey et al., 1998), setting the upper age limits of the deformation affecting the Gangpur rocks.

References

Acharyya, S.K. (2003). A plate tectonic model for Proterozoic crustal evolution of Central Indian Tectonic Zone. ond. Geol. Mag. Spec., vol. **7**, pp. 9–31.

Bandyopadhyay, P.K. and Sengupta, S. (1984). The Chakradharpur granite gneiss complex of west Singhbhum, Bihar. In: *Monograph on Crustal Evolution*. Indian Soc. Earth Sci., pp. 75–90.

Banerjee, P.K. (1968). Revision of stratigraphy, structure and metamorphic history of the Gangpur Series, Sundergarh district, Orissa. Rec. Geol. Surv. India, vol. **95**(2), pp. 328–354.

Banerji, A.K. (1975). On the evolution of the Singhbhum nucleus, eastern India. Q. J. Geol. Min. Met. Soc. India, vol. **47**, pp. 51–60.

Bhattacharya, H.N. (1991). A reappraisal of the depositional environment of the Precambrian metasediments around Ghatsila-Galudih, eastern Singhbhum. J. Geol. Soc. India, vol. **37**, pp. 47–54.

Bhattacharya, D.S. and Sanyal, P. (1988). The Singhbhum orogen—its structure and stratigraphy. In: Mukhopadhyay, D. (ed.), Precambrian of the Eastern Indian Shield. Geol. Soc. India Mem., Bangalore, vol. **8**, pp. 85–111.

Blackburn, W.H. and Srivastava, D.C. (1994). Geochemistry and tectonic significance of the Ongerbira metavolcanic rocks, Singhbhum district, India. Precambrian Res., vol. **67**, pp. 181–206.

References

Bose, M.K. (1990). Growth of Precambrian continental crust—a study of the Singhbhum segment in the eastern Indian shield. In: Naqvi, S.M. (ed.), *Precambrian Continental Crust and Its Economic Resources*. Dev. Precambrian Geol., vol. **8**. Elsevier, Amsterdam, pp. 267–286.

Bose, M.K. (1994). Sedimentation pattern and tectonic evolution of the Proterozoic Singhbhum basin in the eastern Indian shield. Tectonophysics, vol. **231**, pp. 325–346.

Bose, M.K. (2000). Mafic-ultramafic magmatism in the eastern Indian craton—a review. In: Dr. M.S. Krishnan Centenary Commemoration Seminar Volume. Geol. Surv. India Spl. Publ., vol. **55**, pp. 227.

Bose, M.K. and Chakraborti, M.K. (1981). Fossil marginal basin from the Indian shield: a model for the evolution of Singhbhum Precambrian belt, eastern India. Geol. Rundsch., vol. **78**, pp. 633–648.

Bose, P.K., Mazumder, R. and Sarkar, S. (1997). Tidal sand waves and related strom deposits in the transgressive Protoproterozoic Chaibasa Formation, India. Precambrian Res., vol. **84**, pp. 63–81.

Dunn, J.A. (1929). The geology of north Singhbhum including parts of Ranchi and Manbhum districts. Geol. Surv. India Mem., vol. **54**(2), pp. 1–166.

Dunn, J. and Dey, A.K. (1942). Geology and petrology of eastern Singhbhum and surrounding areas. Geol. Surv. India Mem., vol. **69**, pp. 281–456.

Gangopadhyay, P.K. and Samanta, M.K. (1988). Microstructures and quartz c-axis patterns in mylonitic rocks from the Singhbhum shear zone, Bihar. Indian J. Geology, Calcutta, vol. **70**, pp. 107–122.

Gupta, A. and Basu, A. (1985). Structural evolution of Precambrians in parts of North Singhbhum, Bihar. Rec. Geol. Surv. India, vol. **113**(3), pp. 13–24.

Gupta, A. and Basu, A. (1991). Evolutionary trend of the mafic-ultramafic volcanism in the Proterozoic North Singhbhum Mobile Belt. Indian Minerals, vol. **45**, pp. 273–288.

Gupta, A. and Basu, A (2000). North Singhbhum Proterozoic mobile belt, Eastern India—a review. In: Dr. M.S. Krishnan Centenary Commemoration Seminar Volume. Geol. Surv. India Spl. Publ., vol. **55**, pp. 195–226.

Gupta, A., Basu, A. and Ghosh, P.K. (1980). The Proterozoic ultramafic and mafic lavas and tuffs of the Dalma greenstone belt, Singhbhum, eastern India. Can. J. Earth Sci., vol. **17**, pp. 210–231.

Gupta, A., Basu, A. and Ghosh, P.K. (1982). Ultramafic volcaniclastics of the Precambrian Dalma volcanic belt, Singhbhum, eastern India. Geol. Mag., vol. **119**, pp. 505–510.

Gupta, A., Basu, A. and Srivastava, D.C. (1981). Mafic and ultramafic volcanism of Ongerbira greenstone belt, Singhbhum, Bihar. J. Geol. Soc. India, vol. **22**, pp. 593–596.

Iyengar, S.V.P. and Anand Alwar, M. (1965). The Dhanjori eugeosyncline and its bearing on the stratigraphy of Singhbhum, Keonjhar, and Mayurbhanj districts. (D.N. Wadia Commemoration Volume), Min. Geol. Met. Inst. India, pp. 138–162.

Iyengar, S.V.P., Chandy, K.C. and Narayanswamy, R. (1981). Geochronology and Rb-Sr systematics of the igneous rocks of the Simlipal Complex, Orissa. Indian J. Earth Sci., vol. **8**, pp. 61–65.

Iyengar, S.V.P. and Murthy, Y.G.K. (1982). The evolution of the Archaean-Proterozoic crust in parts of Bihar and Orissa, eastern India. Rec. Geol. Surv. India, vol. **112**(3), pp. 1–5.

Johnson, P.T., Dasgupta, D. and Smith, A.D. (1993). Pb-Pb systematic of copper sulphide mineralization, Singhbhum area, Bihar. Indian J. Geol., vol. **65**, pp. 211–213.

Krishnan, M.S. (1937). Geology of Gangpur state. Mem. Geol. Surv. India, vol. **71**, pp. 1–128.

Lal, R.K., Ackermand, D. and Singh, J.B. (1987). Geothermobarometry in Barrovian type of metamorphism of politic schists, Sini, district Singhbhum. In: Saha, A.K. (ed.), Geological Evolution of Peninsular India—Petrological and Structural Aspect (Professor Saurindranath Sen, Commemoration Volume), Recent Res. Geol., vol. **13**. Hindustan Publ. Corpn., New Delhi, pp. 125–142.

Mahadevan, T.M. (2002). Geology of Bihar and Jharkhand. Geol. Soc. India, Bangalore, 563p.

Mahalik, N.K. (1987). Geology of rocks lying between Gangpur Group and Iron-Ore Group of the horse-shoe syncline in north Orissa. Indian J. Earth Sci., vol. **14**, pp. 73–83.

Mallik, A.K. and Sarkar, A. (1994). Geochronology and geochemistry of mafic dykes from the Precambrian of Keonjhar, Orissa. Indian Minerals, vol. **48**, pp. 13–24.

Mazumdar, S.K. (1996). Precambrian geology of peninsular eastern India. Indian Minerals, vol. **50**, pp. 139–174.

Misra, S. (2006). Precambrian chronostratigraphic growth of Singhbhum-Orissa craton, eastern Indian shield: an alternative model. J. Geol. Soc. India, vol. **67**, pp. 356–378.

Misra, S. and Johnson, P.T. (2005). Geochronological constraints on the evolution of the Singhbhum Mobile Belt and associated basic volcanics of eastern Indian shield. Gondwana Res., vol. **8**, pp. 129–142.

Mukhopadhyay, D. (1984). The Singhbhum shear zone and its place in the evolution of the Precambrian mobile belt, north Singhbhum. Indian J. Earth Sci., CEISM Volume, pp. 205–212.

Mukhopadhyay, D. (1988). Precambrian of the eastern Indian shield—perspective of the problem. Geol. Soc. India Mem., vol. **8**, pp. 1–12.

Mukhopadhyay, D., Bhattacharya, T., Chakraborty, T. and Dey, A.K. (1990). Structural pattern in the Precambrian rocks of Sonua-Lotapahar region, North Singhbhum, eastern India. Proc. Indian Acad. Sci. (Earth Planet Sci.), vol. **99**, pp. 249–268.

Mukhopadhyay, D., Ghosh, A.K. and Bhattacharya, S. (1975). A reassessment of structures in the Singhbhum shear zone. Bull. Geol. Min. Met. Soc. India, vol. **48**, pp. 49–67.

Mukhopadhyay, D. and Mukhopadhyay, C. (2008). Cycles of deformation and metamorphism during Mesoproterozoic orogesis in North Singhbhum fold belt, eastern India: International association for Gondwana Res. Conference Series, 5, (Abstract Volume), pp. 170–171

Naha, K. (1965). Metamorphism in relation to stratigraphy, structure and movement in Singhbhum, East India. Q. J. Geol. Min. Met. Soc. India, vol. **37**, pp. 41–85.

Pandey, B.K., Krishna, V. and Chabria, T. (1998). An overview of the geochronological data of the rocks of Chhotanagpur Gneiss-Granulite Complex and adjoining sedimentary sequences, eastern and central India. Proc. Intern. Seminar, *Crust in Eastern and Central India* (UNESCO-IUGS-IGCP 368). Geol. Surv. India, Bhubaneswar, India, (Abstract), pp. 131–135.

Pradhan, A.K. and Srivastava, D.C. (1996). Deformation style of the Singhbhum shear zone and the adjoining rocks in the eastern Indian shield. In: Naha, K. and Ghosh, S.K. (eds.), *Recent Researches in Geology and Geophysics,* vol. **16**. Hindustan Publ., New Delhi, pp. 1–14.

Rao, N.K., Aggarwal, S.K. and Rao, G.U.V. (1979). Lead Isotopic ratios of uraninites and the age of uranium mineralization in Singhbhum shear zone, Bihar. J. Geol. Soc. India, Bangalore, vol. **20**, pp. 124–127.

Ray, K.K. (1990). The Dalma volcanics—a Precambrian analogue of the Mesozoic-Cenozoic suture. Group Discussion on *"Suture Zones, Young and Old"*. Geol. Surv. India, Abstract, pp. 17–21.

Ray, K.K., Ghosh, S.K., Roy, A.K. and Sengupta, S. (1996). Acid volcanic rocks between the Dalma Volcanic Belt and the Chhotanagpur Gneissic Complex, East Singhbhum and Purulia district of Bihar and West Bengal. Indian Minerals, vol. **50**, pp. 1–18.

Roy, A.B. (1966). Interrelation of metamorphism and deformation in central Singhbhum, eastern India. Geol. Mag., vol. **45**, pp. 365–374.

Roy, A., Sarkar, A., Jayakumar, S., Aggrawal, S.K. and Ebihara, M. (2002a). Mid-Proterozoic plume-related thermal event in eastern Indian craton: evidence from trace elements, REE geochemistry and Sr-Nd isotope systematics of basic-ultrabasic intrusives from Dalma volcanic belt. Gondwana Res., vol. **5**, pp. 133–146.

Roy, A., Sarkar, A., Jayakumar, S., Aggrawal, S.K. and Ebihara, M. (2002b). Sm-Nd age and mantle source characteristics of the Dhanjori volcanic rocks, eastern India. Geochem. J., vol. **36**, pp. 503–518.

Saha, A.K. (1994). Crustal evolution of Singhbhum-North Orissa, Eastern India. Geol. Soc. India Mem., vol. **27**, 341p.

Saha, A.K., Ray, S.L. and Sarkar, S.N. (1988). Early history of the Earth: evidence from the eastern Indian shield. In: Mukhopadhyay, D. (ed.), *Precambrian of the Eastern Indian Shield*. Mem. Geol. Soc. India, Bangalore, vol. **8**, pp. 13–37.

Sarkar, A.N. (1982). Precambrian tectonic evolution of eastern India: a model of converging microplates. Tectonophysics, vol. **86**, pp. 363–397.

References

Sarkar, A.N. and Chakraborti, D.K. (1982). One orogenic belt or two? A structural reinterpretation supported by Landsat data products of the Precambrian metamorphites of Singhbhum, eastern India. Photogrammetria, vol. **37**, pp. 185–201.

Sarkar, S.N., Ghosh, D.K. and Lambert, R.J.St. (1985). Rubidium-Strontium and Lead isotopic studies of the Soda granites from Mosabani, Singhbhum Copper Belt, eastern India. Indian J. Earth Sci., vol. **13**, pp. 101–116.

Sarkar, S.C., Gupta, A. and Basu, A. (1992). North Singhbhum Proterozoic mobile belt, eastern India: its character, evolution, and metallogeny. In: Sarkar, S.C. (ed.), *Metallogeny Related to Tectonics of the Proterozoic Mobile Belts*. Oxford & IBH Publ. Co. Pvt Ltd., New Delhi, pp. 271–305.

Sarkar, S.N and Saha, A.K. (1962). A revision of Precambrian stratigraphy and tectonics of Singhbhum and adjacent region. Q. J. Geol. Min. Met. Soc. India, vol. **34**, pp. 97–136.

Sarkar, S.N. and Saha, A.K. (1977). The present status of the Precambrian stratigraphy, tectonics and geochronology of Singhbhum-Keonjhar-Mayurbhanj region, eastern India. Indian J. Earth Sci., S. Ray Volume, pp. 37–65.

Sarkar, S.N. and Saha, A.K. (1983). Structure and tectonics of the Singhbhum-Orissa Iron-Ore Craton, eastern India. In: Sinha-Roy, S. (ed.), *Recent Researches in Geology* (Structure and Tectonics of the Precambrian Rocks), vol. **10**. Hindustan Publishing Corpn., New Delhi, pp. 1–25.

Sarkar, S.N., Saha, A.K., Boelrijk, N.A.L.M. and Hebeda, E.H. (1979). New data on the geochronology of the Older Metamorphic Group and the Singhbhum Granite of Singhbhum-Keonjhar-Mayurbhanj region, eastern India. Indian J. Earth Sci., vol. **6**, pp. 32–51.

Sengupta, S., Acharyya, S.K. and de Smeeth, J.B. (1997). Geochemistry of Archaean volcanic rocks from Iron-Ore Supergroup, Singhbhum, eastern India. Proc. Indian Acad. Sci. (Earth Planet. Sci.), vol. **106**, pp. 327–342.

Sengupta, S. and Chattopadhyay, B. (2004). Singhbhum mobile belt—how fat it fits an ancient orogen. Geol. Surv. India Spl. Publ., vol. **84**, pp. 23–31.

Sengupta, S. and Mukhopadhyay, P.K. (2000). Sequence of Precambrian events in the eastern Indian craton. Geol. Surv. India Spl. Publ., vol. **57**, pp. 49–56.

Sengupta, S., Paul, D.K., Bishui, P.K., Gupta, S.N., Chakrabarti, R. and Sen, P. (1994). Geochemical and Rb-Sr isotopic study of Kuilapal granite and Arkasani granophyre from the eastern Indian craton. Indian Minerals, vol. **48**(1–2), pp. 77–88.

Sengupta, S., Paul, D.K., de Laeter, J.R., McnNaughton, N.J., Bandyopadhyay, B.K. and de Smeth, J.B. (1991). Mid-Archaean evolution of the eastern Indian craton: geochemical and isotopic evidence from Bonai pluton. Precamb. Res., vol. **49**, pp. 23–37.

Srivastava, D.C. and Pradhan, A.K. (1995). Late brittle tectonics in a Precambrian ductile belt: evidence from brittle structures in the Singhbhum shear zone, eastern India. J. Struct. Geol., vol. **17**(3), pp. 385–396.

Verma, R.K., Sharma, A.U.S. and Mukhopadhyay, M. (1984). Gravity field over Singhbhum, its relationship to geology and tectonic history. Tectonophysics, vol. **106**, pp. 87–107.

Vohra, C.P., Dasgupta, S., Paul, D.K., Bishui, P.K., Gupta, S.N. and Guha, S. (1991). Rb-Sr chronology and petrochemistry of granitoids from the southeastern part of the Singhbhum craton, Orissa. J. Geol. Soc. Inc., vol. **38**, pp. 5–22.

Chapter 7
Eastern Ghats Mobile Belt

7.1 Introduction: Geological Setting

The Eastern Ghats Mobile Belt (EGMB) derives its name from the mountain of Eastern Ghats in the east coast of India. Apart from reconnaissance survey by individual geologists (C.S. Middlemiss, T.L. Walker, L.L. Fermor, V. Ball), the Geological Survey of India provided a geological map of the EGMB on 1:50,000 scale in later part of the 20th century and also gave periodical reviews in its Reports (see Ramakrishnan and Vaidyanadhan, 2008). Recently, Ramakrishnan et al. (1998) gave a new geological map (1: 1000,000) with a summary of regional geology of the EGMB.

The Eastern Ghats Mobile Belt is a NE-SW trending arcuate Precambrian fold belt of high-grade rocks, extending over a length of ca. 600 km along the east coast of India from north of Cuttack in Orissa to Nellore in Andhra Pradesh (Fig. 7.1). The belt has a maximum width of 100 km in the northern part and less than 20 km in the south where it is concealed under the Phanerozoic cover. The fold belt is in contact with three cratonic blocks, namely the Singhbhum(-Orissa) craton in the north, the Bastar craton in the west, and the Dharwar craton in the south and southwest. To the east of the fold belt is the Bay of Bengal. The boundary between these cratons and the EGMB is termed the *Transition Zone*, having characteristics common to both (see Bhattacharya and Kar, 2002). The contact of western belt of the EGMB against Bastar craton is gradational. Basic granulite, dykes with chilled margins and incipient or arrested charnockite patches are noticed in this marginal zone (cf. Ramakrishnan and Vaidyanadhan, 2008). Around Deobhog, the contact is marked by a Terrane Boundary Shear zone (TBSZ) by Biswal et al. (2004). Shear sense indicators show top-to NW movement. This contact is described as the Eastern Ghats Boundary Shear zone in the map of Ramakrishnan et al. (1998) and is regarded as continuation of Sileru Shear zone. The contact between EGMB and Singhbhum craton is also marked by shear zones, predominantly Sukinda Thrust. The craton-mobile belt boundary is characterized by the occurrence of alkaline rocks in the Bastar and Singhbhum cratons, like those of Prakasan alkaline province in the Transition zone of Dharwar craton to the south of Godavari Graben (Leelanandam, 1990). Alkaline rocks, mainly nepheline syenites are also reported at the contact zone of EGMB-Singhbhum craton in the north as well as in the western

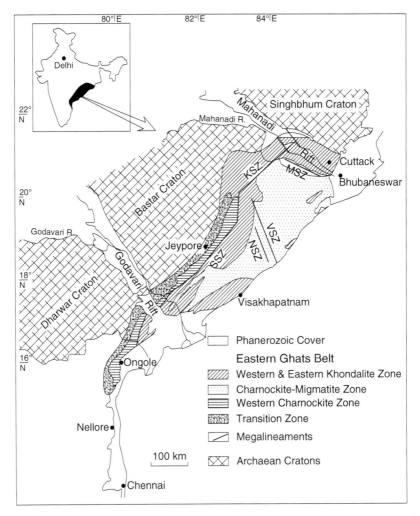

Fig. 7.1 Simplified geological map of the Eastern Ghats Mobile Belt (EGMB) (after Ramakrishnan et al., 1998) with megalineaments after Chetty (1995). MSZ = Mahanadi Shear Zone; NSZ = Nagavalli Shear Zone; SSZ = Sileru Shear Zone; VSZ = Vamsadhara shear Zone. *Inset* shows location of the mobile belt

part of the EGMB, where they are grouped as Khariar Alkaline Complex. These alkaline rocks are deformed and their lenses are seen nearly parallel to the gneissic foliation. The alkaline complex in general has U–Pb and Pb–Pb emplacement ages of 1400–1500 Ma (Upadhya et al., 2006a). The origin of Khariar alkaline complex is attributed to basaltic magma derived from partial melting of enriched lherzolite mantle source within the lithosphere. The basaltic magma fractionated within mantle and gave rise to nepheline syenite (Upadhyay et al., 2006b).

An overview of the EGMB is given by Bhattacharya (1996). The EGMB comprises two major rock groups: one charnockitic and the other khondalitic. The

7.1 Introduction: Geological Setting

charnockitic group consists of mafic to acidic charnockites, hypersthene-bearing granulites, gneisses and leptynites while the khondalitic group includes garnet-sillimanite gneisses, quartzites and calc-silicates. Based on the dominant lithological assemblages, Ramakrishnan et al. (1998) proposed a 4-fold longitudinal division of the EGMB. These divisions (Fig. 7.1) from W to E are: (1) Western Charnockite Zone (WCZ), dominantly charnockite and enderbite with lenses of mafic-ultramafic rocks and minor metapelites. (2) Western Khondalite Zone (WKZ), dominated by metapelite (khondalite) with intercalated quartzite, calc-silicate rocks, marble and high Mg–Al granulites. The metapelitic granulites are intruded by charnockite/enderbite. Several occurrences of massif-type anorthosites are reported from this zone (Bolangir, Turkel and Jugsaipatna). (3) Central Charnockite-Migmatite Zone (CMZ) has dominantly migmatitic gneisses but also includes high Mg–Al granulites and calc-silicate rocks, all of which are intruded by charnockite-enderbite, pophyritic granitoids and massive-type anorthosite (Chilka Lake region), and (4) Eastern Khondalite zone (EKZ), having lithological similarity with WKZ, but without anorthosite. To distinguish the basement upon which the supracrustals (now seen as metamorphosed metasediments of khondalites, calc-silicates, quartzites etc.) were deposited, as the rocks of the mobile belt, is a very difficult task because of the impress of granulite facies metamorphism on both cover and basement rocks in this region. The recrystallized supracrustals (and their unrecognized basement) and occasionally the charnockite-enderbite rocks have been intruded by post-tectonic alkaline/anorthositic/granitoid rocks. These intrusions are in form of numerous plutons of alkaline rocks, anorthosites and granitoids, and have age range between 1450 and 850 Ma (Subba Rao et al., 1989). According to Ramakrishnan et al. (1998), the stated plutons occur in a linear belt, all along the western margin of the EGMB, between the Western Charnockite Zone (WCZ) and the Western Khondalite Zone (WKZ). But, according to Leelanandam (1997), they occur between the cratonic (non-charnockitic) and charnockitic region (WCZ) of the mobile belt. The presence of several alkaline intrusives in the Transition Zone favours the latter proposition. Several exposures of alkaline complexes have been reported from the high-grade EGMB (Leelanandam, 1998), but structural setting has not been worked out to ascribe them to Rift Valley magmatism or Plume tectonic model. For those alkaline complexes, such as the Koraput complex, which are away from the cryptic suture or the craton-mobile belt contact, rift tectonics cannot be envisaged (cf. Bhattacharya and Kar, 2004; Leelanandam et al., 2006).

The high-grade gneiss-granulite ensemble of the EGMB is surrounded all around by a thrust along which the rocks were thrusted up 10–15 km against the cratonic rocks at the Transition Zone. The thrust is, in fact, a crustal-scale ductile shear zone and defines the western boundary of the EGMB. This shear zone against the contact with Bastar craton is named as Sileru shear zone (SSZ) by Chetty and Murthy (1998), where it is found to be 3.5 km wide. According to Chetty (1995), this shear zone is contemporaneous with the emplacement of extensive alkaline magmatism around Kharia, Koraput and Kunavaram. The northeast termination of the thrust is on the north side of the Mahanadi graben, while the SE margin intersects the coastal plain north of Chennai, thus separating the EGMB from the southern Indian Granulite Terrain (SGT). The thrust is demarcated by low gravity anomaly. To the east of the EGMB there is also a shear zone along Angul-Dhankenal, at which Chilka Lake anorthosites are located (Chetty, 1995).

The EGMB is dissected by two prominent rifts with infillings of the Gondwana (Upper Paleozoic) sediments, which are centred with river-courses of Mahanadi in the north and the Godavari in the south. The Godavari graben also contains Proterozoic sediments (Pakhals) that underlie the Gondwanas. These rifts or grabens trend NW-SE and are thus orthogonal to the general trend of the EGMB (see Fig. 7.1). Merely on the basis of scanty geochronological data and apparently different P-T-paths across the Gondwana graben, Mezger and Cosca (1999) and Sengupta et al. (1990) suggested that the EGMB segments on the N and S of the Godavari rift/graben have different thermal history. If this difference is a geological reality, it must be by reasons other than mantle melting or subcrustal heat variation, because the Godavari graben is devoid of any igneous intrusive material during its formation. The absence of alkaline magmatism in particular also rules out the Godavari graben to be an aulocogen or failed arm (cf. Philpots, 1990). Accordingly, the junction-point of the Godavari graben with the Eastern Ghats Mobile Belt does not represent a triple point (Raman and Murthy, 1997). Hence the suggested difference of thermal history in the north and south of Godavari rift cannot be attributed to mantle plume. We shall examine this aspect later when we discuss the tectono-thermal evolution of the EGMB.

7.2 Deformation

The regional geology of the EGMB is less clear, although a general information on structural geology has been given by Sarkar et al. (1981), Halden et al. (1982), Bhattacharya and Gupta (2001), and Chetty (1995). Regional gneissosity generally strikes NE-SW, sub-parallel to the regional trend of the EGMB and dips to the SE. Variations in dip and strike are, however, common. Deformation history of the EGMB has been determined from isolated areas. From the observation that the deformation sequence across the NEGMB is similar, Halden et al. (1982) gave an account of deformational history of the EGMB in the north of the Godavari graben. Structural investigations north of the Godavari rift show four fold phases (D1 to D4) in the EGMB, although the structural elements of different episodes are not uniformly found at different localities of the EGMB. The contact of metapelite and calc-gneiss is taken as the earliest planar structure designated as S0. The first phase, D1, developed reclined and isoclinal folds (F1) with a strong axial-plane foliation (S1) that has become parallel to the compositional layering (S0) in the rocks. Migmatization of rocks of suitable compositions and emplacement of syntectonic granites also occurred during this event. The mineral segregation banding as well as compositional banding in the lithologies including migmatitic quartzo-feldspathic gneisses, metapelites, calc-granulites, mafic granulites and even Opx-bearing granitoids, is taken as S1. The S1 is an axial-plane foliation of rootless and intrafolial folds in the khondalites. In Visakhapatnam area, the pegmatites trend along the foliation or compositional banding of khondalite-charnockite, indicating that these quatzo-feldspathic rocks intruded prior to D2. The D1 was presumably coeval with the granulite facies metamorphism that evidently occurred at great depths of about 25–27 km, considering the estimates of pressure of 8–9 kbar and temperatures of 950 ± 50°C (Paul et al., 1990). The second deformation (D2) also produced isoclinal/reclined folds (F2) which are coaxial with F1, since their axial plane foliation (S2) is parallel to S1. The F2 folds occur on various scales (Bose, 1971; Bhattacharya and Gupta, 2001). Third deformation (D3) produced tight to open folds that are coaxial with F2. The axial plane of F3 folds parallels the shear planes that developed due to westward thrusting of the EGMB against the cratonic nucleus to the west. The shear fabric has either obliterated or transposed earlier foliations in this region. Fourth deformation (D4) produced open, upright folds (F4) transverse to the regional structural grain (Bhattacharya and Gupta, 2001). These folds often show plunges towards SE (cf. Chetty, 1995). The succession of folding indicates that the EGMB underwent a protracted compression, perhaps a thrust-related deformation continuum that can be correlated with the collision with the craton. At places, around Visakhapatnam, one finds NNE kink bands and brittle shears on S1 and S2 foliation, which are considered to be related to D4 that caused retrograde metamorphism (Chetty, 1995). Polyphase deformation is also documented in the southern sector (Sengupta et al., 1999). Here, F1 is a recumbent fold with N-S axial plane, refolded by F2 and F3 (broad warp).

7.3 Geochronology

The isotopic data on the EGMB rocks are scanty and the data also lack proper correlation with tectonic and thermal events. Perraju et al. (1979) recognized a metamorphic event at 3100 Ma, based on Rb–Sr whole rock (granulite) data. Based on U–Pb zircon age, Vinogradov et al. (1964) suggested a metamorphic event at ca. 2600 Ma, which is also supported by some geochronological dates (Rb–Sr whole rock) by Perraju et al. (1979). These ages indicate Archaean ancestory and hence suggestive for the existence of a basement complex (similar to those of the southern high-grade granulite terrain (Chap. 2) on which the Eastern Ghats supracrustals were deposited rather than to suggest protolith ages (> 3.0 Ga old) for the EGMB granulites. Admittedly, supracrustals cannot be laid without a basement. Mezger and Krogstard (1997) dated two distinct zircons from a charnockite gneiss, of which the prismatic zircons gave 3.0 Ga age, interpreted as crystallization/emplacement age of the charnockite, whereas the rounded zircons gave 2.6 Ga (close to lower inrtercept of 3.0 Ga old prismatic zircon) to represent recrystallization or reprecipitation. Recently, Bhattacharya et al. (2001) reported 1.7 Ga old ages for the zircons in the charnockite inclusions which could represent a Pre-F2 granulite facies event or, alternatively, a charnockite magmatism. On the basis of U–Pb method in accessory minerals of the granulites another magmatic/metamorphic event at ca. 900 ± 100 Ma is reported by Grew and Manton (1986), Aftalion et al. (1988) and Paul et al. (1990), which perhaps constrain the last episode of granulite facies metamorphism. A Pan-African thermal event at ~600 Ma has been reported by Aswathanaryan (1964) on the basis of K—Ar biotite age and by Mezger and Cosca (1999) from U—Pb ages in accessory phases in granitoids.

7.4 Petrological Characteristics

Petrologically the EGMB is a gneiss-granulite terrain comprising: (a) metapelitic granulites that include the khondalite (*garnet-sillimanite-perthite-quartz-gneiss*), and Mg—Al granulites *(sapphirine-spinel-cordierite-orthopyroxene-sillimanite-garnet ± quartz)*. The khondalites locally contain quartzite bands and the Mg—Al granulites as lenses within metapelites. The Mg—Al granulites also occur as xenoliths in the mafic granulites and at places are generally asociated with garnet leptynites, (b) leptynite (*plagioclase-quartz-perthite ± garnet*), (c) calc-granulites *(wollastonite-scapolite-calcite-plagioclase-garnet-clinopyroxene)* which mostly occur as bands and lenses within or in association with metapelites, (d) charnockite-enderbite (*orthopyroxene-quartz-feldspar ± garnet gneiss*) and mafic granulite (*orthopyroxene-clinopyroxene-plagioclase-garnet*). True charnockites (perthite >> plagioclase) show intrusive relationship with Opx-bearing orthogneisses (mafic granulites and enderbites). Several bodies of alkaline rocks and associated basic rocks are reported to have intruded the granulite complex. Massive-type anorthosites occur in the northern part of the EGMB (Sarkar et al., 1981). A layered igneous complex, comprising anorthosite-gabbro-pyroxenite-chromitite, is reported from the Kondapalle area in the southern part of the EGMB (Leelanandam, 1967, 1997, 1998).

7.5 Tectono-Thermal Evolution

The EGB granulites are shown to be polymetamorphic. The thermal history is reported to be different on either sides of the Godavari graben. Both Northern and Southern segments of the EGMB, abbreviated as NEGMB and SEGMB have experienced ultra high temperature (UHT) metamorphism, as revealed by the presence of sapphirine (Spr)—quartz and spinel (SPL)-quartz and orthopyroxene (Opx)-sillimanite (Sill)-quartz, but the timing of this extreme event is most confusing and least understood, as can be felt from the following description. It is also intriguing to know from the scanty geochronological data that the SEGMB seems to have escaped reworking during Grenville orogeny (~1000 Ma) as well as Pan-African (~550 Ma) orogeny (cf. Mezger and Cosca, 1999; see also Sengupta et al., 1990). The following account deals with the tectono-thermal evolution separately for the Northern and Southern segments of the EGMB and thereafter the evolution of the entire EGMB will be taken, giving different models.

7.6 Northern Segment

In this segment petrological study of the granulites from near coastal region of EGMB, particularly from Paderu (Lal, 1997), Anantagiri (Sengupta et al., 1990), Rajamundry (Dasgupta et al., 1997), Madhuravada and Vizianagaram (Rao et al., 1995) revealed an early ultra high temperature (UHT) metamorphism with thermal peak at 900–1000°C/8–10 kbar. This was followed by high-pressure isobaric cooling (IBC) as documented by the reaction coronas formed from the reactions involving Spr+Spl+Qz (formed at the thermal peak) (Dasgupta and Sengupta, 1995). There is, however, a serious controversy in respect of the retrograde P-T trajectory, which according to Lal (1997), Bhattacharya and Kar (2004), is clock-wise, whereas Dasgupta et al. (1997) and Sengupta et al. (1990) deduced anticlockwise paths for the investigated localities of the NEGMB. The P-T trajectories by different authors (see Fig. 7.2c) reveal that the EGB granulites are polymetamorphic with an impress of two granulite facies metamorphic events. As revealed in the P-T-t paths by different workers, an early M1 event occurred near 8–10 kbar/950 ± 50°C followed by retrograde path or near isobaric cooling to 750–800°C during which corona and symplectitic textures developed, as in Rayagada area (Shaw and Arima, 1996). Dasgupta et al. (1997) further proposed that the isobarically cooled granulites have been uplifted by isothermal decompression (ITD) path to 5 kbar when late cordierite formed from opx + sill involving reactions. Interesting results were drawn from petrologic study of Chilka Lake granulites, about 200 km north of stated localities of the NEGMB. Here, the UHT metamorphism yield a multi-stage P-T-t path (Sen et al., 1995) with a combination of three isothermal decompression (ITD) paths, intervened by two isobaric cooling (IBC) paths from 8–10 kbar/1000°C down to 4.5 kbar/650°C (Fig. 7.2c). According to Sen et al. (1995), the deduced trajectory indicates a discontinuity after the first cooling, implying re-working of the crust in a separate tectonic event. Rickers et al. (2001) also report multi-stage evolution of the Mg–Al granulites from Anakapalle area, similar to that deduced for Chilka Lake granulites by Sen et al. (1995). The Mg–Al granulites to the NW of Visakhapatnam, on the other hand, show a simple isobaric heating-cooling path (Bose et al., 2000). Interestingly, none of the granulites from EGMB preserved the prograde path of metamorphism to enable one to deduce unambiguous sense of movement of the rocks in the P-T space—whether ACW or CW so as to provide important constraints on tectonic history of orogenic belts. However, there appears a broad similarity in the estimated peak P-T conditions with those deduced by Dasgupta et al. (1997). Textural and compositional characteristics of nearly all the Mg–Al granulites from the EGMB bring out a coherent picture of an early phase of UHT metamorphism ($\sim$1000°C) at 8–10 kbar, followed by near isobaric cooling.

Isotopic record on the granulite facies rocks of the NEGMB denotes Greenville thermal event (Mezger and Cosca, 1999), but Archaean, Mid-Proterozoic and Pan-African dates are also reported (Vinogradov et al., 1964; Paul et al., 1990; Shaw et al., 1997; Sarkar et al., 1998; Dobmeier and Simmat, 2002; Krause et al., 2001). Jarick (2000) dated the UHT metamorphism at Anakapalle to be 1100 Ma for the rocks that occur as xenoliths in basic granulites. However, it is not yet clear

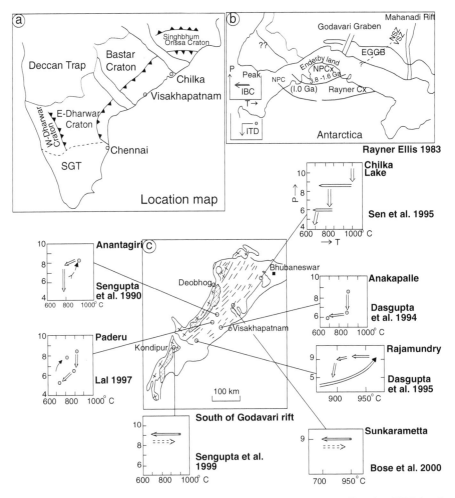

Fig. 7.2 Location map (**a**); the attempted correlation of EGMB with Napier Complex (NPCx) and Rayner Complex (Rayner Cx) of the Enderbyland, Antarctica (**b**); (**c**) shows a synoptic history of P-T paths (see text) in different domains of the Eastern Ghats Mobile Belt. Note that the UHT metamorphism with high pressure IBC and IBC heating-cooling paths are attributed to an earlier event of ~1.6 Ga, whereas ITD path to a later, ~1.0 Ga event (Dasgupta et al., 1994; Mezger and Cosca, 1999; Sengupta et al., 1999; Rickers et al., 2001; and others). The difference in metamorphic evolution in North and South EGMB may not be real (see text)

whether this age is truly representative of the regional UHT metamorphism in the EGMB. The scattered isotopic evidence (Paul et al. 1990; Shaw et al. 1997) suggests a distinct Late Archaean age (2.6–2.8 Ga) for basic magmatism. If the basic magma is the source for high temperature heating of the deeper crustal rocks, the metamorphism would be Late Archaean. However, interpretation of these isotopic data in relation to deformation and petrologic processes has not yet been critically attempted. In the Central Khondalite zone a NE-SW trending megacrystic, syn-D3 granitoid intrudes the khondalites and associated gneisses and granulites (including sapphirine-granulites). Geological setting unambiguously suggests that the granitoid is older than 1400 Ma anorthosite and alkaline intrusions (Sarkar et al., 1981; Ramakrishnan et al., 1998). The granitoid whole-rock Rb—Sr yielded 1890 and 1215 Ma ages as emplacement and overprint ages, respectively (Bhattacharya et al., 2001). This implies that granulite facies metamorphism was between 1890 and 1215 Ma. Again, at Chilka Lake the garnet leptynite, related to axial-plane foliation of F2 in pelitic granulite contains patchy charnockite. The leptynite whole-rock Rb—Sr age is 1913 ± 82 Ma while the zircons in the patchy charnockite gave 1.7 Ga age (Bhattacharya et al., 2001). This zircon age could be taken to represent a Pre-F2 granulite metamorphism (older event), or alternatively a charnockite magmatism. Here, it must be stated that according to the study of Kar (2001), the source rocks of the low Rb—Sr leptynites and high Rb—Sr charnockites must be different. The patchy charnockite in the host leptynite gneisses, considered by Bhattacharya et al. (1994) as relict structure, is interpreted by Dobmeier and Raith (2003) as arrested charnockites (in-situ charnockite). But as stated earlier, Kar (2001) considers them as caught up xenoliths within letynite the foliation (S1) of which is not found in the leptynite having only S2 foliation.

Dasgupta and Sengupta (2003) attribute the UHT metamorphism with high-pressure IBC and IBC heating-cooling path of UHT granulites to an earlier event of about 1.6 Ga and the ITD path to a later event of 1.0 Ga event (see also Rickers et al., 2001; Mezger and Cosca, 1999). It means that there was a granulite event at 1.6 Ga when rocks of NEGMB resided at lower crustal levels. After 600 Ma, i.e. at $\sim$1000 Ma, these granulites were then exhumed by ITD path (superposed on high-P IBC path) to mid- to lower-crustal levels. Partial re-setting during a late overprint at 550$\sim$Ma (Pan-African) has also been suggested by Mezger and Cosca (1999).

7.6.1 Southern Segment

The EGMB rocks to south of the Godavari rift i.e. in the Southern Segment, have only limited information in regard to their thermo-tectonic evolution. In Ongole (Dasgupta et al., 1997) as well as in Kondapalle (Sengupta et al., 1999), where the metapelitic granulites are intruded by layered noritic gabbro-pyroxenite-anorthosite, the granulites also bear the imprint of UHT granulite facies metamorphism, with P-T conditions similar to that documented in the granulites of Northern segment (NEGMB). Also the isobaric cooling trajectory on the retrograde path is comparable with that recorded in the granulites of NEGMB. However, there is no record of any isothermal decompression event, and it is a matter of enquiry as to how these granulites were excavated from deeper levels, about 25 km, where they suffered ultra high temperature (UHT) followed by isobaric cooling (IBC).

Available U—Pb cooling ages of minerals suggest that the areas of Ongole and Kondapalle had UHT metamorphism (granulite facies metamorphism) at 1300 and 1600 Ma, respectively, after which time they presumably did not experience any other granulite facies event.. However, the ^{40}Ar/^{39}Ar in amphibole from near shear zones gave ~1000 Ma (Grenvillian age) which is simply taken to indicate resetting of the radioclock (Crowe et al., 2001).

Resetting of isotope systems is also reported in rocks of the northern segment during M2 event at ~1000 Ma. According to most workers, the UHT metamorphism represents an earlier event of ca. 1600 Ma. The second granulite facies event is understood to have occurred at temperatures around 850°C, which the rocks of NEGMB attained during IBC from UHT event (M1). The source of heat for the UHT metamorphism was obviously the basic/mafic magma that underplated the overlying crustal rocks. The second granulite facies (M2) event in SEGMB is also the result of the isobaric cooling that occurred at around 850°C. Again, the metapelitic granulites south of the Godavari rift show UHT P-T path with isobaric heating and cooling (Sengupta et al., 1999), which is similar to the isobaric heating-cooling path reported from NW of Visakhapatnam in the north of the Godavari rift (Bose et al., 2000). These similarities in two distant areas located in the northern and southern segments are as striking as are the dissimilarities in the P-T-t paths in nearby localities of the northern EGMB, as stated already. Notwithstanding, the isobarically cooled rocks of NEGMB are documented to have been subsequently brought to shallower levels by isothermal decompression (ITD) during an orogney (Pan African) (cf. Sengupta et al., 1999; Dasgupta and Sengupta, 2003). The Southern Segment of the EGMB is considered to have escaped this orogenic event as the rocks retain older ages and do not show isothermal decompression. But it remains to explain as to how the granulites of SEGMB were excavated from lower crustal levels (corresponding to 8 kbar) to be juxtaposed with the granulites of the NEGMB, observing that no later "decompressive" orogeny affected this area (cf. Bhattacharya and Gupta, 2001).

It seems to this author that the entire charnockitic rocks of the EGMB have been exhumed by upthrusting along a ductile shear zone, a suggestion that finds support from the presence of the shear zone around this Proterozoic fold belt (stated earlier). To show the temporal equivalence of the retrograde path of the Northern and Southern segments of the EGMB, one has to accept that the peak of M1 granulite facies metamorphism at 950°C was Mid-Proterozoic (ca. 1600 Ma) in the belt as a whole while the later Grenville event (M2, 1000 Ma) would be at lower temperatures (750–850°C). This implies that following M1, the granulites remained stationary in an isostatically compensated lower crust for about 600 Ma with no uplift. Such a situation, however, appears remote. The multi-stage IBC-ITD paths of the Chilka Lake granulites indicate that they were affected by two or more geological events (Sen et al., 1995). It is almost certain that if the rocks were subjected to high-pressure heating and isobaric cooling they ought to have undergone isothermal decompression to be able to be exposed near the surface. After isobaric cooling if the rocks remained buried in great depths they need an orogeny to be excavated. In the case of the residence for 600 Ma, the 1600 Ma old granulites must have been involved in the Grenville orogeny, failing which their excavation must have been during the Pan-African orogeny. But more geochronological work is required to have a satisfactory answer to this problem.

7.7 Is Godavari Graben a Major Terrain Boundary?

The shear (thrust) zone that surrounds the EGMB all around is also responsible to upthrust the granulites of the mobile belt. It is cut across by the Godavari graben to suggest that crustal extension giving rise to the Godavari rift must be younger than the third deformation (D3) recognized in the region (Bhattacharya and Gupta, 2001). The graben is devoid of any igneous material, but contains sediments of Pakhal Group (Mid-Proterozoic) and Gondwanas (Upper Palaeozoic to Mesozoic). K–Ar date on glauconites from the Pakhals gave 1300 ± 53 to 1188 Ma age (see Raman and Murthy, 1997). These dates indicate that the graben is younger than the UHT metamorphism in the granulites of SEGMB but older than the Grenvillian thermal event in NEGMB. Despite it being younger than the surrounding thrust, the Godavari graben is erroneously regarded as a terrain boundary between the two domains lying on its either side, merely on the basis of scant isotopic data on accessory minerals (Mezger and Cosca, 1999) and partial P-T- paths (Sengupta et al., 1990). If the granitic rocks are missing from the graben and the sediments directly rest upon the granulite basement, as disclosed geophysically by Pande and Rao (2006), there does not seem to be a thermal activity during and after the formation of the Godavari graben. The following points should help the reader on the validity of this proposition.

(1) The retrograde P-T path from the UHT metamorphism (M1 at ca. 950°C) in Southern sector of the EGMB (abbreviated SEGMB) is similar to that deduced for the granulites of NEGMB, although their temporal equivalence is not yet ascertained. Accepting that these two sectors are correlatable, the peak of M1 granulite event at 950°C in NEGMB would also be Mid-Proterozoic (1600–1300 Ma). If so, the second granulite event (M2) would be Grenvillian age ($\sim$1000 Ma) at lower temperature of 750–850°C. In this context, the Ar–Ar age of $\sim$1000 Ma of amphibole from SEGMB can be interpreted to indicate Grenvillian age in the Southern segment. Under the situation, a mechanism very similar to ITD would have to be conceived also for the SEGMB granulites.
(2) If the granulites (charnockites-enderbites) from the Western Charnockite Zone (WCZ) to the north of the Godavari graben are 1.6 Ga old, as suggested by Mezger and Cosca (1999), and hence comparable to the granulites south of Godavari graben, it rejects the proposition that Godavari rift is a N-S divide within the EGMB; The WCZ extends unobstructed to the south of the Godavari graben.
(3) The difference in metamorphic evolution in NEGMB and SEGMB may not be real if an analogy were not drawn with the Enderby Land, East Antarctica. Dasgupta and Sengupta (2003) suggested similarities of P-T paths such that the North sector of the EGMB is correlatable with Rayner complex while Southern sector of the EGMB is correlatable with the Napier complex (Fig. 7.2b).
(4) The UHT event followed by retrograde P-T trajectory for the granulites of SEGMB is similar to that for the granulites of Northern sector of the EGMB, although time equivalence is not yet demonstrated.

(5) If the ITD path, associated with the presumed orogney (Pan-African?) brought the granulites of NEGMB to shallower depths to be exposed by erosion, how were the granulites from Southern sector excavated from the same depth ($\equiv 8$ kbar) to be juxtaposed with the granulites of Northern sector unless the exhumation was similar in place and time.

In summary, the EGMB granulites are found to be polymetamorphic, and the thermal history for the granulites on either side of the Godavari graben begins with the UHT metamorphism (M1) with 900–1000°C and 8–10 kbar, followed by isobaric cooling (M2) at 750–800°C. However, the retrograde path in the granulites north of Godavari is also characterized by the ITD related to a supposed later metamorphic event, not documented in the granulites of the Southern sector. Temperatures in excess of 900°C, retrieved from thermometry on minerals formed after post-F3, suggest that granulite facies conditions continued from M1 to M2 and even M3. Since the mineralogy of M1 event is different from that of M2 or M3, it is debatable whether the granulite facies metamorphism is a continuous one. Sengupta et al. (1999) contended that the ITD event in NEGMB is essentially a re-working of older UHT granulites and hence this event represents a separate event (? Pan-African), not affecting the region south of the Godavari rift. If the ITD path associated with an orogeny brought the granulites of the NEGMB to shallow depth, how the granulites of SEGMB would then be excavated from same depth unless exhumed in a similar way. This and other points stated already do not support the proposition that the Godavari rift represents a major crustal boundary. If the later event related to the uplift of the high-grade terrain could simply be in the form of a passive (i.e. with no associated structures or fabric) thermal pulse then it had only reset the isotopic systems during the Grenvillian orogeny (cf. Bhattacharya and Gupta, 2001). A more precise definition of structures and fabrics is necessary to distinguish it from the earlier event. As stated earlier, the M2 event (ca. 1000 Ma) is inferred from U–Pb cooling ages of zircon, allanite, monazite, sphene etc. (Grew and Manton, 1986). Partial resetting during a late Pan-African thermal overprint at ~550 Ma has also been suggested (Mezger and Cosca, 1999; Mezger et al., 1996).

7.8 Agreed Observations and Facts

In order to have a comprehensive picture and to conceive a more plausible evolutionary model, we should have an inventory of the agreed geological data and facts on the rocks of the EGMB. These are enumerated below:

1. The Eastern Ghats Mobile Belt (EGMB) is a gneiss-granulite tract of intensely deformed granulite facies rocks.
2. The EGMB has dominantly two types of rocks: charnockitic-enderbitic in the western part while khondalites and other orthopyroxene-free rocks occur to the east.

7.9 Evolutionary Models and Discussion   243

3. The rocks trend in NE-SW to NNE-SSW direction and are parallel to the orographic trend.
4. Anorthosites, mafic intrusions and alkaline bodies, including carbonatites clearly post-date all tectonic events and are restricted in the western part of the EGMB.
5. Anorthosites and mafic intrusives are younger than most other rocks in the EGMB Rb—Sr whole rock age of anorthosites and alkaline rocks ranges from 1400 to 1300 Ma (Sarkar et al., 1981; Subba Rao et al., 1989).
6. The EGMB has a peripheral thrust zone (ductile shear zone) which dips towards SE.
7. Extensional shears in EGMB are late and post-date all penetrative structures related to F1 and F2 (Bhattacharya and Gupta, 2001).
8. EGMB has a gravity high and the thrust shows a gravity low
9. The Godavari rift intersects the EGMB and disrupts its gravity profile.
10. The EGMB granulites are polymetamorphic with an early UHT (of uncertain age) with ~950°C, followed by isobaric cooling at temperatures of ~800 ± 50°C (granulite metamorphism M2.
11. South of the Godavari graben the EGMB granulites have no record of either Grenvillian (1000 Ma) or Pan-African (600 Ma) reworking; its rocks retain older ages and do not show isothermal decompression.
12. It is unclear whether the EGMB rocks had an early Archaean or early Proterozoic metamorphic history.
13. U—Pb cooling ages of accessory minerals in the granulites from north of Godavari graben indicate last granulite facies metamorphism to be Grenvillian age (~1000 Ma) (Grew and Manton, 1986; Mezger and Cosca, 1999), but partial re-setting during 550 Ma (Pan-African) is also suggested (Mezger and Cosca, 1999).
14. U—Pb cooling ages of minerals in the granulites from south of Godavari graben indicate granulite facies metamorphism during Mesoproterozoic (1600 and 1300 Ma). Grenvillian age is indicated by Ar—Ar dates of amphibole, which could be interpreted differently.
15. The folding events are a thrust-related deformation continuum correlatable with collision with the cratons during Mesoproterozoic.
16. The EGMB is supposed to have been juxtaposed against the East Antarctica in all Proterozoic reconstructions.

7.9 Evolutionary Models and Discussion

As stated already, the EGMB is a NE-SW trending gneiss-granulite tract, comprising charno-enderbite and mafic granulite of magmatic origin and khondalite with minor calc-gneisses and quartzites as recrystallized metasediments (supracrustals) under granulite-facies conditions. These magmatic and metasedimentary rocks appear in alternate bands each of which is designated by a name (Ramakrishnan et al., 1998), depending on the dominance of the rock type. The western part dominated by the charnockitic rocks is designated the Western Charnockite Zone (WCZ) while the central part of the fold belt dominated by the charnockitic-migmatitic rocks is designated Central Charnockite-Migmatite Zone (CMZ). The khondalite dominated zone in the west, between the two charnockite zones, is designated Western Khondalite Zone (WKZ), while khondalites dominantly occurring in the eastern part of the fold belt are designated Eastern Khondalite Zone (EKZ). It is interesting to note that a given zone also contain rocks from other zone, though in minor amount. The contact between the charnockites and khondalites is obscured. The contact of the EGMB with the cratonic rocks, in particular of Bastar craton, is also a wide zone which has characteristics of both the terrains (i.e. Craton and mobile belt) and hence is called the Transition Zone. The zone occurs along a significant length of the western margin. Geological information on rocks of the Transition zone is not sufficiently known, except from isolated regions (cf. Bhadra et al., 2003).

Field studies have not been able to recognize the basement upon which the precursor of the metasedimentary granulites were deposited, perhaps because both cover and basement have been deformed and thoroughly recrystallized under granulite facies conditions during the Eastern Ghats orogeny. But geochronological data suggest that protoliths between 3.2 and 2.6 Ga are present in the EGMB (Vinogradov et al., 1964; Perraju et al., 1979). The charnockites of the EGMB, yielding a whole-rock Rb–Sr age of 2800–2600 Ma (Perraju et al., 1979) compare well with the charnockites from Madras and Nilgiri Hills, having Rb–Sr isochron age of 2580 ± 80 Ma (Crawford, 1969), as also with the Kabbaldurga charnockites of 2500 Ma age.

The possible basement may have been the Precambrian gneiss laid upon by a supracrustal sequence of shale-silt-sandstone and impure carbonates, as a precursor of metapelite/khondalite, quartzite, and calc-gneisses. Models have been proposed for the evolution of the EGMB and are discussed below.

Model 1 (Ramakrishnan and Vaidyanadhan, 2008)

This model is based on nature and composition of the possible basement and the geochronology of the cover rocks constituting the EGMB. The Proterozoic age supracrustals of khondalite (garnet-sillimanite-graphite gneiss), calc-silicates, quartzites with or without garnet/sillimanite, and the migmatites must have an Archaean basement. The supracrustals and the charnockites of the EGMB occur in parallel bands or longitudinal zones. Gradual contact is often found between the cratons and the EGMB, implying that Archaean gneisses and granites (converted now into charnockite and enderbite of the EGMB) were the possible basement for supracrustal rocks of the EGMB. Geochronological data of the enderbite yielded U–Pb zircon ages of 1700–1720 Ma with ΣNd (T_{DM}) model ages (protolith) of 2.3–2.5 Ga to ssuggest that the basement was mainly a gneiss-granite association. Nd model ages of granitoids of the Eastern Dharwar craton also indicate a provenance age of 2.6–2.8 Ga which correspond to the ages of granitoids of Eastern Dharwar craton and also with the rocks of the Napier Complex in East Antarctica (Yoshida, 2007). From this it is assumed that the basement rocks for the EGMB supracrustals were Late Archaean gneiss-granitoid association.

7.9 Evolutionary Models and Discussion

To understand the geodynamic evolution of the EGMB, we must know that India was in contact with East Antarctica and that the EGMB formed a part of the SWEAT (SouthWest United States and East Antarctica) orogen during Rodinia (1.3–1.0 Ga) supercontinent. Prior to the Rodinia assembly, the Archaean continental crust of the united India and East Antarctica rifted with initiation of alkaline magmatism followed by sedimentation of the Khondalite Group. It is also suggested by the authors that the minor Mg–Al rich sediments have been produced from weathering of mafic volcanics of the rift basin. There was also a likelihood for the occurrence of basic sills and flows that were intercalated with the sedimentary pile. In addition, felsic volcanics and volcani-clastics were also deposited but all are not identifiable within deformed and metamorphosed rocks of the EGMB. The cover rocks and the basement were then deformed (superposedly) and metamorphosed (granulite facies grade) by magmatic underplating, as a result of crustal delamination, producing ultra-high temperature (UHT) rocks. The movment path of these UHT rocks in the P-T space was found to be anticlock-wise with isothermal (ITD) decompression and also isobaric cooling (IBC) in some instances. The mobile belt witnessed intrusion of massive anorthosites and late-tectonic granitoids and charniockites. This is a geo-history of the Granulite belt, which is not interpreted in terms of Phanerozoic-type plate tectonics which so elegantly explains the evolution of most fold belts in various continents of the world.

Model 2 (Upadhyay, 2008)

The geodynamic evolution of the EGMB has been discussed in reference to the origin of the alkaline complex occurring at the contact of the EGMB and the Archaean cratons. According to Upadhyay (2008), there was a rifting during Meso-proterozoic period when Columbia Supercontinent fragmented, resulting into the opening of ocean between eastern India and East Antarctica. Following the rifting, metasedimentary sequence of EGMB was deposited during 1.4–1.2 Ga. The rifted basin closed during the Grenville time (1.0 Ga) due to collision of eastern India and East Antarctica when Rodinia began to assemble. The collision is considered to result into the development of the Grenvillian Eastern Ghats—Rayner Complex orogen whose constituent rocks were metamorphosed to granulite facies conditions, by underplating due to crustal delamination. The EGMB accreted to the cratons and subsequently thermally overprinted during Pan-African (0.5–0.6 Ga) orogeny when the EGMB granulite belt was thrust westward over the cratonic foreland. The author (Upadhyay, loc. cit.) suggests that the present crustal geometry exposed only the eroded Pan-African thrust contact between the Craton and the EGMB, and not the original Grenvillian suture which may be underlying the granulite thrust sheet.

Model 3

The depositional basin was obviously a rift (extensional regime), formed due to ductile stretching and thinning of the crust by subcrustal convection currents. Consequently, the supracrustal rocks were heated by under- and intra-plated mafic/ultramafic melt whose parental magma was possibly derived (by decompression melting) from depleted mantle in Palaeoproterozoic time. The crustal rocks were thereby subjected to deep-seated UHT metamorphism (granulite facies conditions) as if there was a regional scale thermal metamorphism at lower crustal depth. This conclusion, however, depends on the age of UHT metamorphism and that of the emplacement of the magmatic rocks. Fractionation in dry conditions of the parental magma gave rise to gabbro, norite, enderbite and charnockite at lower crustal levels (20–25 km) at which depths these magmatic rocks intruded the basement complex (incl. metasediments). It is perhaps this way that the khondalitic rocks are seen intruded by charnockitic-enderbitic rocks in the EGMB, although other mechanisms may also be considered for charnockite-metapelite association in the EGMB). In this proposed model, the charnockitic rocks appear as products of deep-seated (granulite) metamorphism of quartzo-feldspathic rocks many of which were initially igneous. The ultra high temperature metamorphism (UHT) by magma underplating of the crust not only caused granulite facies metamorphism but also resulted in partial melting of biotite-bearing protoliths, which gave rise to migmatization, formation of Mg–Al granulites (sapphirine-bearing) and emplacement of syntectonic granite. There are no geochronological data on the timing of UHT metamorphism, but Jarick (cited in Dasgupta and Sengupta, 2003), on the basis of U–Pb TIMS study of zircon and a Pb isotopic study of leached feldspar, suggests that the M1 (UHT) metamorphism took place at about 1400 Ma. On the other hand, Dobmeier and Raith (2003), from synthesis of available geochronological data, concluded that M1 event occurred at about 1100 Ma. This discrepancy could be due to partial resetting of mineral ages during cooling and due to overprinting by a later high-grade M2 event. Parallelism of melt fraction with gneissic layering and occurrence of F2 and F3 folds in them imply that the F1 leucosomes were solid throughout the later deformation (Gupta et al., 2000). From the given evidence, it may be suggested that both charnockitic-khondalitic zones exhibit a similar deformation history in which D1 produced reclined to isoclinal folds as well as gneissic layering in the rocks, refolded coaxially by D2 and then superposed by open upright D3 folds plunging eastwards. The main phase of migmatization was syn-F1, but melts are documented to have generated continuously during the granulite facies conditions but their amount decreased successively. It is possible that F1, F2 and F3 could be continuum or they are temporally separated phases, but peak of M1 is reported to be pre- to syn-F1.

The anticlock-wise path (ACW) deduced for the granulites from different areas of the EGMB is consistent with the heating by magma underplating (England and Thompson, 1984). Most of the information of UHT was obtained from sapphirine-spinel- bearing high Mg–Al granulites and calc-silicate granulites (see Dasgupta and Sengupta, 2003). The P-T trajectory, deduced by different authors, reveals that the EGMB granulites are polymetamorphic with an impress of two granulite events. An early M1 occurred near 8–10 kbar/$\sim$950 $\pm$ 50°C., followed by retrograde path or near isobaric cooling to 750–800°C during which coronas and symplectite textures developed. According to Dasgupta et al. (1997) the isobarically cooled granulites have been re-worked by M2 at $P = 8$–8.5 kbar and 850°C which is characterized by near ITD to 5 kbar when late cordierite formed from Opx + Sill involving reaction. The age of M2 is given as 950–1000 Ma (see Grew and Manton, 1986; Mezger and Cosca, 1999).

7.9 Evolutionary Models and Discussion

The three deformation/fold phases have caused sufficient shortening of the crust which finally yielded to shearing and helped excavate the high-pressure granulites to shallower depth to be emplaced on the cratonic rock.

The intrusion of anorthosites and alkaline rocks (~1400–800 Ma) occurring only along the western margin (the Transition Zone) of the EGMB suggests that there was a collision tectonics (Chetty, 1995; Chetty and Murthy, 1998). The prevailing NE-SW trend of the fold belt suggests a dominant E-W compression during the Eastern Ghats Orogeny. The compression is attributed to the collision between the continents of India (eastern part) and the Antarctica, giving rise to this Eastern Ghats fold belt. The Grenvillian isotope signature is taken as the time of collision and amalgamation of Rodinia (Hoffman, 1991; Boger et al., 2001; Fitzsimons, 2000; Kelly et al., 2000). The supercontinent Rodinia is conceived to have assembled due to suturing at the end of Mesoproterozoic (1100–900 Ma). According to SWEAT model (SW US-East Antarctica Model), the EGMB is the site of suturing that linked East India and the Antarctica (see Gopalkrishnan, 1998). According to Powell et al. (1994), the Rodinia fragmented between 830 and 750 Ma and re-assembled into Gondwanaland at ca. 500–550 Ma due to Pan-African orogeny (Rogers, 1996). Owing to the similar thermo-tectonic evolution of EGMB-Antarctica, these continents do not seem to have separated during the fragmentation of Rodinia.

It is commonly believed that the Eastern Ghats orogeny gave rise to large-scale isoclinal folding superposed by another fold phase that occurs all over the EGMB. While the litho-units of EGMB were subjected to UHT metamorphism at lower crustal levels, intrusions of alkaline rocks and anorthosites occurred, mostly in the western margin of the EGMB. The intrusive rocks form an integral part of the high-grade belt of the EGMB. These rocks are frequently located within the Western Charnockite Zone to appear as if they are confined between the cratonic (non-charnockitic) and mobile belt, or as if confined between WCZ and Western Khondalite Zone (WKZ) (cf. Ramakrishnan et al., 1998; Leelanandam, 1998). From their nearly undeformed nature but mildly sheared fabric near shear zones, it is suggested that these rocks were definitely post-D2, but pre- or syn-D3, although they have not been studied for their structural setting. Nor are these alkaline and anorthosites isotopically dated, although a range from 1400 to 800 Ma is given by Subba Rao et al. (1989). Like Adirondack mountains and Stillwater complex, layered anorthosite-gabbro-pyroxenite are found at several places in the EGMB, particularly in Kondapalle area south of the Godavari graben and in Bolangir and Chilka lake in the north of the graben. In these localities, the predominance of orthopyroxene (rather than olivine) as an early cumulate phase in the ultramafic zones indicate high-pressure crystallization of a tholeiitic magma. This parental magma is considered to have been intruded at depth just after the peak of M1 and cooled slowly together with the enclosing charnockites and granulites (cf. Leelanandam, 1998). It may be conceived that consequent upon the magmatic underplating near the crust/mantle interface, the plagioclase concentration as a crystal-laden mush separated from the parental magma, perhaps by filter pressing. The next event that followed culminated into shearing along which the EGMB was upthrusted against the cratons in the west. This shear boundary (Sileru shear zone of Chetty, 1995) is most conspicuous, particularly between the EGMB and the Bastar craton (Gupta et al., 2000; Bhattacharya et al., 2001). The thrusting of the EGMB over the Bastar craton is regarded Pan-African in age. The development of mylonite foliation and its superposed folding are the features of this event. The last episode of deformation is marked by open folds (F4) the axial plane of which is nearly parallel to the shear plane. The overprinting of the weak amphibolite facies metamorphism (M3), suggested by Mezger and Cosca (1999, and references therein) is related to this event recognized as Pan-African (cf. Gupta et al., 2000) during which the granulite-khondalite rock-suite of the EGMB was excavated along a shear zone.

The contrasting P-T-t paths in neighbouring localities of the EGMB of the northern segment to the north of Godavari graben and similarity of P-T-t paths in localities of the belt on either side of the graben should be accepted as an incomplete P-T trajectory deduced from silica-deficient Mg—Al granulites. Therefore these deduced paths may not be able to give a clear picture of the thermo-tectonic evolution of the two segments of the EGMB. From the interpretation of isotopic data and from the age and lithological composition of the Godavari graben, it is doubted that the geohistory of the regions on either side of the graben is different, as discussed earlier. The shear zone bordering the EGMB all around the fold belt does not support the excavation of the North and South segments of the EGMB in different times. This is also evident from the coincidence of the shear plane with the axial planes of the F3 fold that predate the peak of granulite metamorphism (because 900–950°C was obtained on minerals annealed in post-F3). Excavation along the thrust would account for the existing structural and mineralogical data on the EGMB. The development of the mylonitic foliation is related to the upthrusting movement of the EGMB rocks along the shear zone.

7.10 Newer Divisions of the Eastern Ghats Mobile Belt

We know that the shear-bound Eastern Ghats Mobile Belt (EGMB) is entirely made of granulite facies rocks, and attempts have been made to propose its divisions from different point of view. The GSI made the first attempt to prepare a geological map (1: 1 million scale) of the EGMB with four-fold longitudinal sub-divisions, based on the dominant lithological assemblage (Ramakrishnan et al., 1998). To recall, these divisions from west to east are called Western Charnockite Zone (WCZ), Western Khondalite Zone (WKZ), Central Charnockite-Migmatite Zone (CMZ) and Eastern Khondalite Zone (EKZ). These zones are alternately dominated by charnockitic and khondalite rocks as if the pelitic granulite (khondalite) and charnockites were folded. Ramakrishnan et al. (1998) also proposed a Transition Zone (TZ) along the western margin of the EGMB. This Zone contains a mixture of high-grade rocks (similar to those occurring in the east EGMB) and lower grade rocks of the craton and alkaline intrusions. The TZ of Ramakrishnan and others has been now shifted further west by Gupta et al. (2000). The thermo-tectonic evolution of granulites from all the four zones is similar with UHT metamorphism followed by retrograde path of isobaric cooling (IBC) or isothermal decompression (ITD), although prograde path has not been established along this anticlock-wise trajectory suggested by most workers in the EGMB (Dasgupta and Sengupta, 2003).

With limited geological information available on rocks from south of Godavari graben, Mezger and Cosca (1999) suggested on the basis of scant isotopic data that these rocks of EGMB had a different tectono-thermal history as compared to those occurring to the north of the Godavari graben. This author has already given reasons that do not support this division, significant among them is the southern continuation of 1.6 Ga old granulites of the West charnockite Zone across the Godavari graben into Kondapalle-Ongole areas. Again, if the ITD path associated with an orogeny brought the Northern region of the EGMB to shallower depth, it is difficult to understand as to how were the granulites south of the Godavari rift excavated from similar depths to be juxtaposed with the granulites north of the rift, unless they were uplifted together. Also, the suggested difference in metamorphic evolution in these two sectors is not real, because of incomplete P-T-t paths. These and other arguments invalidate the two-fold division of the EGMB across the Godavari graben. The correlatibility of northern EGB with the Rayner complex and of southern EGMB with the Napier complex of the Enderby Land is also questionable because of the similarity in metamorphic evolution of the North and South regions of the EGMB (cf. Dasgupta and Sengupta, 2003). Unless an intensive geochronolgical data is available, these two divisions of the EGMB, even if tentative, cannot be taken seriously. An isotopic age must ensure that it is event specific and is not indicative of a mixed age or even re-setting age.

A different scheme of subdivision of the EGMB has been adopted by Chetty (1995) and Chetty and Murthy (1998) who gave a two-fold division of the EGMB on the basis of interpretation of landsat imagery. According to these authors, the EGMB is divisible into two blocks on either side of the Negavalli-Vamosdhara shear zone (NVSZ) recognized as Negavalli shear zone (NSZ) and Vamasdhara shear zone (VSZ) in the centre of the EGMB (Fig. 7.1). These blocks are called the Archaean Araku Block in the south and the Proterozoic Chilka Block in the north. Although the NVSZ is far from the Godavari graben, the shear zone itself is considered to be an extension of the Napier-Rayner boundary in the East Antarctica (Chetty, 1995). It appears that neither the Godavari graben nor the NVSZ be regarded as a terrain boundary because of inconsistent isotopic data on the granulitic rocks of the EGMB. The age-data are too few to uphold these divisions of the EGMB, either by the NV-shear zone or the Godavari graben. A division based on structural criteria is useful only to separate homogenous/ inhomogeneous domains and surely not for delimiting the antiquity of rocks that were all subjected to granulite facies metamorphism over a vast region.

It is not known if the shear zone (NVSZ) in the central part of the EGMB is coeval with the encircling thrust zone of the EGMB along which the granulites of the Eastern Ghats were thrust upon the craton on the west. This boundary shear zone has received different names, sector-wise. The NW- trending sector is named Sileru shear zone and the E-W trending sector in the north is named Mahanadi shear zone. This Terrain boundary shear zone or thrust defines the border between the Archaean rocks (schist/gneiss) of the eastern margin of the Craton and Proterozoic granulites on the western margin of the Mobile belt. The seismic profile along Alampur-Koniki revealed a major eastern dipping thrust (Kaila et al., 1987). The greater crustal thickness westward underneath near Cuddapah basin, compared to the eastern margin, is the consequence of this upthrusting (Kaila et al., loc. cit). The central shear zone of Nagavalli-Vamosdhara (see Fig. 7.1) is in fact a combination of two NNW-SSE trending shear zones that are considered to coalesce at both ends (Chetty et al. 2003). The coalesced shear zone is thought to be an extension of the Napier-Rayner boundary in the East Antarctica, but palaeomagnetic data are inconsistent with the correlation of the EGMB with the East Antarctica (Torsvik et al., 2001).

Within the southern Indian shield, the nature and extension of the EGMB towards south is also enigmatic. Based on gravity data it is considered that the EGMB continues southwards along the east coast up to Cape Comorin and then swerves to west and north (Subramanyan, 1978). In this way the EGMB crosses the coast between Ongole and Kavali and re-emerges again near Chennai. According to this setting the EGMB is linked to the whole part of the Southern Granulite Terrain (SGT) through Bay of Bengal. However, such a link does not seem possible as the SGT is devoid of Mesoproterozoic event, unlike the EGMB (Braun and Kriegsman, 2002). In contrast to the SGT, the EGMB shows (a) close intermingling of charnockitic rocks and khondalites, (b) greater abundance of khondalite, (c) intense deformation of the belt before UHT metamorphism, and (d) abundance of Mn-formation as pointed out by Naqvi and Rogers (1987, p. 113).

In view of its significance for the continental assembly and break-up of Rodinia and East Gondwanaland, the EGMB has been receiving international attention. Based on Nd model ages and supported by Rb–Sr and Pb isotope data, Rickers et al. (2001) have identified 4 crustal domains in the EGMB (Fig. 7.3). Their Domain-I nearly coincides with the Western Charnockite Zone (WCZ) of Ramakrishnan et al. (1998). But the Domain south of the Godavari Rift (Domain IA of Rickers et al., 2001) is a distinct crustal block formed from juvenile material between 2300 and 1700 Ma. However, Sengupta et al. (1999) report mid-Archaean Nd modal ages for the mafic granulites of the Kondapalle area, which is considered a heat source for the UHT metamorphism over a vast region of the EGMB. Notwithstanding, Rickers et al. (2001) observed a notable difference between the WCZ rocks occurring on either side of the Godavari Rift. They noticed Archaean isotopic signatures in the north as compared to Proterozoic signatures in the south, supporting the earlier notion (Mezger et al., 1996; Sengupta et al., 1999) that the Godavari Rift is a terrain boundary (crustal boundary).

The remaining 3 Domains are in the north of the Godavari graben. The Domain II and III are correlatable with the N and S structural divisions suggested by Chetty and Murthy (1998). Rickers et al. (2001) placed the boundary between Domains II and III along the Nagavalli-Vamosdhara lineament (Chetty and Murthy, ibid) in the central part of the EGMB. The metasediments of Domain II have model ages of 2100–2500 Ma while those of Domain III have Nd model ages between 1800 and 2200 Ma (Fig. 7.3). In Domain II, the Nd model ages for orthogneisses show a gradual variation from areas near the Godavari Rift (3200 Ma) and decreases (1800 Ma) towards NE in beginning of Domain III. The Domain IV of Rickers et al. contains

7.10 Newer Divisions of the Eastern Ghats Mobile Belt

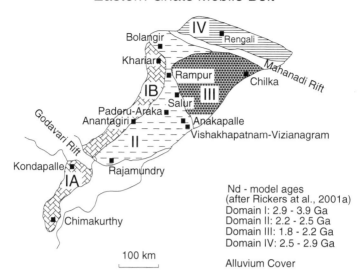

Fig. 7.3 Age domains in the Eastern Ghats Mobile Belt (after Rickers et al., 2001) (see text for details)

metasediments with Nd model ages of 2500–2900 Ma, although Domains II and IV are lithologically similar.

The similar U–Pb mineral ages from orthogneisses at Phulbani (Domain III) and at Angul (Domain IV) suggest Grenvillian thermal event. It is intriguing that the strong Grenvillian impress in Phulbani is not recorded at Chilka Lake in the same Domain. It is also not clear from the subdivisions of Rickers et al. (2001) as to which domain should the anorthosites of Bolangir and Turkel belong, although their geological setting is similar to that in Angul and Dhenkanal in Domain IV.

These divisions after Rickers et al. (2001) are confusing in that the isotopic data overlap and the divisions are at high angles to the regional trend. These divisions based on Nd model ages are, therefore, ambiguous. This author would like to make here a note of caution, particularly for field geologists, on the application and use of Nd model ages on crustal rocks. We infer growth curve of a rock with time from a measured $\sum Nd$ and known $^{147}Sm/^{144}Nd$ ratios. However, if the rock sample was derived from CHUR (Chondrite Uniform Reservoir) the separation event can be fixed by the intersection of this growth curve with the horizontal curve of CHUR. But if the source of the rock sample was DM (Depleted Mantle), the separation time can be obtained as the intersection point of DM and growth curve. However, if the rock sample was derived from a crustal source that had separated from DM at a certain time, then the age will be very much younger than previous cases and difference could be as high as 500 Ma (see Sharma and Pandit, 2003). Moreover, if the Sm/Nd ratios are high, model ages need to be calculated using a two-step evolution. Also, if the orthogneisses are highly evolved, their Sm/Nd ratios would be changed by fractional crystallization. So it is really a question as to how reliable are these model ages on the granulitic rocks that have been metamorphosed at great depth and subjected to melting and fluid activity, thereby disturbing their protolith composition and even isotopic systems. In brief, the model ages cannot be a superior criteria over the one based on lithological assemblage for dividing the granulite facies rocks of the EGMB (cf. Ramakrishnan et al., 1998).

Recently, a new scheme of subdivision of the EGMB has been proposed by Dobmeier and Raith (2003) on mere assumption that the entire granulite terrain of EGMB consists of several crustal segments with distinct geological history. Their evaluation of the existing geological and isotopic data led them to suggest that the EGMB also includes the Nellore-Khammam Schist Belt as well as the lower grade rock-units at the southern margin of the Singhbhum craton. Their four subdivisions of this Eastern Ghats region are called Provinces and are named as follows (Fig. 7.4)

1. Rengali province
2. Jeypore province
3. Krishna province
4. Eastern Ghats province.

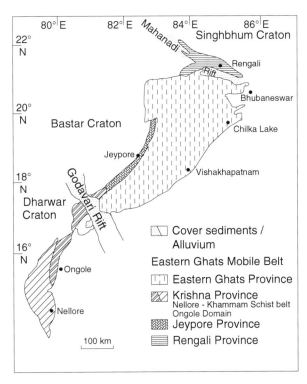

Fig. 7.4 Newer division of Eastern Ghats Mobile Belt into crustal provinces (after Dobmeier and Raith, 2003). See text for details

7.10 Newer Divisions of the Eastern Ghats Mobile Belt

Two of these four Provinces are further divided into Domains separated by mega lineaments and shear zones, following the works of Nash et al. (1996) and Crowe et al. (2001), but ignoring the lineaments recognized in the central part of the EGMB by Chetty and Murthy (1998).

7.10.1 Rengali Province

The Rengali Province is the area with SW-NE trend located between the Bastar/Bhandara and Singhbhum cratons and hence beyond the north of Eastern Ghats Boundary Thrust Zone. This Province is mainly a low- to medium-grade volcano-sedimentary sequence which is similar to the Iron Ore Group and on which basis it was earlier assigned to the Singhbhum craton by Mahalik (1994). On the contrary, Nash et al. (1996) regarded them as a cover sequence to the south-dipping charnockitic-migmatitic gneisses of Angole within EGMB. The Rengali Province is dissected by faults and exposes a block of granulite facies gneisses and mafic granulites—the Badarma Complex, which is also included by Dobmeier and Raith in the Rengali Province. It must be noted that their metamorphic grade and geological history are quite contrasting. The very basis of recognizing Provinces is thus challenged. Moreover, the age and P-T evolution of Rengali Province are poorly constrained.

7.10.2 Jeypore Province

The Jeypore Province represents the northern portion of Western Charnockite Zone of Ramakrishnan et al. (1998), hence dominantly composed of charnockitic-enderbitic rocks. Because of dominance of intermediate rocks with calc-alkaline affinity Dobmeier and Raith (2003) think that these rocks are related to magmatism in an Andean-type active continental margin setting. But it is uncertain what part of the terrain was subducted and why do these granulites give so high pressures coinciding with the pelitic granulites with which they are associated in the EGMB.

7.10.3 Krishna Province

The Krishna Province of Dobmeier and Raith (2003) lies to the south of the Godavari graben. This Province comprises granulites of Ongole (equivalent to southern part of the WCZ of Ramakrishnan et al. (1998) and low- to medium-grade Nellore-Khammam Schist belt which is located between Cuddapah Boundary Shear Zone (CBSZ) and marginal zone of the EGMB. Dobmeier and Raith justify the grouping of these crustal terrains by geochronological data. Following Raman and Murthy (1997), the Nellore-Khammam belt is recognized as the Nellore Schist Belt (NSB) in the south and the Khammam Schist Belt (KSB) in the north. The western upper part of the NSB, consisting of low-grade metasediments, has been named

Udayagiri Domain and the eastern lower unit of medium-grade meta-volcano sedimentary sequence named as Vinajamuru Domain (VD) by Dobmeier and Raith (2003, p. 153). At Chamalpahar in the northern end of VD there are layered mafic complexes that synkinematically intruded migmatitic gneisses and amphibolites (Leelanandam and Narashima Reddy, 1998). Dobmeier and Raith assigned the amphibolites to back-arc setting, following Mukhopadhyay et al. (1994). But Hari Prasad et al. (2000) argued that the protoliths formed in two different tectonic settings, comparable to recent ocean island arc and continental marginal arcs. The boundary between the VD and Ongole Domain (South of Godavari graben) is blurred or uncertain (cf. Dobmeier and Raith, 2003, p. 156). From the Nd model ages of 2.6–2.8 Ga and zircon/monazite ages near 1.6 Ga for UHT metamorphism, Kovach et al. (2001) suggested a separate Ongole Domain that possibly thrusted on Domain 1A of Rickers et al. (2001).

7.10.4 Eastern Ghats Province

On the consideration that the results of Nd isotope mapping by Rickers et al. (2001) do not support the subdivisions of the khondalite lithologies earlier proposed by GSI (see Ramakrishnan et al., 1998), Dobmeier and Raith (2003) carved out the largest Province called the Eastern Ghats Province. This Province includes the Eastern Khondalite Zone (EKZ), the Central Charnockite-Migmatite Zone (CMZ) and a part of the Western Khondalite Zone (WKZ) of Ramakrishnan et al. (1998). The Eastern Ghats Province contains all granulite facies rocks and the sapphirine-bearing assemblages. Lastly, if the metamorphic history of the proposed domains is different, it is a matter of serious enquiry as to when and how the collage of the different domains/blocks had occurred. The models of Rickers et al. (2001) as well as of Dobmeier and Raith (2003) fail to answer this and other questions put forth in the preceding sections.

The different schemes of classification have been compiled by the writer in a comparative way and shown in Table 7.1.

The Eastern Ghats Province is extended by Dobmeier and Raith (2003) to south of Godavari graben, which was earlier recognized as a crustal-scale boundary of the EGMB. The ranges of Nd model ages given by Rickers et al. (2001) were also used for further subdivision of this Province into Domains. However, these divisions reflect regional scale compositional heterogeneity for the EG-Province.

From U–Pb ages of detrital zircons Dobmeier and Raith infer that deposition of sediments (precursor of khondalitic gneisses etc.) did not cease before 1.35 Ga, while an early episode of felsic magmatism (obtained from Sm–Nd whole-rock isochron) at 1.45 Ga (Shaw et al., 1997) was found to be almost coeval with an alkaline complex near Khariar (Sarkar and Paul, 1998; Aftalion et al., 2000). But it is not definitely known whether the alkaline complex intruded the Bastar craton or the Eastern Ghats Province. A second pulse of felsic plutonism is inferred at about 1.2 Ga, on the basis of discordant U–Pb zircon data (Crowe, 2003). Dobmeier and Raith have tried to constrain the early UHT metamorphism (8 kbar/950°C) at

7.10 Newer Divisions of the Eastern Ghats Mobile Belt

Table 7.1 Newer divisions of the Eastern Ghats Mobile Belt with respect to those given by GSI (1998)

Basis	Divisions					References
Lithological assemblage	Transition zone	Western charnockite zone	Western khondalite zone	Central charnockite-migmatite zone	Eastern khondalite zone	Ramakrishnan et al. (1998) GSI view
	(TZ) between cratons and EGMB and has characteristics of both	(WCZ) mainly charnockite-enderbite with lenses of mafic/ultra-mafics and minor metapelites	(WKZ) mainly metapelites intercalated with Qtzites, calcsilicates, marbles and Mg–Al granulites. Contain anorthothosite	(CMZ) mainly migmatite incl. Qz-diatextite, Mg–Al granulites, intrusions of charnockite. CMZ missing S of Godavari rift	(EKZ) similar to WKZ but devoid of anorthosites	
Nd Model age (assumed depleted mantle as source of nearly all EGMB rocks)	← Domain-I, 2.9–3.9 Ga →		← Domain-II, 2.2–2.5 Ga → (south of Nagavalli-Vamsadhara (NV) SZ)			Rickers et al. (2001)
	Domain-IV 2.5–2.9 Ga between Mahanadi SZ and SB-craton in the North			← Domain-III → 1.8–2.2 Ga bounded on three sides by lineaments; NVSz on SW Mahanadi Sz on N and Koraput-Sonapur Sz on W		

Table 7.1 (continued)

Basis	Divisions		References
Distinct Geol. history; limits of mega-lineaments; also geochronology for further sub-divisions into domains	← *Rengali province* → (N of Mahanadi graben)	← *Jeypore province* → (Confined only to N of Mahanadi graben)	Dobmeier and Raith (2003)
		← *Krishna province* → (confined mainly to S of Godavari; consists of igneous complexes (anorthosite incl.) Includes Nellore and Khammam Schist Belts on W)	
		← *Eastern Ghats province* → (recognized on both sides of Godavari but missing to the N of Mahanadi)	

1100 Ma from dating with common Pb method and ^{207}Pb–^{206}Pb SHRIMP ages for euhedral zircons from leucogranite. This is a very significant conclusion along with the longitudinal extent of the Eastern Ghats Province across the Godavari graben. The second granulite facies metamorphism is dated at about 1000 Ma from U–Pb monazite ages from metasedimentary and metaigneous rocks (Aftalion et al., 1988; Mezger and Cosca, 1999) and from Pb–Pb single zircon evaporation ages from high Mg–Al granulites (Jarick, 2000) and EPMA monazite dates (Simmat, 2003). U–Pb age-data of sphene from calc-silicate rocks of the Eastern Ghats Province gave cooling age at about 570–530 Ma, pointing to a Pan-African thermal overprinting of the Khariar Domain. But it should be noted that the Khariar Domain of the Eastern Ghats Province is located near the mega-lineament, called Sileru Shear Zone (Chetty, 1995), or Lecher-Kunavaram-Koraput shear zone (Chetty and Murthy, 1998), or Lakhna thrust (Biswal, 2000), and this Pan-African age could be due to reactivation of this shear zone. The Angul Domain north of Mahanadi graben, however, gave different cooling ages at 930 ± 30 Ma, which led Dobmeier and Raith to suggest that this Domain was tectonically and thermally decoupled from the Eastern Ghats Province prior to Pan-African thermal event. The authors have given a comparative picture of isotopic data and thermo-tectonic events (Dobmeier and Raith, 2003, Fig. 5, p. 154) for their proposed crustal provinces of the Eastern Ghats fold belt, which are likely to be modified by more intense geological and geochronological work on this mobile belt.

References

Aftalion, M., Bowes, D.R., Dash, B. and Dempster, T.J. (1988). Late Proterozoic charnockites in Orissa, India: U–Pb and Rb–Sr isotopic study. J. Geol., vol. 96, pp. 663–676.

Aftalion, M., Bowes, D.R., Dash, B. and Fallick, A.E. (2000). Late Pan-African thermal history in the Eastern Ghats terrane, India from U–Pb and K–Ar isotopic study of the Mid-Proterozoic Khariar alkali syenite, Orissa. Geol. Surv. India Spl. Publ., vol. 57, pp. 26–33.

Aswathanaryan, U. (1964). Isotopic ages from the Eastern Ghats and Cuddapahs of India. J. Geophys. Res., vol. **69**, pp. 3479–3486.

Bhadra, S., Banerjee, M. and Bhattacharya, A. (2003). Tectonic restoration of polychronous mobile belt-craton assembly: constraints from corridor study across the western margin of the Eastern Ghats Belt, India. Mem. Geol. Soc. India, vol. **52**, pp. 109–130.

Bhattacharya, S. (1996). Eastern Ghats granulite terrain of India: an overview. J. Southeast Asian Earth Sci., vol. **14**, pp. 165–174.

Bhattacharya, A. and Gupta, S. (2001). A reappraisal of polymetamorphism in the Eastern Ghats Belt—a view from north of the Godavari rift. Proc. Indian Acad. Sci. (Earth Planet. Sci.), vol. **110**(4), pp. 369–383.

Bhattacharya, S. and Kar, R. (2002). High temperature dehydration melting and decompressive P-T path in a granulite complex from the Eastern Ghats, India. Contrib Mineral Petrol, vol. **143**, pp. 175–191.

Bhattacharya, S. and Kar, R. (2004). Alkaline intrusion in a granulite ensemble in the Eastern Ghats Belt, India: shear zone pathways and pull-apart structure. Proc. Indian Acad. Sci. (Earth Planet Sci.), vol. **113**(1), pp. 37–48.

Bhattacharya, S., Kar, R., Misra, S. and Teixeira, W. (2001). Early Archaean continental crust in the Eastern Ghats Belt, India: isotopic evidence from the charnockite suite. Geol. Mag., vol. **138**, pp. 609–618.

Bhattacharya, S., Sen, S.K. and Acharyya, A. (1994). Structural setting of the Chilka Lake granulite-migmatite-anorthosite suite with emphasis on the time relation of charnockites. Precambrian Res., vol. **66**, pp. 393–409.

Biswal, T.K. (2000). Fold-thrust geometry of the Eastern Ghats Mobile Belt, a structural study from its western margin, Orissa, India. J. African Earth Sci., vol. **31**, pp. 25–33.

Biswal, T.K., Ahuja, H. and Sahu, H.S. (2004). Emplacement kinematics of nepheline syenites from the Terrane Boundary shear zone of the Eastern Ghats Mobile Belt, west Khariar, NW Orissa: Evidence from Meso- and micro-structures. Proc. Indian Acad. Sci. (EPS), vol. **113**, pp. 785–793.

Boger, S.D., Wilson, C.J.L. and Fanning, C.M. (2001). Early Palaeozoic tectonism within the East Antarctica craton: the final suture between east and west Gondwana? Geology, vol. **29**, pp. 463–466.

Bose, M.K. (1971). Petrology of the alkaline suite of Sivamalai, Coimbatore, Tamil Nadu. J. Geol. Soc. India, vol. **12**, pp. 241–261.

Bose, S., Fukuoka, M., Sengupta, S. and Dasgupta, S. (2000). Evolution of high Mg–Al granulites from Sunkarametta, Eastern Ghats, India: evidence for a lower crustal heating-cooling trajectory. J. Metam. Geol., vol. **18**, pp. 223–240.

Braun, I. and Kriegsman, L.M. (2002). Proterozoic crustal evolution of southernmost India and Sri Lanka. In: Yoshida, M., Windley, B.F. and Dasgupta, S. (eds.), Proterozoic East Gondwana: Supercontinent Assembly and Breakup. Geol. Soc. Lond. Spl. Publ., vol. **206**, London.

Chetty, T.R.K. (1995). A correlation of Proterozoic shear zones between Eastern Ghats Belt, India and Enderby Land, East Antarctica, based on LANDSAT imagery. Mem. Geol. Soc. India, vol. **34**, pp. 205–220.

Chetty, T.R.K. and Murthy, D.S.N. (1998). Eluchuru-Kunavaram-Koraput (EKK) shear zone, Eastern Ghats Granulite Terrane, India: a possible Precambrian suture zone. Gondwana Res. Group Mem., vol. **4**, pp. 37–48.

Chetty, T.R.K., Vijay, P., Narayana, B.L. and Girdhar, G.V. (2003). Structure of the Nagavali shear zone, Eastern Ghats Mobile Belt, India: correlation in the East Gondwana reconstruction. Gondwna Res., vol. **6**(2), pp. 215–219.

Crawford, A.R. (1969). Reconnaissance Rb–Sr dating of Precambrian rocks of southern peninsular India. J. Geol. Soc. India, vol. **10**, pp. 117–166.

Crowe, W.A. (2003). Age constraints for magmatism and deformation in the northern Eastern Ghats Belt, India: implications for the development of East Gondwanaland. Precambrian Res., vol. **110**, pp. 1–8.

Crowe, W.A., Cosca, M.A. and Harris, L.B. (2001). ^{40}Ar/^{39}Ar geochronology and Neoproterozoic tectonics along the northern margin of the Eastern Ghats Belt, in north Orissa, India. Precambrian Res., vol. **108**, pp. 237–266.

Dasgupta, S., Ehl, J., Raith, M.M. and Sengupta, P. (1997). Mid-crustal contact metamorphism around the Chimakurthy mafic-ultramafic complex, Eastern Ghats Belt, India. Contrib Mineral Petrol, vol. **129**, pp. 182–197.

Dasgupta, S., Sanyal, S., Sengupta, P. and Fukuoka, M. (1994). Petrology of the granulites from Anakapalle—evidence for Proterozoic decompression in the Eastern Ghats, India. J. Petrol., vol. **35**, pp. 433–459.

Dasgupta, S. and Sengupta, P. (1995). Ultrametamorphism in Precambrian granulite terrains—evidence from Mg-Al granulites and calc-granulites of the Eastern Ghats, India. Geol. J., vol. **30**, pp. 307–318.

Dasgupta, S. and Sengupta, P. (2003). Indo-Antarctic correlation: a perspective from the Eastern Ghats granulite belt, India. In: Yoshida, M., Windley, B.F. and Dasgupta, S. (eds.), Proterozoic East Gondwana: Supercontinent Assembly and Breakup. Geol. Soc. Lond. Spl. Publ., vol. **206**, London, pp. 131–143.

Dasgupta, S. and Sengupta, P., Ehl, J., Raith, M. and Bardhan, S. (1995). Reaction textures in a suite of spinel granulites from the Eastern Ghats Belt, India: evidence for polymetamorphism, a partial petrogenetic grid in the system KFMASH and roles of ZnO and Fe_2O_3. J. Petrol., vol. **36**, pp. 435–461.

Dobmeier, C. and Raith, M.M. (2003). Crustal architecture and evolution of the Eastern Ghats Belt and adjacent regions of India. In: Yoshida, M., Windley, B.F. and Dasgupta, S. (eds.), *Proterozoic East Gondwana: Supercontinent Assembly and Breakup*. Geol. Soc. Lond. Spl. Publ., vol. **206**, London, pp. 145–168.

Dobmeier, C and Simmat, R. (2002). Post-Grenvillian transpression in the Chilka Lake area, Eastern Ghats Belt—implications for the geological evolution of peninsular India. Precambrian Res., vol. **113**, pp. 243–268.

England, P.C. and Thompson, A.B. (1984). Pressure-temperature-time paths of regional metamorphism. I. Heat transfer during the evolution of regions of overthickened continental crust. J. Petrol., vol. **25**, pp. 894–928.

Fitzsimons, I.C.W. (2000). Grenville-age basement provinces in East Antarctica: evidence for three separate collisional orogens. Geology, vol. **28**, pp. 879–882.

Gopalkrishnan, K. (1998). Extensions of Eastern Ghats Mobile Belt, India—a geological enigma. Geol. Surv. India Spl. Publ., vol. **44**, pp. 22–38.

Grew, E.S. and Manton, W.I. (1986). A new correlation of sapphirine granulites in the Indo-Antarctic metamorphic terrain: Late Proterozoic dates from the Eastern Ghats Province of India. Precambrian Res., vol. **33**, pp. 123–137.

GSI (1998). Geological and Mineral Map of Tamil Nadu and Pondicherry. Scale 1 : 500,000 with Explanatory Brochure, 38 p. GSI Publ. Kolkata.

Gupta, S., Bhattacharya, A., Raith, M. and Nanda, J.K. (2000). Contrasting pressure–temperature–deformational history across a vestigial craton-mobile belt assembly: the western margin of the Eastern Ghats Belt at Deobogh, India. J. Metam. Geol., vol. **18**, pp. 683–697.

Halden, N.M., Bowes, D.R. and Dash, B. (1982). Structural evolution of migmatites in granulite facies terrane: Precambrian crystalline complex of Angul, Orissa, India. Trans. R. Soc. Edinburgh: Earth Sciences, vol. **73**, pp. 109–118.

Hari Prasad, B., Okudaira, T., Hayasaka, Y., Yoshida, M, and Divi, R.S. (2000). Petrology and geochemistry of amphibolites from the Nellore-Khammam Schist Belt, SE India. J. Geol. Soc. India, vol. **56**, pp. 67–78.

Hoffman, P.F. (1991). Did the breakup of Laurentia turn Gondwanaland inside-out? Science, vol. **42**, pp. 1409–1412.

Jarick, J. (2000). Die thermotecktonometamorphe Entwicklung des Eastern Ghats Belt, Indien—ein Test der SWEAT Hypothese. Ph.D. thesis, J.W. Goethe Universitaet, Frankfurt.

Kaila, K.L., Tewari, H.C., Roy Chowdhury, K., Rao, V.K., Sridhar, A.R. and Mall, D.M. (1987). Crustal structure of the northern part of the Proterozoic Cuddapah basin of India from deep seismic soundings and gravity data. Tectonophysics, vol. **140**, pp. 1–12.

Kar, R. (2001). Patchy charnockites from Jenapore, Eastern Ghats granulite belt, India: structural and petrochemical evidences attesting to their relict nature. Proc. Indian Acad. Sci. (Earth Planet. Sci.), vol. **110**, pp. 337–350.

Kelly, N.M., Clark, G.L., Carson, C.J. and White, R.W. (2000). Thrusting in the lower crust: evidence from the Oygarden Islands, Kemp Land, East Antarctica. Geol. Mag., vol. **137**, pp. 219–234.

Kovach, V.P., Simmat, R., Rickers, K., Berezhnaya, N.G., Salnikova, E.B., Dobmeier, C., Raith, M.M., Yakovleva, S.Z. and Kotov, A.B. (2001). The Western charnockite zone of the Eastern Ghats Belt, India: An independent crustal province of Late Archaean (2.8 Ga) and Paleoproterozoic (1.7–1.6 Ga) terrains. Gondwana Res., vol. **4**, pp. 666–667.

Krause, O., Dobmeier, C., Raith, M.M. and Mezger, K. (2001). Age of emplacement of massif-type anorthosites in the Eastern Ghats Belt, India: constraints from U-Pb zircon dating and structural studies. Precambrian Res., vol. **109**, pp. 25–38.

Lal, R.K. (1997). Internally consistent calibrations for geothermobarometry of high-grade Mg–Al rich rocks in the system $MgO-Al_2O_3-SiO_2$ and their application to sapphirine-spinel granulites of Eastern Ghats, India and Enderby Land, Antarctica. Proc. Indian Acad. Sci. (Earth Planet Sci.), vol. **103**, pp. 91–113.

Leelanandam, C. (1967). Occurrence of anorthosites from the charnockitic area of Kondapalle, Andhra Pradesh. Bull. Geol. Soc. India, vol. **4**, pp. 5–7.

Leelanandam, C. (1990). The anorthosite complexes and Proterozoic mobile belt of peninsular India: a review. In: Naqvi, S.M. (ed.), Precambrian Continental Crust and Its Mineral Resources. Dev. Precambrian Geol., vol. **8**. Elsevier, Amsterdam, pp. 409–435.

Leelanandam, C. (1997). The Kondapalle layered complex, Andhra Pradesh, India: a synoptic overview. Gondwana Res., vol. **1**, pp. 95–114.

Leelanandam, C. (1998). Alkaline magmatism in the Eastern Ghats Belt—a critique. Geol. Surv. India Spl. Publ., vol. **44**, pp. 170–179.

Leelanandam, C., Burke, K., Ashwal, L.D. and Webb, S.J. (2006). Proterozoic mountain building in peninsular India: an analysis based primarily on alkaline rocks distribution. Geol. Mag., vol. **43**, pp. 195–212.

Leelanandam, C. and Narashima Reddy, M. (1998). Precambrian anorthosites from peninsular India—problems and perspectives. Geol. Surv. India Spl. Publ., vol. **44**, pp. 152–169.

Mahalik, N.K. (1994). Geology of the contact between the Eastern Ghats Belt and the north Orissa craton, India. J. Geol. Soc. India, vol. **44**, pp. 41–51.

Mezger, K. and Cosca, M.A. (1999). The thermal history of the Eastern Ghats Belt (India), as revealed by U–Pb and $^{40}Ar/^{39}Ar$ dating of metamorphic and magmatic minerals: implications for the SWEAT correlation. Precambrian Res., vol. **94**, pp. 251–271.

Mezger, K., Cosca, M.A. and Raith, M.M. (1996). Thermal history of the Eastern Ghats Belt (India) deduced from U–Pb and $^{40}Ar/^{39}Ar$ dating of metamorphic minerals. (Abstract) V.M. Goldschmidt Conference, 1, p. 401.

Mezger, K. and Krogstard, E.J. (1997). Interpretation of discordant U–Pb zircon ages: An evaluation. J. Metam. Geol., vol. **15**, pp. 127–140.

Mukhopadhyay, I., Ray, J. and Guha, S.B. (1994). Amphibolitic rocks around Kotturu, Khammam district, Andhra Pradesh: structural and petrological aspects. Indian J. Geochem., vol. **9**, pp. 39–53.

Naqvi, S.M. and Rogers, J.J.W. (1987). Precambrian Geology of India. Oxford University Press, New York, 223p.

Nash, C.R., Rankun, L.R., Leeming, P.M. and Harris, L.B. (1996). Delineation of lithostructural domains in northern Orissa (India) from landsat thematic mapper imagery. Tectonophysics, vol. **260**, pp. 245–257.

Pande, O.P. and Rao, V.K. (2006). Missing granitic crust (?) in the Godavari graben of southeastern India. J. Geol. Soc. India, vol. **67**(3), pp. 307–319.

Paul, D.K., Ray Barman, T., McNaughton, N.J., Fletscher, I.R., Potts, P.J., Ramakrishnan, M. and Augustine, P.F. (1990). Archaean-Proterozoic evolution of Indian charnockites: isotopic and geochemical evidence from granulites of the Eastern Ghats Belt. J. Geol., vol. **98**, pp. 253–263.

Perraju, P., Kovach, A. and Svingor, E. (1979). Rubidium-strontium ages of some rocks from parts of the Eastern Ghats in Orissa and Andhra Pradesh, India. J. Geol. Soc. India, vol. **20**, pp. 290–296.

Philpots, A. (1990). Principles of Igneous and Metamorphic Petrology. Prentice Hall, New Jersey, 498p.

Powell, C.McA., Li, Z.X., Mcellhinny, M.W., Meert, J.G. and Park, J.K. (1994). Palaeomagnetic constraints on the timing of the Neoproterozoic breakup of Rodinia and the Cambrian formation of Gondwana. Geology, vol. **22**, pp. 889–892.

Ramakrishnan, M., Nanda, J.K. and Augustine, P.F. (1998). Geological evolution of the Proterozoic Eastern Ghats Mobile Belt. Geol. Surv. India Spl. Publ., vol. **44**, pp. 1–21.

Ramakrishnan, M. and Vaidyanadhan, R. (2008). *Geology of India,* vol. **1**. Geological Society India, Bangalore, 556p.

Raman, P.K. and Murthy, V.N. (1997). Geology of Andhra Pradesh. Geological Society of India, Bangalore, 245p.

Rao, A.T., Kamineni, D.C., Arima, M. and Yoshida, M. (1995). Mineral chemistry and metamorphic P-T conditions of a new occurrence of sapphirine granulites near Madhuravada in the Eastern Ghats, India. In: Yoshida, M., Santosh, M. and Rao, A.T. (eds.), *India as a Fragment of East Gondwana*. Gondwana Res. Group Mem., vol. **2**, Gondwana, pp. 109–121.

References

Rickers, K., Mezger, K. and Raith, M.M. (2001). Evolution of the continental crust in the Proterozoic Eastern Ghats Belt, India and new constraints for Rodinia reconstruction: implications from Sm—Nd, Rb—Sr and Pb—Pb isotopes. Precambrian Res., vol. **112**, pp. 183–212.

Rogers, J.J.W. (1996). History of 3 billion years old continents. J. Geol., vol. **104**, pp. 91–107.

Sarkar, A., Bhanumathi, L. and Balasubramanyan, M.N. (1981). Petrology, geochemistry, and geochronology of the Chilka Lake igneous complex, Orissa State, India. Lithos, vol. **14**, pp. 93–111.

Sarkar, A., Pati, U.C., Panda, P.K., Patra, P.C., Kundu, H.K. and Ghosh, S. (1998). Late-Archaean charnockitic rocks from the northern marginal zones of the Eastern Ghats Belt: a geochronolgical study (Abstract). International Seminar on *Precambrian Crust in Eastern and Central India,* held in October 29–30, Bhubaneswar.

Sarkar, A. and Paul, D.K. (1998). Geochronology of the Eastern Ghats Precambrian Mobile Belt—a review. Geol. Surv. India Spl. Publ., vol. **44**, pp. 51–86.

Sen, S.K., Bhattacharya, S. and Acharyya, A. (1995). A multi-stage pressure-temperature record in the Chilka Lake granulites: the epitome of metamorphic evolution in the Eastern Ghats, India. J. Metam. Geol., vol. **13**, pp. 287–298.

Sengupta, P., Dasgupta, S., Bhattacharya, P.K., Fukoka, M., Chakraborti, S. and Bhowmik, S. (1990). Petrotectonic imprints in the sapphirine granulites from Anantagiri, Eastern Ghats Mobile Belt, India. J. Petrol., vol. **31**, pp. 971–996.

Sengupta, P., Dasgupta, S., Raith, M., Bhui, U.K. and Ehl, J. (1999). Ultra-high temperature metamorphism of metapelitic granulites from Kondapalle, Eastern Ghats Belt: implications for the Indo-Antarctic correlation. J. Petrol., vol. **40**, pp. 1065–1087.

Sharma, R.S. and Pandit, M.K. (2003). Evolution of early continental crust. Curr. Sci., vol. **84**(8), pp. 995–1001.

Shaw, R.K. and Arima, M. (1996). High-temperature metamorphic imprint from calc-granulites of Rayagada, Eastern Ghats, India: implications of isobaric cooling path. Contrib Mineral Petrol, vol. **126**, pp. 169–180.

Shaw, R.K., Arima, M., Kagami, H., Fanning, C.M., Shiraishi, K. and Motoyoshi, Y. (1997). Proterozoic events in the Eastern Ghats granulite belt, India: evidence from Rb—Sr, Sm—Nd systematics and SHRIMP dating. J. Geol., vol. **105**, pp. 645–656.

Simmat, R. (2003). Indentifizierung hochgradig metamorpher Provinzen des Eastern Ghats Belt in Indien anhand EMS-Studie von Monazit-Altersmustern. Ph.D. thesis, University of Bonn, Germany.

Subba Rao, T.V., Bhaskar Rao, Y.J., Sivaraman, T.V. and Gopalan, K. (1989). Rb—Sr age and petrology of Elchuru alkaline complex: implication to alkaline magmatism in the Eastern Ghats Mobile Belt. Mem. Geol. Soc. India, vol. **15**, pp. 207–223.

Subramanyan, C. (1978). On the relation of gravity anomalies to geotctonics of Precambrian terrains of south Indian shield. J. Geol. Soc. India, vol. **19**, pp. 241–263.

Torsvik, T.H., Carter, L.M., Ashwal, L.D., Bhusan, S.K., Pandit, M.K. and Jamtveit, B. (2001). Rodinia refined or obscured: palaeomgnetism of the Malani igneous suite (NW India). Precambrian Res., vol. **108**, pp. 319–333.

Upadhyay, D. (2008). Alkaline magmatism along the southeastern margin of the Indian shield: Implications for regional geodynamics and constraints on craton-Eastern Ghats Belt suturing. Precambrian Res., vol. **162**, pp. 59–62.

Upadhyay, D., Raith, M.M., Mezger, K., Bhattacharya, A. and Kinny, P.D. (2006b). Mesoproterozoic rifting and Pan-African continental collision in SE India: evidence from the Khariar alkaline complex. Contrib Mineral Petrol, vol. **151**, pp. 434–456.

Upadhyay, D., Raith, M.M., Mezger, K. and Hammerschmidt, K. (2006a). Mesoproterozoic rift-related alkaline magmatism at Eluchuru, Prakasam alkaline complex, SE India. Lithos, vol. **89**, pp. 447–477.

Vinogradov, A., Tugarinov, A., Zhykov, C., Stapnikova, N., Bibikova, E. and Korre, K. (1964). *Geochronology of the Indian Precambrian.* Report 22nd International Geological Congress, Part 10, pp. 553–567.

Yoshida, M. (2007). Geochronological data evaluation: Implications for the Ptroterozoic tectonics of East Gondwana. Gondwana Res., vol. **12**, pp. 228–241.

Chapter 8
Pandyan Mobile Belt

8.1 Introduction: Geological Setting

Pandyan Mobile Belt (PMB) is the name given by Ramakrishnan (1993, 1988) to the Southern Granulite Terrain (SGT) situated to the south of the E-W trending Palghat-Cauvery Shear Zone (PCSZ) (Fig. 8.1). The name Pandyan is adopted after the legendary dynasty that ruled this part of South India in the historical past. Interestingly, the SGT has been defined variously by different workers. According to Fermor (1936), this terrain is a part of the large "Charnockite Province" located to the south of the orthopyroxene-in (Opx-in) isograd, delineated along a line straddling the join Mangalore-Mysore-Bangalore-Chennai (Pichamuthu, 1965). It would not be justifiable to encompass the entire granulite province under the Pandyan mobile belt since this would include both Archaean granulites (Northern Terrain, see below) and Neoproterozoic granulites (south of the PCSZ), each with their distinctive differences in geological setting (Ramakrishnan, 1998). Vemban et al. (1977) pointed out this difference which was later emphasized and elaborated by Drury et al. (1984).

A reference to Fig. 8.1 reveals that the SGT (south of Opx-in isograd) is dissected into different granulite blocks by structural discontinuities, lineament or shear zones which are taken as boundaries of the differerent granulite blocks. The most conspicuous is the E-W trending Palghat-Cauvery shear zone (PCSZ), first mapped as Cauvery Fault by ONGC and recognized by Grady (1971) and Vemban et al. (1977) as a boundary between two different geological domains. Drury and Holt (1980) established this boundary from Landsat imagery and named it as Noyil-Cauvery shear zone with Moyar Shear zone (MSZ) and Bhavani Shear zone (BSZ) as its subsidiary (Fig. 8.1). On this basis, the southern Indian granulite terrain to the south of the Opx-in isograd is divided into two distinct crustal blocks (Drury et al., 1984; see also Meissner et al., 2002), viz. the Northern Block (NB) or the Northern Granulite Terrain (NBSGT); and the Southern Block (SB) or Southern Granulite Block (SBSGT), separated by the Palghat-Cauvery Shear zone, PCSZ (see inset in Fig. 8.1). These two Blocks correspond to the Northern Marginal Zone and Central Madurai Zone of Ramakrishnan (1988).

Realizing the grand scale of this shear zone, the term Cauvery Shear zone (CSZ) system was used in maps by GSI and ISRO (1994; Gopalkrishnan, 1996 and

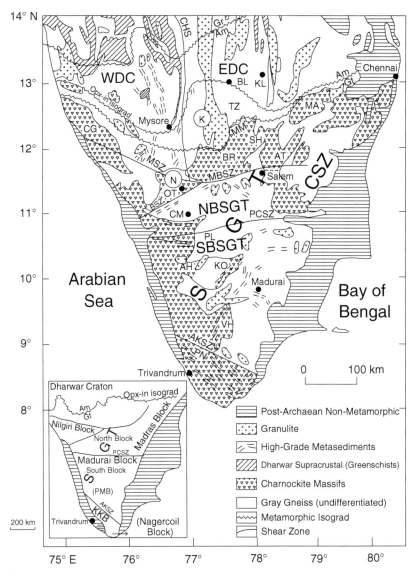

Fig. 8.1 Simplified Geological map of the southern India (after GSI and ISRO, 1994), showing the major geological domains, the Western Dharwar Craton (WDC), Eastern Dharwar Craton (EDC), and Southern Granulite Terrain (SGT) along with the Cauveri Shear Zone System (CSZ). Abbreviations: AKSZ = Achankovil Shear Zone; AH = Anamalai Hills; AT = Attur; BS = Bhavani Shear zone; BL = Bangalore; BR = Biligirirangan; CHS = Chitradurga Shear Zone; CG = Coorg; CM = Coimbatore; EDC = East Dharwar Craton; K = Kabbaldurga; KL = Kolar; KKB = Kerala Khondalite Belt; MS = Moyar Shear zone; N = Nilgiri; OT= Ooty; PCSZ = Palgahat Shear Zone; PL = Pollachi; PMB = Pandyan Mobile Belt; SGT = Southern Granulite Terrain; SH = Shevaroy Hills; WDC = West Dharwar Craton; GR-Am = Isograd between Greenschist and Amphibolite Facies; Am-Gt = Isograd between Amphibolite and Granulite Facies; TZ = Transition Zone of amphibolite and granulite facies. *Inset* shows various identified crustal blocks

references therein). The E-W trending crustal-scale shear zones, namely MSZ, BSZ, and PCSZ in Southern Granulite Terrain have been taken to represent major lineaments that are associated with significant regional strike-slip movements and delineated by major river valleys. The BSZ and MSZ enclose the *Nilgiri Block* or Nilgiri granulite massif. These two shear zones join near Bhavani Sagar and extend eastward as Moyar-Bhavani Shear Zone (MBSZ). The SBSGT is the Madurai Block of some workers (see Ramakrishnan, 1998). This block is separated on its south from another granulite block, the Kerala Khondalite Belt (KKB) or Trivandrum Block by the NW-SE trending lineament called the Achankovil Shear Zone (AKSZ) (Santosh, 1996) (see Fig. 8.2). The Madurai Block located between the PCSZ and the AKSZ was designated earlier by the name Pandyan Mobile Belt (PMB) by Ramakrishnan (1993). But the present nomenclature of the PMB is the granulite block south of the PCSZ, being Neoproterozoic in age (see in Ramakrishnan and Vaidyanadhan, 2008).

All these above-stated blocks of the SGT have granulite facies rocks but they differ in regard to their relative abundances of litho-units, deformational history, ages, character of magmatism and exhumation history. The following section gives a brief account of the geological attributes for each granulite blocks so that the reader can appreciate mutual relationship of the granulite blocks as well as the geological status of the Pandyan Mobile belt and its evolution in light of the temporal and spatial data of the SGT.

8.2 Geological Attributes of the Granulite Blocks

As mentioned earlier, the southern Indian granulite terrain occurs to the south of the Opx-in isograd. This terrain is dissected by Proterozoic shear zones that separate the granulite rocks into four regions or blocks, viz. Madras Block, Nilgiri Block, Madurai Block, and Trivandrum Block (se inset Fig. 8.1). In all these blocks of the SGT, charnockites are conspicuously exposed. Two distinct types of charnockites can be recognized in the SGT: (1) massive charnockite and (2) Incipient or arrested chatrnockite. The massive charnockite occurs in the Nilgiri and Madras Blocks.

Historically, the *Madras granulites* are significant, being the first to be investigated. However, the charnockites vary in composition from adamellite, monzonite, diorite to granite. These hypersthene-bearing rocks were used for the tombstone of Job Charnock, the founder of modern Kolkata (formerly Calcutta) and as a mark of respect to him, they were named Charnockite by Thomas Holland (1900). In the area around Madras, the charnockites are interbedded with garnet-cordierite-sillimanite gneisses, graphite-sillimanite gneiss, pyroxenite, and hornblende granulite and also contain intrusion of norite (hypersthene gabbro). The charnockites were first considered igneous origin but later they were found to be metamorphic, recrystallized in deep-seated conditions into granulite facies. The acid charnockites form St. Thomas Mount near Chennai (formerly Madras) yielded 2580 ± 45 Ma (Crawford, 1969; see also Bernard-Griffiths et al., 1987; Unnikrishnan-Warrior et al., 1995). The P-T

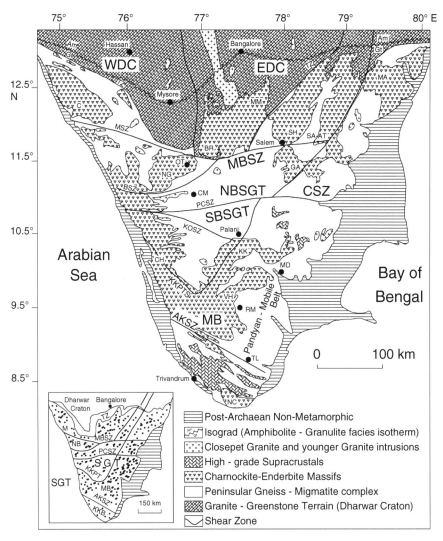

Fig. 8.2 Simplified Geological map of part of the southern Indian shield (based on GSI and ISRO, 1994), showing the location of the Southern Granulite Terrain (SGT) to the south of the Am/Gt (Amphibolite-granulite facies) isograd (drawn after Pichamuthu, 1965). To the north of the isograd are the Western and Eastern Dharwar Cratons (WDC and EDC). The Highland massifs are: Bilgirirangan (BR), Coorg (C), Cardamom (CH), Madras (MA), Kodaikanal (KK), Malai Mahadeswara (MM), Nilgiri (NG), Shevroy (SH), Varshunadu (VH). The prominent shear zones: The Cauvery Shear Zone System (CSZ), comprising Moyar (MS), Bhavani (BS), Palghat-Cauvery (PCS), Salem-Attur (SA-At). Others are: Gangavalli (GA), Karur-Kambam-Painvu-Trichur (KKPT) and Achankovil Shear Zones (AKSZ), Korur-Odanchitram Shear Zone (KOSZ). The different crustal blocks recognized in the SGT *sensu lato* are: the Northern Block (NBSGT) and Southern Block (SBSGT), Madurai Block (MB). The block between the Palghat Shear Zone (PCS) and the Achankovil Shear zone (AKS) is designated Pandyan Mobile Belt (PMB). To the south of the AKS is the Kerala Khondalite Belt (KKB). Other abbreviations are: CM = Coimbatore; NC = Nagercoil; OT = Ooty; RM = Rajapalayam; TL = Tirunelveli. *Inset* shows the SGT, the KKB and the different shear zones traversing the SGT. Other abbreviations as in Fig. 8.1

estimates by Weaver et al. (1978) for the charnockites from near Pallavaram gave 720–840°C and 9–10 kbar.

The charnockites of the *Transition zone* from southern Karnataka, especially at Kabbal quarry, are associated with gneisses. Pichamuthu (1953, 1960) observed this association of incipient charnockite-gneiss and proposed that the charnockites were the products of metamorphic recrystallization. Impressed by patches and streaks of charnockite, unchanged foliation from gneiss to charnockite, and by fluid inclusion studies, some workers (Janardhan et al., 1982; Newton, 1987; Santosh, 2003) suggested that the charnockites were formed by influx of CO_2- rich fluids from mantle.

Another group of geologists, considered the presence of quartzo-feldspathic veins and pods at the site of gneiss-charnockite association, together with the results of experimental studies on granite system as well as P-T estimates and geochemical-geochronological data. On these considerations, a new model of dehydration melting was proposed for the origin of charnockite. According to this model, vapour-absent melting of biotite and/or hornblende is responsible to produce granite melt and hypersthene in rocks of suitable composition (e.g. tonalite gneiss). In general, all along the Transition Zone, the amphibolite facies gneisses (Peninsular Gneiss) have been transformed isobarically ($\sim$5 kbar/750°C) into foliated to massive charnockite. The incipient charnockites from Kabbal have been dated at 2500 Ma by zircon, monazite and allanite U–Pb geochronology (Buhl et al., 1983; Grew and Menton, 1986). Zircon ion probe studies from Kabbal indicate that the protolith is 2965 $\pm$ 14 Ma old, and was transformed into charnockite at 2500 Ma ago (Friend and Nutman, 1992; Hansen et al., 1987). Besides Kabbal, there are several incipient charnockite regions in this Zone (Fig. 8.1), for example around Krishnagiri-Dharmapuri area (Allen et al., 1983), northern part of the Malai Mahadeswara hills (MM Hills), Bilgiri Rangan hills (B-R Hills) (see Raith et al., 1990). In all these localities, the charnockites retain the dominant N-S to NNE-SSW structural trend that characterize the Dharwar craton located to the north of the Transition zone.

In the *Nilgiri Block*, the granulites are mainly enderbites, although minor charnockites are also found. The granulites are massive to foliated, and are characterized by hypersthene-plagioclase-quartz with or without garnet and relics of biotite. The foliation trends about N65°E with steep dips. There are also numerous concordant lenses of pyroxenite and gabbro which are almost undeformed and show sharp contacts with the enderbitic granulites. Besides, a set of dolerite and picrite dykes are also found. The granulites of the Nilgiri have been sheared on its north and south by Moyar and Bhavani shear zones (Srikantappa et al., 2003). The shearing has developed mylonitic texture in these granulites, without affecting the granulite facies assemblages. However, minor hydration during shearing has developed greenish amphiboles and biotite from the breakdown of pyroxene. The general opinion about the Nilgiri granulite is that they represent deep crust and by their uplift the Nilgiri massif juxtaposed against the amphibolite facies terrain (DC terrain) in the north. The P-T data for the garnet-granulites from Nilgiri yielded 730°C/7 kb in SW part and 750°C/9 kb in the NE toward the Moyar shear zone, indicating a differential uplift of the Nilgiri massif (Raith et al., 1983). It has also been proposed by Raith et al. (1999), that the Nilgiri terrain is an allochthonous unit which was thrust upon the Dharwar craton in a Paleoproterozoic collision tctonics. On the other hand, Drury et al. (1984) correlated the Nilgiri charnockites with those of the Biligiri Rangan Hills (B-R Hills) across the Moyar shear zone. In contrast, Peucat et al. (1989) consider the Nilgiri granulites as syn-accretional type in contrast to the B-R Hills granulite massif which is presumed to be post-accretion. On the basis of geochemical characteristic, the charnockitic rocks of the Nilgiri are inferred to have igneous protoliths of tonalite to granodiorite composition (Condie and Allen, 1984; Srikantappa et al., 1988). The whole-rock Rb−Sr, Pb−Pb, Sm−Nd and U−Pb zircon ages for the enderbite granulites of the Nilgiri Hills suggest that the granulite metamorphism occurred at about 2.5 Ga. The granulite metamorphism, according to Buhl (1987) has closely followed the emplacement of the protolith. A four-point Pb−Pb whole-rock isochron for the Nilgiri charnockites yielded an age of 2578 ± 63 Ma (Peucat et al., 1989). The garnet-bearing charnockites from Ootacamund showed a Sm−Nd isochron age of 2496 ± 15 Ma (Jayananda et al., 1995, 2003). The BR-Hills charnockites have also been dated at 3.4 Ga by U−Pb zircon geochronology (Buhl, 1987), with protolith ages ranging from 3.0 to 3.4 Ga (Bhaskar Rao et al., 1993, Table 1). The 2.5 Ga old granulites with garnet-pyroxene-plagioclase assemblages are also found in Shevaroy hills and Bhavani Sagar. Large exposures of charnockites also occur east of Salem, particularly in the hills of Kollimalai and Pachaimalai hills where they occur amidst gneisses and metasediments (see Fig. 8.1).

All along the PCSZ, especially in the vicinity of the Moyar-Bhavani shear zones, the charnockite gneisses are associated with extensively developed supracrustals of Satyamangalam Group which consists of metapelites, calc-silicates, marbles, quartzites, amphibolites and granulites. Toward SE, the Satyamangalam rocks are associated with dismembered layered complex of anorthosite-gabbro-pyroxenite-chromitites, such as at Sittampundi and Bhavani complexes. U−Pb zircon geochronology (Ghosh et al., 2004) and Sm−Nd whole-rock isochron (Bhaskar Rao et al., 1996) gave coinciding ages of ca. 3000 Ma for the Satyamangalam layered complexes, suggesting their temporal equivalence with the Sargur Group rocks (see Chap. 2). The concordant charnockite gneisses within these supracrustals gave an age of 2.52 Ga (Ghosh et al., 1998, 2004), which indicates involvement of the Satyamangalam Group in the Early Proterozoic (or Late Archaean) charnockite event, so convincingly documented in the Transition Zone.

These data conclusively suggest that the granulite metamorphic event in the different sectors or Blocks to the north of Moyar-Bhavani shear zone occurred at about 2500 Ma (early Proterozoic).

The granulite area between the MBSZ and the PCSZ has charnockite-enderbite granulites interbedded with BIF, cal-granulites, quartzites and minor mafic to ultramafic rocks as at Chennamalai (Fig. 8.1), described already. South of Chennamalai, there are nepheline syenites of Sivamalai in which numerous enclaves of charnockites can be observed. In the vicinity of the MBSZ, there are pink granites (Sankari) which contain enclaves of rocks of Satyamangalam Group. Rb−Sr whole-rock age of 390 ± 40 Ma is obtained for pegmatoidal variety of the Sankari Granite.

The granulites of *Madurai Block* occur south of the PCSZ (Fig. 8.2). The block comprises the Annamalai-Kodaikanal ranges that are made up of massive to gneissic charnockites with minor bands of metasediments (mainly khondalites), conspicuously exposed between Dindigul and Madurai. In addition to these rocks, there are also gneisses, anorthosites Oddanchatran and Kadavur (Janardhan, 1996), and sapphirine-bearing granulites that occur at the margins of Palani-Kodaikanal Highlands (Lal, 1993; Mohan and Windley, 1993; Raith et al., 1997). Toward east, the basic granulites dominate as at Palani. The metasedimentary rocks of Annamalai-Palani ranges in the east are intruded by the Proterozoic anorthosites, mentioned above. The anorthosite intrusion contains large inclusions of deformed granulites and quartzites. The post-emplacement deformation is seen in the stretched and boudinaged inclusions. Symplectite minerals garnet-plagioclase-quart around pyroxene yielded 5–7 kb and 725°C (see Subramanian and Selvan, 1981).

South of Palani, the granulites are spinel-bearing metapelites that are interlayered with charnockite gneisses. These Al-Mg rich granulites are considered to have formed by anatexis at 5 kb/740°C under $P_{H2O} \ll P_{total}$ conditions (Harris et al., 1982). Some authors also consider that the magmatism of large anorthosites and ultramafic bodies and to some extent the abundant Late Proterozoic sodic- and potash–granites in the Madurai Block may have been responsible for causing the ultra high temperature (UHT) metamorphism, seen at places where sapphirine-bearing Mg–Al rich assemblages have formed (Braun and Kriegsman, 2002; Grew, 1982). In the southern Madurai Block, massive to streaky charnockites are found. The streaky appearance is attributed to volatile-deficient dehydration melting of biotite at high temperatures (Brown and Raith, 1996). The charnockites of Madurai Block yielded Sm–Nd ages in the range 2.2–2.5 Ga (Harris et al., 1994). The Kodaikanal charnockites gave model Nd ages of 2.2–3.0 Ga. The U–Pb zircon age from acid charnockite gave an age of 2115 ± 8 Ma for the core and 1838 ± 31 Ma for the rim of the mineral (Jayananda et al., 1995). Recently, Pan-African ages have been reported from the Madurai Block. Rb–Sr age of 550 ± 15 Ma is reported from Madurai gneisses (Hensen et al., 199), and Pb–Pb zircon age by evaporation method gave an age of ∼600 Ma (Bartlett et al., 1998; see also Bhaskar Rao et al., 2003). Mineral ages from the charnockites and Pb–Pb ages of zircon overgrowth are found to range between 550 and 500 Ma (Jayananda et al., 1995; Bartlett et al., 1998). U–Pb zircon geochronology from charnockites of Kodaikanal-Cardimom Hills gave 580 Ma (Miller et al., 1997). All these geochronological data convincingly show that the Madurai block has imprints of Pan-African tectono-thermal event (orogey), unlike the terrain located north of the PCSZ (see above).

Geothermobarometry data point to paleopressures of 4–5 kbar (∼12–15 km depth) in the high-grade Karnataka craton; 9 kbar (∼27 km depth) in the Nilgiri and the BR Hills; and 8 kbar (∼24 km depth) in the Madurai block in an early phase followed by a 4.5 kbar and 700°C. P-T-t paths have a dominant isothermal decompresssive history in general (cf. Bhaskara Rao et al., 2003).

To the south of the Madurai Block and across the AKSZ is the *Trivandrum Block* (also called Kerala Khondalite Belt, KKB) of granulite facies rocks (Fig. 8.2). It is characterized by large exposures of folded khondalite and charnockite. The khondalites are predominant in the northern part, called the *Ponmudi Unit*, while charnockites are abundant in the southern part, called the *Nagercoil Unit* (Santosh, 1996). Both rock-types show migmatitic structures in which quartz-feldspar leucosome alternate with garnet-sillimanite bearing bands and hypersthene-plagioclase bearing bands or layers. The metasedimentary-dominated Ponmudi Unit shows spectacular in-situ charnockitization of gneisses and bleaching of charnockites. The arrested charnockites occur as patches, veins and rafts, all devoid of orientation. Arrested growth is identified by cross-cutting relation of charnockite with the gneissic foliation. The charnockitization seems to have been brought about by dehydration melting reactions, because granitic leucosome is always in proximity of the arrested charnockites and because inverse modal relationship is observed between biotite and hypersthene (Ravindra Kumar and Raghavan, 1992). In Trivandrum Block, anatexis of the gneisses is found to pre-date the charnockite formation since patches and veins of charnockites overprint both melanosome and leucosome as at Ponmudi (Santosh et al., 1990). P-T calculations showed that the incipient charniockitization at Ponmudi occurred at 5.5 kbar and 700–750°C (Santosh et al., 1990). The southern end of the Trivandrum Block (i.e. KKB) around Kanyakumari has migmatites, khondalite and charnockite. P-T data for charnockite and khondalite of Nagercoil massif are 670–720°C and 5 ± 1 kbar (Harris et al., 1982), while Srikantappa et al. (1985) report 750 ± 50°C and 6 ± 1 kbar. Nand Kumar et al. (1991) gave a variable range of 626–833°C and 4–6 kbar. All these values clearly suggest that the Trivandrum Block is formed of low- to medium-pressure granulites (see Chacko et al., 1987). There are also various stages of conversion of Opx-bearing charnockites to biotite gneisses (retrogression) which also occurred under low-pressure conditions (Ravindra Kumar, 2004). Other rocks in the Trivandrum Block are cordierite-gneisses and some calc-silicates with diopside-wollastonite-scapolite-garnet-quartz assemblages, often interleaved with charnockites and gneisses. At fringes of the Ponmudi Unit of the Trivandrum Block, there are massive charnockite-enderbites (Nagercoil Unit). Geochemically, the massive charnockites-enderbites resemble C-type magmas of Kilpatric and Ellis (1992). In contrast to the arrested charnockites whose composition is close to granitoids, Chacko et al. (1992) divided the massive charnockites of south Kerala into two types—felsic and intermediate—which are distinguishable by the contents of K-feldspar and mafic minerals. Chacko et al. argued that the massive charnockite varieties are the source of ultra high temperature (UHT) metamorphism. Like the Madurai Block, the Trivandrum Block hosts extensive alkali granite and granite/syenite of 550 Ma age, indicating anorogenic magmatism related to Pan-African event. In summary, the Madurai Block and Trivandrum Block have undergone similar tectono-thermal evolution in the Pan-African times (Jayananda et al., 1995). And this justifies to combine them as one terrain, or the Pandyan Mobile Belt after Ramakrishnan (see Ramakrishnan and Vaidyanadhan, 2008). Palaeoproterozoic ages of ∼2.0–1.8 Ga, obtained from single zircon Pb evaporation ages, CHIME zircon and monazite ages have been interpreted as maximum ages of sedimentation or high-grade metamorphism (Santosh et al., 2006).

8.2 Geological Attributes of the Granulite Blocks

Deformation in the southern Indian granulite terrain as a whole records evidence of more than one fold phases and shear zones, particularly in the granulites of Karnataka and Tamil Nadu. The granulite massifs of BR-Hills, Shevaroy and Kalravan Hills to the north of the Moyar-Bhavani Shear zones are characterized by N-S to NNE-SSW trend (Dharwar trend). This trend is cut across by two distinct sets of lineaments—NNW and WNW. As a result, the NNE-trending foliation is rotated along the Moyar shear zone with E-W, ESE and ENE trends (Jain et al., 2003). According to the Geological Survey of India (Sugavanam and Vidyadharan, 1988), the granulite terrain is characterized by isoclinal and asymmetrical F1 folds with NNE-SSW axial planes, observed in both 2.6–2.5 Ga old charnockites and supracrustals. The superposed F2 folds with WNW-ESE trend affected migmatites, granulites and 2450 Ma old granites. The F3 are shear folds with N-S trend. Regional warps on WNW-ESE axis mark the F4 deformation and their interference with earlier folds produced domes and basin structures. In the southern extreme part of the Madurai Block, the deformation history involves an earlier folding (F1) about NW-SE axis. The F2 folds have refolded the F1 about NNW-SSE axis. The F2 generally plunges toward NW. Sometimes large folds are produced by the fold interference. Because of the cross folding the foliations show abrupt changes both in strike and dip (Narayanswami and Lakshmi, 1967). The superposed deformation is not observed in the granite intrusions that cut across the foliation and banding of the granulitic rocks and migmatites of the Madurai Block. Near the southern boundary of the PCSZ, the D1 deformation in granulites produced isoclinal folds (F1) and a regional foliation (gneissosity) as a result of the transposition of the original layering in the rocks. The second deformation (D2) in the granulites produced F2 folds with variable fold axes. Their axial planes are subvertical and strike E-W to ESE-WNW. The F2 folds are less tight than the F1 folds. According to Mukhopadhyay et al. (2003), the superposition of F2 on F1 folds produced interference patterns in the gneisses and granulites. The refolding has caused bending of mineral lineation which gave rise to different orientations on the two limbs—subhorizontal on the northern limb and northwesterly plunges on the southern limb. The sheared granulites are often retrograded to amphibolite facies gneisses with quartz-feldspar-hornblende-biotite assemblages. In these rocks, two sets of minor folds have formed wherein isoclinal minor folds with axial planes parallel to the regional gneissosity are refolded by later folds with N-S to NE-SW trending axial planes (cf. Mukhopadhyay et al., 2001, 2003). The sheared rocks are characterized by north-plunging folds with N-S axial trance. Shearing is characteristically seen in the Moyar and Bhavani shear zones. Here, the N-S fabric of the Dharwar craton (lower grade terrain) have been rotated to near E-W orientation along these shear zones which evidently developed later than the regional N-S fabric and granulite facies metamorphism of 2.5 Ga age.

Besides these shear zones, there are a few more shear zones, recently identified by the southern Indian workers (see Bhaskar Rao et al., 2003; Chetty et al., 2003). Amongst these are Karu-Kambam-Painvu-Trichur (KKPT) shear zone, Karur-Odanchatram shear zone (KOSZ), Salem-Attur shear zone (SASZ) can be mentioned (Fig. 8.2). In the SGT, some shear zones are located at the interface of two compositionally different rocks, for example charnockite and metasediments. The differing fabric and competence can facilitate rock-movement at the interface during any stage of deformation, developing a shear surface such as the curved KKPTSZ. Such shear zones do not have large-scale strike-slip movement. Attenuation of a limb of a regional fold could also give rise to a shear zone. The AKSZ may be cited as an example of this type of shear zone (cf. Ghosh and deWit, 2004).

8.3 Pandyan Mobile Belt

The Pandyan mobile belt (PMB), according to Ramakrishnan (1993), is the geological domain between the PCSZ in the north and the AKSZ in the south. Impressed by swirling structural pattern in the Madurai Block, Similar to Limpopo belt in South Africa, and by the general occurrence of fold belts either at the peripehery of a continent or sandwiched between two continents, Ramakrishnan carved out his Pandyan mobile belt from the segmented Southern Granulite Terrain. A few years later, Ramakrishnan (2003) enlarged the domain of his mobile belt and included areas of granulites on both north and south margins of his initially proposed Pandyan mobile belt, perhaps on the consideration of meaningful geochronological data available over almost entire SGT. Ramaskrishnan also incorporated the granulite region north of the MBSZ.

Ramakrishnan (2003) divided the enlarged Pandyan Mobile Belt into three zones, analogous to the division of the Limpopo belt. These zones are: (1) Northern Marginal Zone (NMZ), (2) Central Zone (CZ), and (3) Southern Zone (SZ). According to Ramakrishnan, the NMZ lies just north of the E-W trending Moyar-Bhavani Shear zone (MBSZ) into which the N-S cratonic grains abruptly swereve. This zone is characterized by the charnockite massifs of Shevaroy, Javada, BR-Hills and Nilgiri. The Central Zone, CZ (2) is located between the MBSZ and Noyil-Cauvery shear zone (i.e. PCSZ). This zone consists of mainly high-grade relics of older greenstone (Sargur Group), namely Satyamangalam Group, along with supracrustals enclaves (metasediments and carbonates of Sankari dome) and layered basic complex of Sittampundi. This layered complex has dismembered layered basic intrusions (layered gabbro and leucogabbro) within quartzo-feldspathic gneisses surrounding the Sankari (Sankagiri) dome. The gneisses contain enclaves of amphibolites, marble, calc-silicate rock and BIF. The complex is considered to have derived from LREE depleted mantle sources, by fractional crystallization. Sm–Nd whole-rock isochron gives 2935 ± 60 Ma for the complex. Still younger ages of ~2000 and 730 Ma recorded in the complex are due to later thermal imprints (cf. Bhaskar Rao et al., 1996).

The area lying to the south of the PCSZ is the Southern Zone (SZ). It contains charnockite-khondalite rocks. The SZ of the PMB, according to Ramakrishnan (2003), contains two sub-units on either side of the AKSZ. First is the Kerala Khondalite Belt (KKB) to the south of the AKSZ and is predominantly formed of khondalite-leptynite suite with subordinate charnockite. Second sub-unit is the

Madurai Block to the north of the AKSZ. It consists of quartzo-feldspathic gneisses; and charnockites, like that of the central zone of the Limpopo belt. The Achankovil unit is characterized by the occurrence of cordierite gneisses, marbles and quartzites and pink granite—all of which are nearly absent in other units of the KKB (Cenki et al., 2004).

Age data on the charnockite and metasediments of the Southern Zone of the PMB unambiguously denote Pan-African tecto-thermal event. Jayananda et al. (1995) gave zircon evaporation ages of 652–560 Ma from charnockites near Palani area of the Madurai Block. U–Pb zircon geochronology from Varshanadu, Kodaikanal-Cardimom Hills gave 580 Ma (Miller et al., 1997). Also Madurai gneisses yield Rb–Sr ages of 550 ± 15 Ma (Hansen et al., 1995). The Pan-African event is also noticed in the carbonatites, syenites, and alkali granites of the Madurai Block (Deans and Powell, 1968; Subramanian, 1983; Santosh et al., 1989). More recently, Collins and Santosh (2004) showed from SHRIMP age data on zircons from Gangavarpatti quartzite (adjacent to sapphirine-bearing metapelites from Madurai Block), that zircon core gave age of 695 ± 16 Ma and zircon rim yielded 508 ± 9 Ma. The zircon rim age denotes the time of high-grade (granulite facies) metamorphism. A similar age is also obtained as the lower intercept age (550 Ma) of a zircon from felsic gneiss in the vicinity of the Cauvery Shear zone, giving upper U–Pb SHRIMP zircon age at 2600 Ma (upper intercept). Again, the biotites from Karur and Kodaikanal granulites in the domain of Madurai Block to the south of the PCSZ yielded 600–500 Ma age, although the granulite rocks gave T_{DM} age of 2.8 Ga or slightly less.

In the rocks of Trivandrum Block (i.e. KKB), no ages older than 1.8 Ga are recorded (Bartlett et al., 1995). Recent dating of zircons from granites affected by charnockitization in KKB yielded 548 ± 2 and 526 ± 2 Ma (Ghosh and de Wit, 2004). Also a monazite from a small khondalite enclave within a charnockite massif gave 525 ± 2 Ma (Ghosh and de Wit, loc. cit.). Again, the Sm–Nd isotopic data in minerals of cordierite-bearing charnockites in south Kerala yielded a mineral isochron age of 539 ± 20 Ma (Santosh et al., 1992).

These results clearly suggest that the entire region (Madurai Block and KKB) to the south of the PCSZ witnessed the charnockite formation during the Pan-African tectono-thermal event (see Bhaskar Rao, 2004). The charnockitization occurred in suitable compositions with variable prtolith ages. Based on these geochronological data, discussed above, Ramakrishnan defined his Pandyan Mobile Belt as the Southern Granulite Terrain lying south of the PCSZ (see Ramakrishnan and Vaidyanadhan, 2008). In other words, the Madurai Block and KKB (or Trivandrum Block) formed a single granulite terrain (The Pandyan Mobile Belt sensu stricto), despite the presence of the Achankovil shear zone between them. The AKSZ is not believed to be a shear zone but is considered to represent an attenuated limb of a large fold (Ghosh and de Wit, 2004). Within and across the AKSZ, both strike and dip of foliation are conformable. The isoclinal folding in the charnockite-khondalite rocks of the KKB together with the layer-parallel stretching and pinch-and-swell structure indicate that the AKSZ is an attenuated limb of a fold and not a simple shear. K–Ar date of ~490 Ma on phlogopite from ultramafic rocks within AKSZ may suggest the age of AKSZ. This clearly shows that the AKSZ cannot be a terrane boundary. The record of the Pan-African event is also found in low-pressure granulite terranes in the Gondwanaland (see Harris et al., 1994).

Geochronological data do not support the amalgamation of the Pan-African terrain south of the PCSZ with the granulite terrain in the north. Age data indicate that the Northern Block (CZ and MNZ of Ramakrishnan, 2003) has undergone the granulite metamorphism in the Late Archaean/Early Proterozoic time (2.6–2.5 Ga). The high-grade rocks to the north of the Moyar-Bhavani-Attur shear zone (NMZ of Ramakrishnan, 2003) appear to be part of the Dharwar craton (Archaean age). This terrain contains granulite massifs of BR-Hills, Male Mahadeswara, Shevaroy and Kalravan Hills, and Coorg granulites—all characterized by NNE-SSW to N-S trend (Dharwar trend). All these areas preserve the 2.5 Ga old granulite metamorphism as documented in the Transition zone bordering the amphibolite facies rocks to the south of the Dharwar craton. The Nilgiri massif also consists of 2.5 Ga old granulites, although it is separated from the Dharwar craton (DC) by Moyar shear zone. The whole-rock Rb–Sr, Pb–Pb, Sm–Nd and U–Pb zircon ages for the enderbite granulites of the Nilgiri Hills suggest that the granulite facies metamorphism occurred at about 2.5 Ga ago. Also, a four-point Pb–Pb whole-rock isochron for the Nilgiri charnockites yielded an age of 2578 ± 63 Ma (Peucat et al., 1989). The BR-Hills charnockites have also been dated at 2.5 Ga by different isotopic methods, with protolith ages ranging from 2.9 to 3.0 Ga (cf. Bhaskar Rao et al., 2003, Table 1). A concordant charnockite gneiss within the supracrustals of Satyamangalam Group (at least 3.0 Ga old; Bhaskar Rao et al., 1996) gave an age of 2.52 Ga (Ghosh et al., 1998), indicating the involvement of these trocks in the Early Proterozoic charnockite event, coeval with that in the Transition Zone.

The above discussion conclusively suggests that the granulite metamorphic event in the terrane north of the MBSZ occurred at about 2500 Ma ago. It implies that the addition of this terrain (NMZ of Ramakrishnan, 2003) to the Pan-African terrane south of the PCSZ is geochronologivcally not valid. The continuation of the N-S fabric of the Dharwar schist belts beyond the Moyar-Bhavani shear zones (Naha et al., 1993) and the continuation of the pressure gradient for 7 kbar in the north near Dharmapuri and 10 kbar near the MBSZ (Raith et al., 1983) suggest that the Dharwar greenstone-granite terrane, the Transition Zone, and the Granulite terrane up to MBSZ (Northern Zone of the Pandyan Mobile Belt of Ramakrishnan, 2003) form a single terrane—the Dharwar Crustal Province. As such, the NMZ of Ramakrishnan (2003) has to be excluded from being a part of the Mobile belt. The continuity of DC (Dharwar craton) and granulite facies rocks north of the MBSZ is also supported by a continuity of the gravity anomalies across the Opx-in isograd (Reddi et al., 2003).

The N-S fabric of the Dharwar craton is rotated to near E-W orientation along the Moyar-Bhavani sheatr zones. Therefore, the MBSZ is certainly later than the 2.5 Ga old granulite metamorphism. The cross-cutting relationship between the N-S regional foliation and the E-W trending metamorphic isograds indicate that the regional metamorphism (related to 2.5 Ga old Dharwar orogeny) outlasted the major deformation that produced the regional fabric in the basement and cover rocks of the Dharwar Metamorphic Province. If the charnockite-granulites of 2.5 Ga age in the BR-Hills are considered to have been displaced dextrally in relation to the Nilgiri Hills (Drury et al, 1984; Chetty et al., 2003), the Moyar (Bhavani) shear zone has to be later than 2.5 Ga granulite event. Therefore, the MBSZ cannot be regarded as an Archaean Terrane boundary between the Dharwar craton and the Nilgiri granulite Block (cf. Raith et al., 1999). Again, there is a controversy in regard to the sense of shear. The Moyar shear zone is dextral, according to Drury et al. (1984), but Jain et al. (2003) show that the Bhavani shar zone is a sinistral shear zone. According to Naha and Srinivasan (1996), the Moyar shear zone has a predominant dip-slip transport rather than a strike-slip movement. But it seems, as shown by more recent studies, that the MBSZ has both steep and shallow plunging lineations and therefore both vertical and horizontal movements (cf. Chetty et al., 2003; Jain et al., 2003; D'Cruz et al., 2000).

The granulite domain between the MB shear zone and PCSZ is the Central Zone (CZ) of the Pandyan Mobile Belt of Ramakrishnan (2003). This zone contains migmatite gneisses that are structurally concordant with the 3.0 Ga old supracrustal rocks of (i) Sittampundi and Bhavani layered complexes, (ii) Satyamangalam Group metasediments, and (iii) Palaeoproterozoic charnockite gneiss. This zone also contains Neoproterozoic (700–550 Ma) mafic-ultramafic and granite intrusions (Bhaskra Rao et al., 1996). The granulites east of Sittampundi gave Sm−Nd whole-rock isochrones of 2935 ± 60 Ma, while Sm−Nd isochron for minerals (garnet and hornblende) and whole-rock yielded 730 Ma. In this terrane, there are several small domains that have prescrved, although scantly, the Early Proterozoic granulite metamorphism (2.2–2.3 Ga; Bhaskar Rao et al., 2004), but biotites at other places and along the CSZ give Neoproterozoic ages. This suggests Late Archaean magmatic complex was overprinted by the Pan-African tectono-metamorphic event (Bhaskar Rao et al., 1996; Meissner et al., 2002). The Pan-African ages in this zone are found all along the boundary shear zones—the MBSZ and PCSZ, except the domains around Salem and Dharmapuri. If the Sm−Nd model ages relate to the time of extraction of magmatic protolith from the mantle, then the charnockite parents here and across the Moyar-Bhavani shear zone (MBSZ) were formed nearly the same time. If the granulite metamorphism in the Central Zone and the Zone across the MBSZ also formed during Early Proterozoic, there is no justification to separate the two isogradic terrains, nothwithstanding the Pan-African overprinting in certain areas of the Central Zone. This means that the MBSZ is neither a Palaeoproterozoic terrane boundary nor a subduction zone, but is a shear zone that has developed after the 2.5 Ga-old granulite metamorphism during Dharwar orogeny. This is clearly manifested in the deflection of the N-S structural domain (Dharwar trend) and their alignment parallel to the MB shear zone.

The preceding discussion clearly rules out the extension of the Pandyan Mobile Belt beyond the Palghat Cauvery shear Zone. The PCSZ marks the craton-mobile belt boundary, and the Pandyan Mobile Belt occupies the entire granulite domain south of the PCSZ. This is the concept of the Pandyan Mobile Belt, according to the revised opinion of Ramakrishnan (2004).

8.4 Agreed Observations and Facts

Before we discuss the evolutionary models of the Pandyan Mobile Belt (PMB), we summarize below the agreed observations and facts about the rocks of this Neoproterozoic belt. These are:

1. The Pandyan Mobile belt occurs to the south of the Palghat-Cauvery Shear Zone (PCSZ), including a large terrain comprising Madurai Block and Trivandrum Block (KKB). The PMB is made of granulite facies rocks of Pan-African age (700–500 Ma).
2. The PCSZ marks the Dharwar craton—Mobile belt boundary and hence is a Terrane boundary between Late Archaean terrain in the north and Neoproterozoic terrain in the south of PCSZ.
3. The Moyar-Bhavani shear zone (MBSZ) cannot be a terrane boundary because both Terrains on either side of it (North of the PCSZ) are isogradic (granulite facies rocks), have the same metamorphic age (2.5 Ga), excepting sporadic younger intrusives (Overprinting). Both Late Archaean terrains are characterized by N-S Dharwar trend that extend through the Transition Zone, MBSZ and further south up to PCSZ (Bhaskar Rao et al., 1996; Ramakrishnan, 2004).
4. The N-S structural grains (Dharwar trend) show E-W rotation along the MBSZ but Dharwar trend reappears across the shear zone and continue southward showing swirls within the Madurai sector of the PMB (Ramakrishnan, 2004).
5. As an orogenic belt, the PMB must have evolved by collision of an "unknown craton" against the Dharwar craton during Neoproterozoic time.
6. The charnockite protoliths of the PMB granulites are in the age range of 2.4–2.1 Ga but sometimes the granulites record Archaean ages (Bhaskar Rao et al., 2004).
7. Mineral and whole-rock isochron yield 700–500 Ma for the PMB granulites (Bartlett et al., 1998; Jayananda et al., 1995), indicating involvement of the PMB protoliths in a tectono-thermal event related to the Pan-African orogeny.
8. The Novil-Cauvery shear zone (i.e. PCSZ) appears to be a suture zone (? Subduction zone) and has southern dip, supported by seismic tomography (Rai et al., 1993).
9. The exhumation of the Southern Granulite Terrain (SGT) is post-PanAfrican, caused by northerly and also westerly tilt of the Indian Peninsula, perhaps episodically (Ramakrishnan, 2004).

8.5 Evolutionary Models and Discussion

With the basic informations given above, the different evolutionary models of the Pandyan Mobile Belt are discussed in the following pages.

8.5 Evolutionary Models and Discussion

Model 1. Subduction-Collision Model (Ramakrishnan, 1993, 2003, 2004)

This model is based on the consideration that the Neoproterozoic terrain south of the PCSZ is a fold belt and that the Dharwar craton is continuous with the 2.5 Ga old granulite facies terrains to the north of the PCSZ. The fold belt is constituted of both Madurai Block and Trivandrum Block (or KKB), having similar lithologies and Pan-African ages, revealed by the Sm–Nd and Rb–Sr whole-rock and mineral isochrons (Ghosh et al., 1998; Bhaskar Rao et al., 2004), from Pb-isotopes in zircons in leucosomes of the migmatite gneisses (603 ± 14 Ma), from newer zircon growth (560 Ma), from monazite in garnet-sillimanite gneiss (791 ± 1 Ma), and from charnockitization age of about 500 Ma given by a charnockite granite with protolith age of 800 Ma. The granulite facies metamorphism in the Kodaikanal and Palani areas also gave Pan-African ages around 560 Ma (Bartlett et al., 1995; Jayananda and Peucat, 1996). The Pan-African event is also documented in a series of intrusives of alkali granites, syenites and granites, as stated already.

Since the Sm–Nd isotope data on minerals of cordierite-bearing charnockite in southern Kerala yield mineral isochron of 539 ± 20 Ma (Santosh et al., 1992), similar to that in Madurai Block (Deans and Powell, 1968; Subramanian, 1983; Santosh et al., 1989), the Achankovil shear zone (AKSZ) occurring between the Madurai Block and Trivandrum block (i.e. KKB) does not represent a terrane boundary. According to Ghosh and de Wit (2004), the AKSZ represents an attenuated limb of a fold. This finds support from the lithotectonic map of the Pandyan Mobile Belt (Fig. 8.3), based on field data and interpretation of Landsat images that show coaxial folding of both granaulites and gneiss/migmatites, producing hook-shaped antiforms and synforms. This author has delineated the axial traces of the superposed folds, F1 and F2, which are cut across by the AKSZ (see Fig. 8.3). The migmatite outcrop (upper left part of Fig. 8.3) is parallel to the axial trace of F2 fold, indicating that migmatization was syn-F2 in the Pandyan Mobile Belt. The KKB rocks (Trivandrum Block) are seen oriented along this shear zone. The AKSZ is considered 550 Ma old ductile shear zone with left-lateral shear sense and occurring parallel to the Tenmalai-Gutan shear zone on the south of the belt. Contrary to the sinistral shear sense, Sacks et al. (1997), who mapped a part of this shear zone, suggest dextral movement. On the other hand, Radhakrishna (2004, p. 116) does not recognize any shear features along this lineament. To sum up, the AKSZ is not a suture zone or a discrete surface to mark a tectonic break between the blocks across it. Both Madurai and Trivandrum Blocks, south of the PCSZ are therefore part of a single terrain of Neoproterozoic age.

In contrast, the domain to the north of the PCSZ is clearly Late Archaean to Early Proterozoic granulite facies terrain. This terrain, especially north of the MBSZ is undisputedly a Late Archaean terrain, characterized by granulite massifs of Nilgiri, BR-Hills, Shevaroy Hills etc., as discussed earlier. The terrain between the MBSZ and the PCSZ also contains migmatite gneisses, structurally concordant with the supracrustals as well as with the Palaeoproterozozic charnockite gneiss, as stated already. However, there are a few small domains, although preserving Late Archaean or Palaeoproterozoic Rb–Sr ages, which show Pan-African overprinting

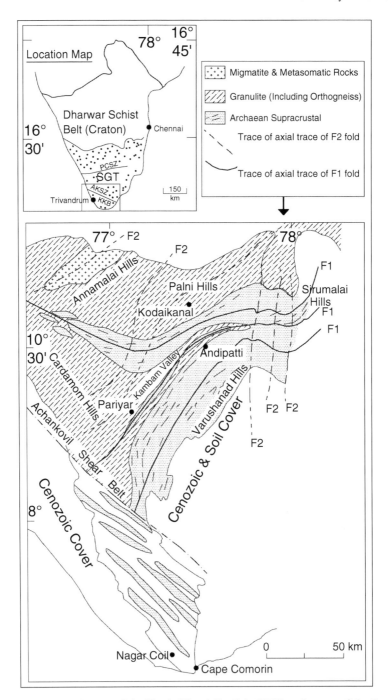

Fig. 8.3 Litho-tectonic map of the Kerala Khondalite Belt (KKB) and a part of the Southern Granulite Terrain (i.e. Madurai Block), redrawn from T. Radhakrishna, 2004. Note that the gneiss/migmatite and granulite are cofolded by superposed deformation F1 and F2 with axial traces (delineated by the writer) which are cut across by the Achankovil shear zone, see text for details

in the mineral of the Archaean protoliths (Bhaskar Rao et al., 1996; Ghosh et al., 1998) could be a thermal overprinting, presumably due to heating by numerous Neoprotoerozoic plutonic rocks (alkali granites, syenites and even carbonatites) that intruded the terrain during the globally recognized Pan-African orogeny. However, the charnockite gneisses of this domain (Central Zone of Ramakrishnan, 2003) are coeval with the 2.5 Ga old granulite belt to the north of the MBSZ, which was recrystallized 2.5 Ga ago during the Dharwar orogeny (Late Archaean/Early Proterozoic). The late Achaean charnockites are characterized by positive Eu anomalies, HREE depletion and near chondritiic values of $\Sigma_{Nd(O)}$ and $^{87}Sr/^{86}Sr$ at 2.5 Ga (Tomson et al., 2006), suggesting their derivation by partial melting of garnet-hornblende bearing mafic parent (mantle component or amphibolite). On the other hand, sources of Neoproterozoic charnockites (south of PCSZ) show greater recycling of crustal components, i.e. intracrustal melting as revealed by negative Eu anomalies, lower degree of HREE depletion and undepleted Y (cf. Tomson et al., 2006).

From the above discussion, it is suggested that the PCSZ is a major terrain boundary separating the charnockite region of Palaeoproterozoic in the north from the Pan-African (Neoproterozoic) charnockite domain (Pandyan Mobile Belt) in the south (Drury et al., 1984; Harris et al., 1994; Jayananda and Peucat, 1996; Yoshida and Santosh, 1995, and others). Since the PMB is located at the southern margin of the Dharwar craton, it is considered to have formed by collision tectonics, involving two oppositely converging plates. The continental block which is assumed to have collided against the Dharwar craton, to raise the Pandyan mobile belt, could be Mozambique craton from South Africa or some uncertain region of Antarctica (Ramakrishnan, 2004).

The present disposition of the metamorphic isograd suggest that in the Dharwar orogeny, the Dharwar craton with its cover sediments subducted southward under an "ancient continent". The southward subduction is also supported by seismic tomography (Srinagesh and Rai, 1996). In the subduction-collision process, the Dharwar crust was thickened and both Archaean basement and its supracrustals were deformed. The resulting increase of temperature as a consequence of crustal thickening gave rise to the progressive regional metamorphism from greenschist facies to amphibolite facies and finally to granulite facies. The cross-cutting relationship between the N-S regional foliation and the E-W trending metamorphic isograds indicates that the regional metamorphism (related to the 2.5 Ga old Dharwar orogeny) outlasted the major deformation that produced the Dharwar trend in the basement and cover rocks of the Dharwar Metamorphic Province. During the progressive metamorphism, the rocks at depth also underwent partial melting and generated granites (Closepet and other coeval granite plutons) of the same age. During this decompressive stage, charnockitization of Peninsular gneiss occurred, spectacularly seen at Kabbal and other places.

Besides metamorphism and magmatism, another internal process that occurred late in the ascending crustal segment of the Dharwar Province has been the development of post-metamorphic E-W shears or lineaments (e.g. Moyar, Bhavani shear zones) in the granulite terrain. Not only is the Dharwar trend deflected but the 2.5 Ga old granulites have also been retrograded and mylonitized to varying degree along these shear, developing shear fabric parallel to the above mentioned shears (cf. Jain et al., 2003). The deep-seated granulite rocks are inferred to have been uplifted along these shears. The P-T data for the garnet-graulite from Nilgiri yielded 730°C/7 kbar in the SW and 750°C/9 kbar along the NE toward the Moyar shear zone. These results suggest that the Nilgiri massif tilted northward during its differential uplift along these shears. It is perhaps this tilting deduced from the P-T data that led Raith (2004) to conceive the Nilgiri massif as an allochthonous massif, thrust northward onto the Dharwar terrain during Palaeoproterozsoic (2.3–2.1 Ga). The northward tilting of the Nilgiri massif could be associated with the northward tilt of the southern Indian shield during Neoproterozoic. It is indeed a matter of enquiry whether the tilting was during Palaeoproterozoic or Neoproterozoic time. The available evidence is taken by Raith et al. (1983) to suggest a differential uplift of the Nilgiri massif. From the proposition that the BR-Hills granulites have a correlation with the Nilgiri granulites (Drury et al., 1984; Chetty et al., 2003), a strike-slip movement is implied along the Moyar shear zone. If the Dharwar crust was segmented by these shears, different granulite blocks may have uplifted differentially from same or different depths. In this situation, the shears need not be of the same age, but necessarily developed in post-granulite metamorphic event, in view of their geological setting, described earlier. Drury and Holt (1980) considered the MBSZ as a part of the major Palghat-Cauvery shear zone system. Considering that the Palghat-Cauvery shear zone (PCSZ) separates the Late-Archaean granulite terrain from the Neoproterozoic Pandyan Mobile Belt, the PCSZ cannot be as old as the Moyar-Bhavani shear zones that separate the 2.5 Ga old granulites on its either side and affected them soon after their metamorphism. Recent geochronological work has given an age constraint on the shearing event. Rb–Sr dating of micas and Sm–Nd dating of garnet from mylonites of the MBSZ (Bhaskar Rao et al., 2003; Meissner et al., 2002) suggest that these shears are Neoproterozoic (700–500 Ma). But laser probe $^{40}Ar/^{39}Ar$ age data of 1250 Ma on pseudotachylites from the MBSZ indicate a Mesoproterozoic and hence older age than the PCSZ. As suggested by the age data of minerals from mylonites, these shear zones (MBSZ) have been reactivated during Neoproterozoic when the Pandyan mobile belt evolved.

Subsequent to the 2.5 Ga old Dharwar orogeny, there are ample evidence for the occurrence of Pan-African tectono-thermal event (orogeny) during which the PMB is considered to have evolved. The Pan-African orogeny was a global event and nearly all the continents of a supercontinent manifest its record in their rocks. The Indian shield also preserves the Pan-African related tectono-thermal event. The best example is the Pandyan Mobile Belt at the southern border of the Dharwar craton. Based on geometric configuration and best-fit of continents, the Pandyan Mobile belt appears to intersect the Grenville-type belt in Antarctica and may even be truncated by the Mozambique belt of eastern Africa (cf. Ramakrishnan, 2003). According to this configuration in the supercontinent (Rodinia), the "ancient continent" that may have collided with the Dharwar craton during the Pan-African time orogeny was Kalahari in Africa or one in Antarctica.

During continental collision, it is believed that the Dharwar craton subducted southward under the "ancient continent". This is inferred from the southern dip of

8.5 Evolutionary Models and Discussion

the PCSZ which may represent a surface expression of continental subduction and collision of 2.5 old Dharwar Metamorphic Province against a southern continent in the pre-Neoproterozoic continental assembly. The subduction was in all probability an A-subduction in which crust-mantle delamination (or crustal decoupling) occurred whereby hot mantle magma (formed as a result of decompression melting) underplated the crust to generate high and also ultra high temperature metamorphism, seen in several areas of PMB (see Grew, 1982; Mohan and Windley, 1993). The clockwise decompression path with ITD (isothermal decompression) loop found in Gangavarpatti and Kodaikanal (Mohan and Windley, 1993) does not, however, support the underplating model for the evolution of the PMB. The emplacement of plutons of alkali affinity during the Pan-African event in southern India indicates a period of extensional tectonics in Gondwanaland at that time. Such a tectonic setting could have provided an appropriate stress regime for channelized influx of CO_2- rich fluid during incipient charnockitization and for enrichment of carbonic fluids (Santosh, 2003). The exposure of the rocks of the Pandyan Mobile Belt is the outcome of northerly and westerly tilt of the Indian peninsula, perhaps during India-Asia collisional events.

Model 2. Accretion Model (Radhakrishna and Naqvi, 1986)

One group of geoscientists believes that the southern Indian shield, particularly the Southern Granulite Terrain (SGT), resulted from accretion of two crustal blocks along the PCSZ. The PCSZ is considered to represent a cryptic suture or collision zone (Gopalkrishnan, 1996) or an accretion zone along which two blocks/cratons have been welded (Radhakrishna and Naqvi, 1986). The model considers welding of the once far-off located SGT with Dharwar craton. To the background of this model is the division of the Dharwar Metamorphic Province into charnockitic and non-charnockitic Provinces by Fermor (1936). This dividing line is the present Opx-in isograd which is also known as the Fermor Line. Although the charnockite/non-charnockite division is germane to the idea of Terrane Accretion Model, the emphasis shifted to the shear zones as the possible site of the terrain accretion. In this model, the shear zones are considered sutures along which continental blocks are assumed to have welded or fused.

The accretion model is, however, beset with serious problems of uncertainty of the direction of crustal movement and the absence of squeezed (deformed) belt of supracrustals that may have been at the site of welding of the once far-off located continental mass. The model also appears weak if the recent conclusions on petrogenesis of the Late Archaean Closepet Granite (Friend and Nutman, 1991) are considered.

Moyen et al. (2001) recognized three zones in the Closepet Granite, which from north to south are: Intrusive zone, Transfer zone and Root zone. The Root zone granite extends from Palghat-Cauvery shear zone in the south up to Opx-in isograd in the north and shows a mixing of mantle-derived magma with the crustal-derived magma. These deep zone granites are heterogenous and contain enclaves and show migmatization at the contact with the Peninsular gneiss. The granites from the Transfer zone extend all through the amphibolite facies terrane from Kabbal to Kalyandungri and have similar granite type as that of root zone granite, except having fewer enclaves and less diffused contact with the country rocks of the Peninsular gneiss. The Intrusive zone granite has very few xenoliths but no thermal effects on the host rocks. These granite types from three zones show geochemical and isotopic similarity that led Moyen et al. (2001) to conclude that the Closepet granite is a continuous body from granulite facies to the greenschist facies (see also Jayananda et al., 1994). If so, the Model 3 is not supported by the Petrological-geochemical data on the Closepet Granite extending from Kabbal to Palghat. Furthermore, the petrological and geochronological data on the granulite massifs suggest uplift and not horizontal movement of the granulite blocks. The Nilgiris modeled as an upthrust-allochthonous terrain to be placed onto Dharwar craton (Raith et al., 1990). As stated earlier, there is no unanimity on the movement direction along the shear zones. Interestingly, the Palghat-Cauvery Shear zone is stated to be non-traceable to the east of the SGT (Mukhopadhyay et al., 2003). Recently, Chetty et al. (2003) consider that transpressional orogenesis (Dharwar) in the region was associated with thrusting and complex strike-slip movement, with a regionally consistent sense of shearing. This proposition explains the steep and gentle plunges of the mineral stretching lineations. But, here again, the shearing is not necessarily Dharwar age, in view of the arguments stated previously. In all probability, the shear zones are post-Dharwarian age but appeared to have played a significant role in causing differential movements of the crustal blocks, vertically as well as laterally. If the stated terrane-accretion model is accepted one needs to answer as to what happened to the intervening rocks (unmetamorphosed or already metamorphosed) that may have been there before the SGT moved from far off place and welded with the Dharwar craton. The problem of the SGT, whether an accreted block or a mobile belt formed due to the convergence of a "missing plate" with the Dharwar craton, can be solved by more geochronological and structural work in the critical areas of the southern Indian shield. It is a matter of debate as to how much geophysical investigations of the present crust-mantle configuration could help in unraveling the events that occurred around 2500 million years ago in this part of the Indian shield unless it remained completely inert since then.

The problem of the SGT terrain south of the PCSZ—whether an accreted block or a mobile belt formed due to the collision of "missing plate" against the Dharwar craton can be solved by more structural and geochronological work in critical areas of the southern Indian shield. It is also a matter of debate as to how much geophysical investigations of the present crust-mantle configuration would help in unraveling the events as old as 2500 Ma in this part of the Indian shield, even if the craton remained completely inert since that time.

Model 3. Reworked Ganulite Terrain Sans Mobile Belt

As an alternative to Models 1 and 2, this paradigm presumes that the granulite terrain was one single crustal block of granulite rocks, and the terrain south of the PCSZ is wholly re-Worked Archaean terrain by the superimposed Pan-African tectono-thermal event; i.e. PMB is not a mobile belt. The geological observation that the N-S fabric of the Dharwar craton extends through MBSZ up to PCSZ, and perhaps beyond the AKSZ where the fabric is deflected by post-metamorphic shearing in rocks of different competence. This Dharwar Crustal Province formed one large crustal unit during Late Archaean Dharwar orogeny (2.5 Ga ago). Both Archaean basement and suptracrustals (dominantly greenstones) were superposedly deformed and regionally metamorphosed, giving rise to the E-W trending isograds delineated between greenschist, amhibolite and granulite facies, so extensively developed at the southern margin of the Dharwar craton. Ramiengar et al. (1978) emphasized that the E-W trending belt of charnockites and granulites to the south of the Opx-in isograd does not have any lithostratigraphic significance because the unbroken greenstone and granulite transitions in southern India is orthogonal to the regional structural grain. This means that the SGT is a continuous terrain with the Dharwar craton *sensu stricto*. The regional N-S fabric in the affected rocks indicates that the compressive forces generated by the collision of lithospheric plates in the Archaean were in E-W direction. But the metamorphic isograds also trend E-W, implying that the compressive stresses due to subduction should be N-S and not E-W. This and other reasons require an alternative model for the evolution of SGT.

Considering the dominance of granulites to the south of Opx-in isograd and the southward progression of regional metamorphism in Dharwar orogeny (2.5 Ga), the granulites of the SGT do not seem to belong to a different Terrane, as they are believed to have been exposed by northward tilting of the Indian shield. This proposition of northward tilting finds support from the geophysical studies which show charnockite-granulite rocks (similar to those in the SGT) below the greenstone terrains of the Dharwar craton. This means that the SGT is not a laterally welded terrane, but a lower crustal component that got exposed through tilting of an Archaean continental crust. Conversely, it could be argued that if the northward tilting had not occurred, the SGT would not have such a vast exposure by erosion alone. Rifting of the various continents and related magmatic activities during the Pan-African orogeny are documented in the various regions globally. And its effects in the Indian shield are seen in a series of intrusions of alkaline granites, syenites and granites of 700–550 Ma age (including the Sankari Durg granite near the MBSZ) and also in the re-working of the terrain south of the PCSZ, which document wide spread Pan-African ages. It is interesting to note that the rocks of the Madurai and Trivandrum Blocks (both forming Pandyan mobile belt of Ramakrishnan, 2004), making up the terrain south of the PCSZ, have protolith ages of 3.0 and less, but not very different from the protolith ages from the terrain north of the PCSZ. The Neoproterozoic ages, given by the constituent minerals in rocks from the granulite terrain to the south of PCSZ may be taken to indicate a through rcrystallization during Pan-African overprinting. The tectno-thermal overprinting is also documented occasionally in the granulite facies domain between the MBSZ and PCSZ, although not convincingly in the terrain north of the MBSZ. This may be taken to infer that the Pan-African imprinting was weakening northward.

The PCSZ becomes a clear tectonic divide between the Archaean cratonic crust and the Neoproterozoic granulites of the Pandyan mobile belt. The PCSZ seems to have resulted by collision of a craton (Kalahari in Africa or one in Antarctica) with the Dhawar craton (including the Archaean SGT) during Pan-African time (550–700 Ma). Accordingly one cannot expect Late Archaean or Early Proterozoic deformation in the PCSZ (cf. Chetty and Bhaskar Rao, 2004). The PCSZ is the most plausible terrane boundary in the SGT. Earlier the PCSZ was also proposed as a zone of continental collision (Drury et al., 1984) in which the block south of PCSZ represents large stacks paragneisses, tectonically intercalated in between meta-igneous granulite massifs (represented by charnockite massifs). The first order assumptions of this tectonic model are: (a) magmatic origin of the charnockite massifs south of PCSZ, (b) tectonic contacts between the paragneisses and swathes and the charnockite massifs, and (c) charnockite emplacement pre-dating the tectonic stacking along shear zones. Principal support to this model comes from isotope data (Harris et al., 1994), magnetic anomaly data (Reddi et al., 2003), seismic tomography (Rai et al., 1993) and metamorphic P-T data (Ravindra Kumar and Chacko, 1994). The model of terrane boundary along PCSZ thus appears valid and any other terrane boundary such as along Karur-Kambam-Painvu-Trichur shear zone (KKPTSZ), proposed recently by Ghosh and de Wit (2004), lacks support from different aspects of geology and geophysics. A mere similarity or dissimilarity of lithology along a discrete surface cannot be regarded as convincing criteria of delineating terrane boundary as proposed recently by Ghosh and de Wit (2004). However, the PCSZ is in all probability a terrain boundary between the Dharwar craton (including the Archaean domain of SGT) and the Pandyan Mobile belt that may have accreted subsequent to its recrystallization in Neoproterozoic. The exhumation of the southern Indian shield should be post-Pan-African. The exposure of this vast granulite terrain owes largely to the northward and also eastward tilt of the Indian peninsula in different episodes (Mahadevan, 2004).

We have to collect more geological and geochronological data to accept or reject the proposition of the fold belt versus a cratonic terrain to the south of the Palghat-Cauvery shear zone.

References

Allen, P., Condie, K.C. and Narayana, B.L. (1983). The Archaean low- to high-grade transition in the Krishnagiri-Dharmapuri, Tamil Nadu, southern India. In: Naqvi, S.M. and Rogers, J.J.W. (eds.), Precambrian of South India. Mem. Geol. Soc. India, vol. **4**, Bangalore, pp. 450–462.

Bartlett, J.M., Dougherty-Page, J.S., Harris, N.B.W., Hawkesworth, C.J. and Santosh, M. (1998). The application of single zircon evaporation and Nd model ages to the interpretation of polymetamorphic terrains: an example from the Proterozoic mobile belt of south India. Contrib Mineral Petrol, vol. **131**, pp. 181–195.

Bartlett, J.M., Harris, N.B.W., Hawkesworth, C.J. and Santosh, M. (1995). New isotope constraints on the crustal evolution of south India and Pan-African granulite meta-morphism. Mem. Geol. Soc. India, vol. **34**, pp. 391–397.

Bernard-Griffiths, J., John, B.M. and Sen, S.K. (1987). Sm—Nd isotope and REE geo-chemistry of Madras granulites, India: an introductory statement. Precambrian Res., vol. **37**, pp. 343–355.

Bhaskar Rao, Y.J. (2004). Precambrian terrains of South India. In: *Tectonics and Evolution of the Precambrian Southern Granulite Terrain, India and Gondwana Correlations*. International Workshop, N.G.R.I., Hyderabad, Abstracts and Geological Excursion Guide, pp. 107–115.

Bhaskar Rao, Y.J., Chetty, T.R.K., Janardhan, A.S. and Gopalan, K. (1996). Sm—Nd and Rb—Sr ages and P-T history of the Archaean Sittampundi and Bhavani layered complexes in the Cauvery shear zone. Contrib Mineral Petrol, vol. **125**, pp. 237–250.

Bhaskar Rao, Y.J., Janardhan, A.S., Vijaya Kumar, T., Narayana, B.L., Dayal, A.M., Taylor, P.N. and Chetty, T.R.K. (1993). Sm—Nd model ages and Rb—Sr systematics of charnockites and gneisses across the Cauvery shear zone, southern India: implications for the

Archaean-Neoproterozoic Terrane boundary in Southern Granulite Terrain. Mem. Geol. Soc. India, vol. **50**, pp. 297–317.

Bhaskar Rao, Y.J., Janardhan, A.S., Vijaya Kumar, T., Narayana, B.L., Dayal, A.M., Taylor, P.N. and Chetty, T.R.K. (2003). Sm–Nd model ages and Rb–Sr isotopic systematics of charnockites and gneisses across the Cauvery shear zone, southern India: implications for the Archaean-Neoproterozoic terrain boundary in the Southern Granulite Terrain. Mem. Geol. Soc. India, vol. **50**, pp. 297–317.

Bhaskar Rao, Y.J., Vijaya Kumar, T., Thomson, J.K., Chetty, T.R.K. and Dayal, A.M. (2004). Sm–Nd and Rb–Sr ages and isotopic constraints on Precambrian terrain assembly and reworking around the Cauvery shear zone system, the Southern Granulite Terrain, India (Abstract), International workshop on *Tectonics and Evolution of the Precambrian Southern Granulite Terrain, India and Gondwanian Correlations*, Feb. 2004, National Geophysical Research Institute, Hyderabad, pp. 50–55.

Braun, I. and Kriegsman, L.M. (2002). Proterozoic crustal evolution of southernmost India and Sri Lanka. In: Yoshida, M., Windley, B.F. and Dasgupta, S. (eds.), Proterozoic East Gondwana Supercontinent Assembly and Breakup. Geol. Soc. Lond. Spl. Publ., London

Brown, M. and Raith, M. (1996). First evidence of ultra high temperature decompression from the granulite province of southern India. J. Geol. Soc. Lond., vol. **153**, pp. 819–822.

Buhl, D. (1987). U–Pb und Rb–Sr Altersbestimmungen und Untersuchungen zum Strontium Isotopenaustausch und Granuliten Studiens. Ph.D. Thesis, University of Munster, Germany (unpublished).

Buhl, D., Grauert, B. and Raith, M. (1983). U–Pb zircon dating of Archaean rocks from the South Indian Craton: results from the amphibolite to granulite facies transitions zone at Kabbal Quarry, southern Karnataka. Fortsch. Mineral., vol. **61**, pp. 43–45.

Cenki, B., Braun, L. and Brocker, M. (2004). Evolution of continental crust in the Kerala Khondalite Belt, southernmost India: evidence from Nd isotope mapping, U–Pb and Rb–Sr geochronology. Precambrian Res., vol. **134**, pp. 37–56.

Chacko, T., Ravindra Kumar, G.R., Meen, J.K. and Rogers, J.J.W. (1992). Geochemistry of high-grade supracrustal rocks from the Kerala Khondalite Belt and adjacent massif charnockites, south India. Precambrian Res., vol. **55**, pp. 469–489.

Chacko, T., Ravindra Kumar, G.R. and Newton, R.C. (1987). Metamorphic P-T conditions of the Kerala (South India) Khondalite Belt: a granulite facies supracrustals terrain. J. Geol., vol. **95**, pp. 343–358.

Chetty, T.R.K. and Bhaskar Rao, Y.J. (2004). Strain variations and deformation styles across the Cauvery shear zone, southern Granulite Terrain, India. (Abstract), International workshop on *Tectonics and Evolution of the Precambrian Southern Granulite Terrain, India and Gondwanian Correlations*, Feb. 2004, National Geophysical Research Institute, Hyderabad, pp. 47–49.

Chetty, T.R.K., Bhaskar Rao, Y.J. and Narayana, B.L. (2003). A structural cross-section along Krishnagiri-Palani corridor, Southern Granulite Terrain of India. In: Ramakrishnan, M. (ed.), *Tectonics of Southern Granulite Terrain, Kuppam-Palani Geotransect*. Mem. Geol. Soc. India, vol. **50**, Bangalore, pp. 253–277.

Collins, A.S. and Santosh, M. (2004). New protolith provenance, crystallization and metamorphic U–Pb zircon SHRIMP ages from southern India. (Abstract), International workshop on *Tectonics and Evolution of the Precambrian Southern Granulite Terrain, India and Gondwanian Correlations*, Feb. 2004, National Geophysical Research Institute, Hyderabad, pp. 73–76.

Condie, K.C. and Allen, P. (1984). Origin of Archaean charnockites from southern India. In: Kroener, A., Hanson, G.N. and Godwin, A.M. (eds.), Archaean Geochemistry. Springer Verlag, Berlin, pp. 183–203.

Crawford, A.R. (1969). Reconnaissance Rb–Sr dating of Precambrian rocks of southern Peninsular India. J. Geol. Soc. India, vol. **10**, pp. 117–166.

Deans, T. and Powell, J.L. (1968). Trace elements and strontium isotopes in carbonatites, fluorites and limestones from India and Pakistan. Nature, vol. **218**, pp. 750–752.

Drury, S.A., Harris, N.B.W., Holt, R.W., Reeves-Smith, G.J. and Wightman, R.T. (1984). Precambrian tectonics and crustal evolution in south India. J. Geol., vol. **92**, pp. 3–20.

Drury, S.A. and Holt, R.W. (1980). The tectonic framework of the south Indian craton: a reconnaissance involving landsat imagery. Tectonophysics, vol. **65**, pp. T1–T5.

D'Cruz, E., Nair, P.K.R. and Prasanna Kumar, V. (2000). Palghat gap—a dextral shear zone from the south Indian granulite terrain. Gondwana Res., vol. **3**, pp. 21–32.

Fermor, L.L. (1936). An attempt at correlation of ancient schistose formations of peninsular India. Mem. Geol. Surv. India, vol. **70**, pp. 1–52.

Friend, C.L. and Nutman, A.P. (1991). SHRIMP U–Pb geochronolgy of the Closepet granite and peninsular gneiss, Karnataka, India. J. Geol. Soc. India, vol. **38**, pp. 357–368.

Friend, C.L. and Nutman, A.P. (1992). Response of zircon U–Pb isotopes and whole-rock geochemistry to CO_2- fluid induced granulite facies metamorphism, Karnataka, south India. Contrib Mineral Petrol, vol. **111**, pp. 299–310.

GSI and ISRO. (1994). Project Vasundhara: Generalized Geological Map (Scale 1: 2 Million). Geol. Surv. India & Indian Space Res. Orgn., Bangalore.

Ghosh, J.G., Zartman, R.E. and de Wit, M.J. (1998). Re-evaluation of tectonic framework of southernmost India: new U–Pb geochronological and structural data, and their implication for Gondwana reconstruction. In: Almond, J. and others (eds.), Gondwana 10, Event Stratigraphy of Gondwana. J. African Earth Sci., Elsevier, vol. **27-1A**, (Abstract) 86p.

Ghosh, J.G. and de Wit, M. (2004). Tectonic relation between the Dharwar craton and the Southern Granulite Terrain in India (Abstract), International workshop on Tectonics and Evolution of the Precambrian Southern Granulite Terrain, India and Gondwanian Correlations, Feb. 2004, National Geophysical Research Institute, Hyderabad, pp. 56–58.

Ghosh, J.G., de Wit, M.J. and Zartman, R.E. (2004). Age and tectonic evolution of Neoproterozoic ductile shear zones in the Southern Granulite Terrain of India, with implications for Gondwana studies. Tectonophysics, vol. **23**, pp. 130–144.

Gopalkrishnan, K. (1996). An overview of Southern Granulite Terrain, India. Geol. Surv. India Spl. Publ., vol. **55**, pp. 85–96.

Grady, J.C. (1971). Deep main faults of south India. J. Geol. Soc. India, vol. **12**, pp. 56–62.

Grew, E.S. (1982). Sapphirine, kornerupine and sillimanite + orthopyroxene in the charnockite regions of south India. J. Geol. Soc. India, vol. **23(10)**, pp. 469–505.

Grew, E.S. and Manton, W.L. (1986). A new correlation of sapphirine granulites in the Indo-Antarctic metamorphic terrain: Late Proterozoic dates from the Eastern Ghats Province of India. Precambrian Res., vol. **33**, pp. 123–137.

Hansen, E.C., Janardhan, A.S., Newton, R.C., Prame, W.K.B.N. and Ravindra Kumar, G.R. (1987). Arrested charnockite formation of south India and Sri Lanka. Contrib Mineral Petrol, vol. **97**, pp. 225–244.

Hansen, E.C., Newton, R.C., Janardhan, A.S. and Lindenberg, S. (1995). Differentiation of Late Archaean crust in Eastern Dharwar craton, Krishnagiri-Salem area, south India. J. Geol., vol. **103**, pp. 629–651.

Harris, N.B.W., Holt, R.W. and Drury, S.A. (1982). Geobarometry and geothermometry and the late Archaean geotherm from the granulite facies terrain in south India. J. Geol., vol. **90**, pp. 509–527.

Harris, N.B.W., Santosh, M. and Taylor, P. (1994). Crustal evolution of south India: constraints from Nd isotopes. J. Geol., vol. **102**, pp. 139–150.

Holland, T.H. (1900). The charnockite series, a group of Archaean hypersthenic rocks in Peninsular India. Mem. Geol. Surv. India, vol. **102**, pp. 139–150.

Jain, A.K., Singh, S. and Manickavasagam, R.M. (2003). Intracontinental shear zones in the Southern Granulite Terrain: their kinematics and evolution. In: Ramakrishnan, M. (ed.), *Tectonics of Southern Granulite Terrain: Kuppam-Palani Geotransect*. Mem. Geol. Soc. India, vol. **50**, Bangalore, pp. 225–253.

References

Janardhan, A.S. (1996). The Oddanchatran anorthosite body, Madurai block, southern India. In: Santosh, M. and Yoshida, M. (eds.), The Archaean and Proterozoic Terrains in Southern India within East Gondwana. Mem. Gondwana Res. Group, Japan, vol. **3**, pp. 385–390.

Janardhan, A.S., Newton, R.C. and Hensen, E.C. (1982). The transformation of amphibolite facies gneiss to charnockite in southern Karnataka and northern Tamil Nadu, India. Contrib Mineral Petrol, vol. **79**, pp. 130–149.

Jayananda, M., Kano, T., Harish Kumar, S.B., Mohan, A., Shadakshra Swamy, N. and Mahabaleswar, B. (2003). Thermal history of the 2.5 Ga juvenile continental crust in the Kuppam-Karimangalam area, Eastern Dharwar Craton, southern India. In: Ramakrishnan, M. (ed.), Tectonics of Southern Granulite Terrain: Kuppam-Palani Geotransect. Mem. Geol. Soc. India, vol. **50**, Bangalore, pp. 255–287.

Jayananda, J.J., Martin, H., Peucat, J.J. and Mahabaleswar, B. (1995). Geochronologic and isotopic constraints on granulite formation in the Kodaikanal area, southern India. In: Yoshida, M. and Santosh, M. (eds.), India and Antarctica During the Precambrian. Mem. Geol. Soc. India, vol. **34**, Bangalore, pp. 373–390.

Jayananda, M. and Peucat, J.J. (1996). Geochronological framework of south India. In: Santosh, M. and Yoshida, M. (eds.), *The Archaean and Proterozoic Terrains in Southern India Within East Gondwana*. Mem. Gondwana Res. Group, Japan, vol. **3**, pp. 53–75.

Jayananda, M., Peucat, J.J., Martin, H. and Mahabaleswar, B. (1994). Magma mixing in plutonic environment: geochemical and isotopic evidence from the Closepet batholith, southern India. Curr. Sci., vol. **66**, pp. 928–935.

Kilpatric, J.A. and Ellis, D.J. (1992). C-type magmas: igneous charnockite and their extrusive equivalents. Trans. Roy. Soc. Edinburgh, vol. **83**, pp. 155–164.

Lal, R.K. (1993). Internally consistent recalibrations of mineral equilibria for geothermobarometry involving garnet-orthopyroxene-plagioclase-quartz assemblages and their applications to southern Indian granulites. J. Metam. Geol., vol. **11**, pp. 855–866.

Mahadevan, T.N. (2004). Continent evolution through time: new insights from southern Indian shield (Abstract). International workshop on *Tectonics and Evolution of the Precambrian Southern Granulite Terrain, India and Gondwanian Correlations*, Feb. 2004, National Geophysical Research Institute, Hyderabad, pp. 37–38.

Meissner, B., Deters, P., Srikantappa, C. and Kohler, H. (2002). Geochronological evolution of the Moyar, Bhavani and Palghat shear zones of southern India: implications for Gondwana correlations. Precambrian Res., vol. **114**, pp. 149–175.

Miller, J.S., Santosh, M., Pressley, R.A., Clements, A.S. and Rogers, J.J.W. (1997). A Pan-African thermal event in southern India. J. Southeast Asian Earth Sci., vol. **14**, pp. 127–136.

Mohan, A. and Windley, B.F. (1993). Crustal trajectory of sapphirine-bearing granulites from Gangavarpatti, south India: evidence for an isothermal decompression path. J. Metam. Geol., vol. **11**, pp. 867–878.

Moyen, J.R., Martin, H., and Jayananda, M. (2001). Multi-element geochemical modeling of crust mantle interactions during Late Archaean crustal growth: the Closepet granite (South India). Precambrian Res., vol. **112**, pp. 87–105.

Mukhopadhyay, D., Senthil Kumar, P., Srinivasan, R. and Bhattacharya, T. (2003). Nature of the Palghat-Cauvery lineament in the region south of Namakkal, Tamil Nadu: implications for the terrain assembly in the South Indian Granulite Province. In: Ramakrishnan, M. (ed.), Tectonics of Southern Granulite Terrain: Kuppam-Palani Geotransect. Mem. Geol. Soc. India, vol. **50**, Bangalore, pp. 279–296.

Mukhopadhyay, D., Sentil Kumar, P., Srinivasan, R., Bhattacharya, T. and Sengupta, P. (2001). Tectonics of the eastern sector of the Palghat-Cauvery lineament near Namakkal, Tamil Nadu. Deep Continental Studies in India, DST New Letter, vol. **11**, pp. 9–13.

Naha, K., and Srinivasan, R. (1996). Nature of the Moyar and Bhavani shear zones, with a note on its implication on the tectonics of the southern Indian Precambrian shield. Proc. Indian Acad. Sci. (Earth Planet. Sci.), vol. **105**, pp. 173–189.

Naha, K., Srinivasan, R. and Jayaram, S. (1993). Structural relations of charnockites of the Archaean Dharwar craton, southern Indian Precambrian shield. Proc. Indian Acad. Sci. (Earth Planet. Sci.), vol. **105**, pp. 173–189.

Nand Kumar, V., Santosh, M. and Yoshida, M. (1991). Decompression granulites of southern Kerala, south India: microstructures and mineral chemistry. J. Geosci., Osaka City University, vol. **34(6)**, pp. 119–145.

Narayanswami, S. and Lakshmi, P. (1967). Charnockitic rocks of Tinnelvelley district, Madras. J. Geol. Soc. India, vol. **8**, pp. 35–50.

Newton, R.C. (1987). Petrologic aspects of the Precambrian granulite facies terrains bearing on their origin. In: Kroener, A. (ed.), Proterozoic Lithospheric Evolution. Am. Geophys. Union Geodynamic Series, Washington D.C., 17, pp. 11–26.

Peucat, J.J., Vidal, P., Bernard-Griffiths, J. and Condie, K.C. (1989). Sr, Nd, and Pb isotopic systems in the Archaean low- to high-grade transition zone of southern India: syn-accretion vs. post-accretion granulites. J. Geol., vol. **97**, pp. 537–550.

Pichamuthu, C.S. (1953). The Charnockite Problem. Mysore Geol. Assoc., Spl. Publ., Bangalore.

Pichamuthu, C.S. (1960). Charnockite in the making. Nature, vol. **188**, pp.135–136.

Pichamuthu, C.S. (1965). Regional metamorphism and charnockitization in Mysore state, India. Indian Mineral., vol. **6**, pp. 46–49.

Radhakrishna, T. (2004). The Achankovil shear zone. Abstract in Intern. Workshop on Tectonics and Evolution of the Precambrian Southern Granulite Terrain, India & Gondwana Correlations, National Geophysical Research Institute, Hyderabad, pp. 116–118.

Radhakrishna, B.P. and Naqvi, S.M. (1986). Precambrian continental crust of India and its evolution. J. Geol., vol. **94**, pp. 145–166.

Rai, S.S., Srinagesh, D. and Gaur, V.K. (1993). Granulite evolution in southern India—a seismic tomographic perspective. Mem. Geol. Soc. India, vol. **25**, pp. 235–264.

Raith, M. (2004). Nature and tectonic significance of the Moyar and Bhavani shear zones (Abstract), International workshop on *Tectonics and Evolution of the Precambrian Southern Granulite Terrain, India and Gondwanian Correlations*, Feb. 2004, National Geophysical Research Institute, Hyderabad, p.46.

Raith, M., Karmakar, S., and Brown, M. (1997). Ultrahigh-temperature metamorphism and multistage decompressional evolution of sapphirine granulites from the Palani Hill Ranges, southern India. J. Metam. Geol., vol. **15**, pp. 379–399.

Raith, M., Raase, P., Ackermand, D. and Lal, R. K. (1983). Regional geothermobarometry in the granulitefacies terrane of South Inida. Trans. Roy. Soc. Edinburgh (Earth Sciences), vol. **73**, pp. 221–244.

Raith, M., Srikantappa, C., Ashamanjeri, K. and Spiering, B. (1990). The granulite terrane of the Nilgiri hills (south India): characterization of high-grade metamorphism. In: Vielzeuf, D. and Vidal, Ph. (eds.), *Granulites and Crustal Evolution*. NATO ASI Series, 311. Kluwer Academic, Dordrecht, pp. 339–365.

Raith, M., Srikantappa, C., Buhl, D. and Koehler, H. (1999). The Nilgiri enderbites, south India: nature and age constraints on protolith formation, high-grade metamorphism and cooling history. Precambrian Res., vol. **98**, pp. 129–150.

Ramakrishnan, M. (1988). Tectonic evolution of the Archaean high-grade terrain of south India (Abstract), workshop on The Deep Continental Crust of South India, Jan. 1988, Field excursion guide, pp. 118–119.

Ramakrishnan, M. (1993). Tectonic evolution of the granulite terrains of south India. Mem. Geol. Soc. India, vol. **25**, pp. 35–44.

Ramakrishnan, M. (2003). Craton-mobile belt relations in Southern Granulite Terrain. Mem. Geol. Soc. India, vol. **50**, pp. 1–24.

Ramakrishnan, M. (2004). Evolution of Pandyan Mobile Belt in relation to Dharwar craton (Abstract), International workshop on *Tectonics and evolution of the Precambrian Southern Granulite Terrain, India and Gondwanian correlations*, Feb. 2004, National Geophysical Research Institute, Hyderabad, pp. 39–40.

Ramakrishnan, M. and Vaidyanadhan, R. (2008). Geology of India, vol. **1**. Geol. Soc. India, Bangalore, 556p.

Ramiengar, A.S., Ramakrishnan, M. and Vishwanathan, M.N. (1978). Charnockite gneiss complex relationship in southern Karanataka. J. Geol. Soc. India, vol. **17**, pp. 411–419.

Ravindra Kumar, G.R. (2004). Kerala Khondalite Belt: major rock types (Abstract), International workshop on Tectonics and Evolution of the Precambrian Southern Granulite Terrain, India and Gondwanian Correlations, Feb. 2004, National Geophysical Research Institute, Hyderabad, pp. 117–122.

Ravindra Kumar, G.R. and Chacko, T. (1994). Geothermobarometry of mafic granulites and metapelites from Palghat Gap region, south India: petrologic evidence for isothermal decompression and rapid cooling. J. Metam. Geol., vol. **12**, pp. 479–492.

Ravindra Kumar, G.R. and Raghavan, V. (1992). The incipient charnockites in transition zone, khondalite zone and granulite zone of south India: controlling factors and contrasting mechanisms. J. Geol. Soc. India, vol. **39**, pp. 293–302.

Reddi, P.R., Rajendra Prasad, B., Vijaya Rao, V., Sain, K., Prasad Rao, P., Khare, P. and Reddy, M.S. (2003). Deep seismic reflection and refraction/wide-angle reflection studies along Kuppam-Palani transect in the Southern Granulite Terrain of India. In: Ramakrishnan, M. (ed.), *Tectonics of Southern Granulite Terrain: Kuppam-Palani Geotransect*. Mem. Geol. Soc. India, vol. **50**, Bangalore, pp. 279–106.

Sacks, P.E., Nambiar, C.G. and Walters, L.J. (1997). Dextral Pan-African shear along the southwestern edge of the Achankovil shear belt, south India, constraints on Gondwana Reconstruction. J. Geol., vol. **105**, pp. 275–284.

Santosh, M. (1996). The Trivandrum and Nagercoil blocks. In: Santosh, M. and Yoshida, M. (eds.), *The Archaean and Proterozoic Terrains of Southern India Within Gondwana*. Gondwana Res. Group Mem., vol. **3**, Field Sci. Publ., Osaka, pp. 243–277.

Santosh, M. (2003). Granulites and fluid: a petrologic paradigm. Mem. Geol. Soc. India, vol. **52**, pp. 289–311.

Santosh, M., Collins, A.S., Tamashiro, I., Koshimoto, S., Tsutsumi, Y. and Yokoyoma, K. (2006). The timing of ultra high temperature metamorphism in southern India: U-Th-Pb electron microprobe ages from zircon and monazite in sapphirine-bearing granulites. Gondwana Res., vol. **10**, pp. 128–155.

Santosh, M., Harris, N.B.W., Jackson, D.H. and Mattey, D.P. (1990). Dehydration and incipient charnockite formation: a phase equilibria and fluid inclusion study from South India. J. Geol., vol. **98**, pp. 915–926.

Santosh, M., Kagami, H., Yoshida, M. and Nand Kumar, V. (1992). Pan-African charnockite formation in East Gondwana: geochronological (Sm–Nd and Rb–Sr) and petrogenetic constraints. Bull. Indian Geol. Assoc., vol. **25**, pp. 1–10.

Santosh, M, Iyer, S.S., Vasconcellos, M.B.A. and Enzweiler, J. (1989). Late Precambrian alkaline plutons in southwest India: geochronological and rare-earth element constraints on Pan-African magmatism. Lithos, vol. **24**, pp. 65–79.

Srikantappa, C., Ashamanjari, K.G. and Raith, M. (1988). Petrology and geochemistry of the high pressure Nilgiri granulite terrain, southern India. J. Geol. Soc. India, vol. **31**, pp. 147–148.

Srikantappa, C., Raith, M. and Spiering, B. (1985). Progressive charnockitization of a leptynite-khondalite suite in southern Kerala, India—evidence for the formation of charnockite through decrease in fluid pressure? J. Geol. Soc. India, vol. **26**, pp. 849–872.

Srikantappa, C., Srinivas, G., Basavarajappa, H.T., Prakash Narashimha, K.N., Basavalinga, B. (2003). Metamorphic evolution and fluid regime in the deep continental crust along the N-S geotransect from Vedlar to Dharmapuram, southern India. In: Ramakrishnan, M. (ed.), *Tectonics of Southern Granulite Terrain, Kuppam-Paalani Geotransect*. Mem. Geol. Soc. India, vol. **50**, Bangalore, pp. 319–373.

Srinagesh, D and Rai, S.S. (1996). Teleseismic tomographic evidence for contrasting crust and upper mantle in south Indian Archaean terrains. Phys. Earth Planet. Int., vol. **97**, pp. 27–41.

Subramanian, V. (1983). Geology and geochemistry of the carbonatities from Tamil Nadu, India. Ph.D. thesis, Indian Inst. Sci., Bangalore (unpublished).

Subramanian, K.S. and Selvan, T.A. (1981). *Geology of Tamil Nadu and Pondicherry*. Geol. Soc. India, Bangalore, 192p.

Sugavanam, E.B. and Vidyadharan, K.T. (1988). Structural patterns in high-grade terrain in parts of Tamil Nadu and Karanataka. In: *Deep Continental Crust of South India*, Workshop Volume. Geol. Soc. India, Bangalore, pp. 153–154.

Tomson, J.K., Bhaskar Rao, Y.J., Vijaya Kumar, T. and Mallikarjuna Rao, J. (2006). Charnockite gneisses across the Archaean-Proterozoic terrane boundary in the Southern Indian Granulite Terrain: constraints from minor-trance element geochemistry and Sr-Nd isotopic systematics. Gondwana Res., vol. **10**, pp. 115–127.

Unnikrishnan-Warrior, C., Santosh, M. and Yoshida, M. (1995). First report of Pan-African Sm—Nd and Rb—Sr mineral isochron ages from regional charnockites of southern India. Geol. Mag., vol. **132**, pp. 253–260.

Vemban, N.A., Subramanian, K.S., Gopalkrishnan, K. and Venkata Rao, V. (1977). Major faults, dislocations/lineaments of Tamil Nadu. Geol. Surv. India Misc. Publ., vol. **31**, pp. 53–56.

Weaver, B.L., Tarney, J., Windley, B.F., Sugavanam, E.B. and Venkata Rao, V. (1978). Madras granulites: geochemistry and P-T conditions of crystallization. In: Windley, B.F. and Naqvi, S.M. (eds.), *Archaen Geochemistry*. Elsevier, Amsterdam, pp. 177–204.

Yoshida, M. and Santosh, M. (1995). India and Antarctica during the Precambrian. Geol. Soc. India Mem., vol. **34**, 412p.

Postscript

Study of the Himalaya and Proterozoic fold belts has progressed to a point that no book can completely satisfy the reader. This book is one more attempt in this direction, covering the fold belts and the cratons of the Indian shield. It tries to probe the subject into a greater depth, giving all about their evolution mechanism, from beginning to the end processes. The different evolutionary models for each fold belt have been evaluated on the basis of available geological and geochronological database and critically interpreted. Not inadvertently, the book contains minimum coverage to a case history on a theme/topic, since its main aim is to emphasize and appreciate the *concepts* that have developed from time to time regarding the evolution of the fold belts of India. The most significant, perhaps major, may be the still unanswered questions that have emerged with ever increasing clarity from the cumulative data presented here in an unbiased manner. This writer could not probe deeper into the topic of the book because of his own constraints in integrating the results of different branches of geosciences with geophysics. Hence one can find much lacuna in the work relating to the fold belts of India. The reader may even say with Sherlock Holmes (Sherlock Holmes, *The Adventure of the Copper Beeches)* "You have erred perhaps in attempting to put colour and life into each of your statements instead of confining yourself to the task of placing upon record that severe reasoning from cause to the effect which is really the only notable feature about the thing. You have degraded what should have been a course of lectures into a series of tales". To all these critiques, the author makes an appeal to send their comments and suggestions for improvement of this book.

Appendix

The Geological Time Scale

Era	Period	Epoch	Ma
Cenozoic	Quaternary	Holocene	0.01
		Pleistocene	1.64
	Tertiary	Pliocene	5.2
		Miocene	23.5
		Oligocene	35.5
		Eocene	56.5
		Palaeocene	65
Mesozoic	Cretaceous		146
	Jurassic		208
	Triassic		245
Palaeozoic	Permian		290
	Carboniferous		363
	Devonian		409
	Silurian		439
	Ordovician		510
	Cambrian		570
Precambrian		*Eon*	
		Proterozoic	2500
		Archaean	

Ages from Harland, W.B., Armstrong, R.L, Cox, A.V., Craig, L.E., Smith, A.G. and Smith, D.G. (1990). *A Geological Time Scale 1989*. Cambridge University Press, Cambridge.

About the Book

This book, Cratons and Fold Belts of India is a unique attempt at presenting updated geological characteristics of the cratons and the evolution of the Himalaya and Proterozoic fold belts of the Indian shield. The author has evaluated the different evolutionary models for each fold belt in light of the available geological and geochronological information. Shortcomings in the various existing evolutionary models are discussed, and a geodynamic model is presented for each fold belt. The book is self-contained—it includes an introduction to the processes of mountain building, especially plate tectonics theory with its application to the evolution of the Himalaya as an illustrative example—so that the reader can better appreciate the novel approach to the evolution of the fold belts. Furthermore, a separate chapter is devoted to the cratons of the Indian shield, covering all aspects of their geology, petrology, deformation, and geochronology. The author eschews a detailed account of the fold belts for a clear description of all the concepts that go into building models. It is primarily written for graduate students, teachers and for those curious geoscientists who aspire to know *all* about the Indian shield.

About the Author

Ram S. Sharma has a vast teaching experience at Banaras Hindu University and several other universities of India and abroad. Ram S. Sharma was born in 1937. He obtained his Bachelor's degree from Rajasthan University, Jaipur and his Masters in Geology from Banaras Hindu University, Varanasi. He obtained his D.Phil. from the University of Basel, Switzerland, for his work on the mineralogy and fabrics of migmatitic rocks of the Swiss Alps, under the supervision of E. Wenk. After obtaining his doctorate, he taught at Banaras Hindu University, Varanasi for over 30 years. During this period, Sharma was also visiting faculty at various universities in India and abroad, including the Curtin University of Technology at Perth, Western Australia. He has been an Alexander von Humboldt Fellow at Karlsruhe (Germany) where he collaborated with E. Althaus in 1976 and 1984, and a Commonwealth Fellow at Leicester, UK in 1983, where he collaborated with B.F. Windley. His research focus has been on metamorphic petrology and fabric studies of crystalline rocks of Archaean gneissic complex and Aravalli mountain belt of Rajasthan. He has published over a hundred refereed works in more than thirty journals. His geological experience includes field work in the classic areas of Lewisian and Highlands of Scotland, Tauern Window in Austrian Alps, Root zone of the Pennine Nappe (Swiss Alps), Adirondack (USA), Albany (Western Australia), Alden (Eastern Siberia), Central Crystallines of Higher Himalaya (India), Southern Granulite Terrain of India, as well as various other locations in Germany, Italy, France, and Greece. Professor Ram Sharma is a Fellow of the Indian Academy of Sciences (FASc), Bangalore, and of the Indian National Science Academy (FNA), New Delhi. Professor Sharma is presently an INSA Honorary Scientist.

Index

A

Accretion prism, **17**, 54, **161**, 202
Ahar river granite, **90**, 92, **93**
Amgaon gneiss, **58**, 60, 62, 186, 187, 192, 193, 194, 195, 197, 202
Amgaon group, **58**, 59, 65, 192, 196, 197
Ampferer subduction, **137**
Anjana granite, 159, **160**, 162
Annamalai-Palani hills, **269**
Aravalli mountain belt, 41, **143–171**
Aravalli supergroup, 41, 85, 87, 90, 143, **145**, 146, 147, 148, 153, 154, 156, 157, 158, 160, 166, 168, 169, 170, 171
Archaean terrains, **3–7**, 67, 276, 277, 282
Arc-trench gap, **17**
Arkasani granophyre, **71**, 215, 219, 226
Asian block, **118**, 122
Aulocogen, **27–28**, 160, 234

B

Bababudan group, **47**, 48
Back-arc basin, 7, **20–22**, 32, 55
Bandanwara granulite, 91, **94**
Banded gneissic complex, 41, 83, **84–85**, 87, 88, 92–94, 143, 145, 201
Banded iron formation (BIF), 4, **50**, 67, 69
Bangong-Nuijiang suture, **121**
Barambara granite, 180, **182**
Barotiya-Sendra belt, **149**, 165
Bastar craton, 42, 43, **58–65**, 177, 178, 179, 181, 182, 185, 186, 187, 188, 195, 198, 200, 202, 203, 231, 232, 233, 238, 244, 248, 252, 254
Berach granite, 41, 83, 84–85, 86, 87, **90**, **92**, 95, 96, 97, 98, 147, 148, 153, 165, 167, 171, 202
Betul belt, **183**, 185, 199
Bhilwara supergroup, **85**, **86**, 149, 171
Bhim-Rajgarh belt, **149**

Bhopalpatnam granulite belt, **64**
Bihar mica belt, 78, 79, **80**
Bilaspur-Raipur granulite, 61, **62, 63**, 190
Blueschists, 33, 35, 38, **124**, 134
Bonai granite, 68, **69**, 71, 73, 74, 209, 215, 220
Bundelkhand granite massif, 83, 92, **94–98**, 169, 181

C

Central Indian tectonic zone (CITZ), 58, 61, 63, 81, 177, 178, 179, 183, 188, **189–191**, 195, 200, 204, 219
Chaibasa formation, **70**, 209, 210, 211, 213, 218, 219
Chakradharpur granite gneiss, 68, **69**, 212, 215
Champaner group, 146, **147**, 148
Channel flow model, **129**, 130
Charnockite, 3, 46, 47, 48, 58, 60, 64, 76, 77, 79, 89, 94, 154, 156, 157, 169, 185, 187, 213, 231, 232, 233, 235, 236, 239, 244, 247, 263, 264, **265**, 266, 267, 268, 269, 270, 271, 272, 273, 274, 275, 276, 277, 279, 281, 283, 284
Chhotanagpur granite gneiss complex, 42, **72–82**, 101, 210, 218, 219, 221, 224
Chilka lake granulite, **237**, 240
Chinwali-Pilwa-Arath granulite, **151**
Closepet granite, 42, 46, **48**, 52, 53, 57, 266, 281, 282
Continental arc, 19, 20, **21**, 22, 53, 117, 163, 222
Continuum model, **133**
Craton
  Bastar, 42, 43, **58–65**, 177, 178, 179, 181, 182, 185, 186, 187, 188, 195, 198, 200, 202, 203, 231, 232, 233, 238, 244, 248, 252, 254

299

Chhotanagpur granite-gneiss complex, 42, 72, **75–82**, 101, 210, 218, 219, 221, 224
Dharwar, 42, **43–58**, 64, 201, 231, 232, 238, 244, 252, 264, 267, 268, 271, 274, 276, 277, 279, 280, 282, 283, 284
Meghalaya, 41, 43, **98–102**
Rajasthan-Bundelkhand, 42, 43, **83**, 120, 177, 202
Singhbhum, 42, 43, **65–72**, 80, 82, 177, 191, 210, 211, 212, 213, 215, 216, 218, 219, 221, 222, 223, 224, 231, 232, 252, 253

D

Dalma group, **209**, 211
Daltonganj-Hazaribagh belt, 76, **79**
Darjin group, **71**, 226
Dauki fault, 99, **101**
Derwal granite, 94, **148**, 153, 159, 160, 162
Dhanjori group, **70**, 71, 209
Dhanjori volcanics, 70, **71**, 210, 216
Dharwar batholith, **48**, 49, 53, 57
Dharwar craton, 42, **43–58**, 64, 201, 231, 232, 238, 244, 252, 264, 267, 268, 271, 274, 276, 277, 279, 280, 282, 283, 284
Dharwar greenstone belt, 50, **51**, 52
Dharwar schist belt, 46, 47, **49, 50–52**, 57, 67, 274, 278
Dharwar supergroup, **45**, 47, 50, 51
Disang thrust, 99, **101**
Dongargarh granite, 178, **193**, 194
Dudhi adamellite-granite, **184**
Dumka granulite, **77**, 78

E

Eastern Dharwar craton, **46**, 51, 244, 264, 266
Eastern Ghats mobile belt, 8, 42, 43, 58, 66, **231–257**
Eastern Ghats province, 252, **254–257**
Ensialic orogenesis, **136–138**, 171, 223, 224
Erinpura granite, 84, **151**, 153
Evolution of
  Aravalli-Delhi fold belt, **164**
  Central Indian fold belt, 61, **177–204**
  Dongargarh fold belt, 8, 42, 65, 178, 179, **191–195**, 196, 198, 202, 204
  Eastern Ghats mobile belt, **231–257**
  Himalayan mountain belt, **118**, 122–126
  Mahakoshal fold belt, **181–182**
  Pandyan mobile belt, **263–284**
  Sakoli fold belt, **197–198**
  Satpura (Sausar) fold belt, **186–189**
  Singhbhum fold belt, **218–220**

F

Fore-arc basin, 17, **18**, 32, 163
Foreland basin, 20, **30**, 34, 57, 133

G

Gangpur group, **226**
Garo-Golapar hills, **100**
Geosynclinal theory, **9–12**
Ghatsila formation, **211**
Gingla granite, 89, **90**
Godavari rift (graben), 42, 58, 61, **234**, 238, 239, 240, 241, 242, 249, 250, 255
Godhra granite, **147**
Graben, 28, 43, 64, 65, 70, 149, 231, 233, 234, 236, 238, **241–242**, 248, 249, 254, 256, 257
Great boundary fault (GBF), **83**, 95, 144, 171
Greenstone belts, 3, **4**, 7, 45, 46, 48, 50, 51, 52, 53, 57, 58, 67, 72, 181
Gyangarh-Thana area, **156**

H

Hercynian belt, **2**
  See also Variscan orogeny
Higher Himalayan crystallines (HHC), 118, 119, **120**, 121, 122, 126, 127, 129
Himalaya, 9, 11, 30, 33, 34, 99, 100, **118–134**
Himalayan frontal thrust, **119**, 133
Himalayan metamorphic belt (HMB), 121, **125, 126**, 127, 129
Hindoli group, 85, **86**, 147, 149, 171
Hutti-Muski schist belt, **52**

I

Indentation model, **133**
Indian plate, 101, 119, **121–128**, 132, 133
Indian shield, **2**, 8, 28, **41–102**, 120, 132, 136, 177, 201, 215, 223, 250, 266, 280, 281, 282, 283
Indus-Tsangpo suture zone (ITSZ), 118, 119, **120**, 122, 123, 124, 125, 130, 132, 133
Inverted metamorphism, 121, 125, **126**, 127, 129
Iron-ore group, 66, **69–70**, 220
Island arc, 1, 4, 6, 7, 13, 14, 15, 16, 18, **19**, 20, 21, 31, 32, 36, 55, 117, 121, 123, 124, 163, 164, 181, 190

Index

J
Jamda-Koira basin, **69**, 70
Jeypore province, 252, **253**, 256

K
Kabbal quarry, **267**
Kakoxili suture, **121**
Karakoram granite batholith, **121**
Karimnagar granulite belt, **43**, 65
Kavital granitoid, **52**
Kerala khondalite belt, 264, 265, 266, **269**, 272, 278
Khairagarh group, 58, 65, 178, **192**, 193, 195, 200, 201, 202
Khariar alkaline complex, **232**
Khetri belt, 91, **149**, 151, 162
Kodaikanal-Cardimom hills, **269**, 273
Kohistan arc, **121**
Kolar belt, **48**
Kondagaon granulite, 64, **65**
Krishna province, 252, **253–254**, 256
Kuilapal granite, 210, 211, 216, **219**, 220, 221
Kuraicha gneiss, **97**

L
Ladakh batholith, **121**
Lalitpur leucogranitoid, **97**
Lesser Himalayan crystalline, 119, 120, 122, **126**, 127, 133
Leucogranite, 88, 93, 122, 126, 127, 128, 129, **131**, 257
Lithosphere, 1, 2, 5, **13**, 14, 17, 20, 22, 23, 24, 29, 30, 31, 32, 36, 51, 117, 120, 124, 130, 132, 137, 138, 164, 167, 168, 169, 197, 218, 221, 223, 232
Lithospheric plates, 12, **13**, 15, 23, 31, 136, 137, 283
Lunavada group, 146, **147**, 148

M
Madras granulite, **265**
Madurai block, 264, **265**, 266, 269, 270, 271, 272, 273, 276, 278
Mahakoshal group, 96, **178**, **179**, 180, 181, 189, 190, 200, 202
Mahendragarh area, **151**
Mahoba granitoid, **96**
Main boundary thrust (MBT), 98, 99, 119, **120**, 123, 127
Main central thrust (MCT), 118, 119, **120**, 121, 123, **126, 127**, 129
Makrohar granulite, 61, 62, **63**, 189
Malanjkhand granite (batholith), **194**
Mangalwar complex, **86**, 149, 171
Marginal basin, 4, 5, 7, 20, **21**, 22, 51, 53, 54, 161, 163, 222
Mayurbhanj granite, 66, 68, **69**, 71, 210, 212, 215, 217, 219
Meghalaya craton, 41, 43, **98–102**
Mewar gneiss, **86, 87**, 97
Mikir Hills massif, **98**
Millie granite, **100**
Mishmi Hills complex, **101**
Monghyr orogeny, **78**
More Valley granulite terrain, **77**
Mountain building processes, **9–12, 13**, 117, 136

N
Nandgaon group, 58, 65, **192**
Newania carbonatite, **93**
Newer dolerite, 210, 212, 217, **223**, 224, 225
Nilgiri block (massif), 264, 265, **267**
North Delhi fold belt, **149**, 151, 152, 158, 160
North Singhbhum mobile belt, *see* Singhbhum fold belt

O
Ocean ridge, 4, **13**, 14, 15, 16, 24, 25, 31, 36, 78
Offscrapping, **18**
Older greenstone belt, **54**
Older Metamorphic group, 66, **67, 68**, 72, 75, 210
Older Metamorphic tonalite gneiss, 66, **67**
Ongerbira volcanics, 71–72, 210, **212**, 213, 219
Ophiolite complex, **124**
Opx-in isograd, 44, **46**, 263, 264, 265, 274, 281, 283
Orogenesis, 1, 2, 9, 22, **30**, 36, 68, 89, 118, 136–138, 158, 163, 166, 171, 223, 224, 282
Orogenic belts, **1**, 2, 3, 8, 10, 16, 28, 30, 33, 34, 35, 38, 39, 42, 82, 117, 118, 134, 135, 189, 237, 276
Orogenic phase, **11–12**
Orogeny, 2, **30–35**, 47, 50, 51, 57, 63, 78, 86, 93, 94, 120, 135, 136, 138, 145, 148–149, 151–153, 166, 170, 171, 185, 236, 240, 242, 244, 247, 249, 280
Outer swell, **16–17**
Owen transform fault, **133**

P
Paired metamorphic belts, 19, **33**
Pandyan mobile belt, 8, 42, 43, 53, **263–284**

Peninsular gneiss, **45**, 46, 47, 48, 49, 50, 51, 52, 53, 57, 58, 266, 267, 279, 281
Petrotectonic indicators, **134**
Plate margins
  conservative, 13, **14–16**
  constructive, **13–14**, 15
  convergent, 12, 13, **14**, 31, 124
Plate tectonic cycle, **31**
Plate tectonic theory, **13**, 16, 136
Proterozoic fold belts, 2, 8, 16, 35, 42, 49, 83, 87, 118, **134, 136**, 144, 153–154, 158, 159, 162, 163, 178, 198, 199
Proterozoic fold belts of India
  Aravalli fold belt, 8, **143–149**, 158, 159, 160, 161, 164, 165, 166, 167, 168, 223
  Delhi fold belt, 95, 143, 144, 145, 148, **149**, 150, 151, 152, 153, 157, 158, 159, 160, 161, 162, 163, 164, 165, 166, 167, 169, 170, 171, 191
  Dongargarh fold belt, 8, 42, 65, 178, 179, **191–195**, 196, 197, 198, 199, 202, 204
  Eastern Ghats mobile belt, 8, 42, 43, 58, 66, **231–257**
  Mahakoshal fold belt, 42, 177, 178, **179–182**, 188, 200, 202, 204
  Pandyan mobile belt, 8, 42, 43, 53, **263–284**
  Sakoli fold belt, 42, 177, 178, 179, 188, 193, **195–198**, 199, 203, 204
  Satpura (Sausar) fold belt, **182–185**, 198
  Singhbhum fold belt, 42, 75, 81, 82, **209–226**
Proterozoic terrains, **7–9**, 134, 177
Pull-apart basins, **28–29**, 149, 159, 160

Q

Quangtong block, **121**
Quetta-Chaman fault, **133**

R

Raialo series, 84, 146, **148**, 149
Rajasthan-Bundelkhand craton, 42, **83**, 120, 177, 202
Rajgir-Kharagpur belt, **79**
Rajmahal trap, **101**
  See also Sylhet trap
Rakhabdev ultramafics, **149**, 159
Ramakona-Katangi granulite (RKG), 61, **62**, 187, 190, 199
Ranakpur metabasalt, **152**, 170
Rengali province, **252, 253**, 256

Rifting, 20, 21, **22–30**, 36, 37, 41, 51, 52, 53, 54, 57, 65, 92, 93, 135, 137, 148, 152, 157, 159, 160, 163, 164, 165, 167, 168, 180, 181, 194, 197, 200, 202, 215, 216, 217, 222, 223, 224, 225, 245, 283

S

Sakoli group, **195, 196**, 197, 198, 199
Sakoli orogeny, **196**, 197, 198
Saltora area, **77**
Sandmata granulite complex, **89**
Sankari Durg granite, **283**
Sargur group, **45**, 46, 47, 48, 268, 272
Satpura fold belt (SFB), 42, 61, 65, 81, 82, 177, 178, 179, 180, 181, **182–185**, 186–189, 190, 191, 195, 196, 197, 198, 200, 203, 204, 209
Satpura orogeny, 60, 78, 147, 182, **184**, 185, 186, 187, 189, 190, 191, 196, 197, 199, 203, 204, 209, 217, 220, 226
  See also Sausar orogeny
Satyamangalam group, **268**, 272, **274**, 275
Sausar group, 60, 64, 178, **182, 183**, 185, 186, 187, 189, 191, 196, 199, 200, 201, 202, 203
Sausar orogeny, **64**, 185, 186, 189
Sawadri group, **88**
Sea-floor spreading, **12**, 13, 20
Sendra granite, **152**, 161
Shear zone
  Achankovil, 42, 264, **265**, 266, 273, 277, 278
  Bhavani, 43, 44, 46, **263**, 264, 265, 266, 267, 268, 271, 272, 274, 275, 276, 279, 280
  Central Indian, 58, **60**, 61, 62, 177, 178, 183, 189, 190, 191, 193, 194, 201, 202
  Chitradurga, **46**, 47, 264
  Moyar, **263**, 264, 267, 268, 271, 274, 275, 280
  Negavalli-Vamosdhara, **249**
  Northern, 80, 185, 209, **222**, 225
  Noyil-Cauvery, 263, **272**
  Palghat-Cauvery, 53, **263**, 276, 280, 281, 282, 284
  Sileru, 231, 232, **233**, 247, 249
  Singhbhum, 66, **71**, 210, 211, 212, 213, 214, 216, 217, 219, 220, 221, 225, 226
Shillong group, **100**
Shillong plateau, 42, **98**, 99, 101

Shyok ophiolite zone, **124**
Singhbhum craton (SC), 42, 43, 61, **65–72**, 80, 82, 177, 191, **209**, 210, 211, 212, 213, 215, 216, 217, 218, **219, 220**, 221, 222, 223, 224, 225, 231, 232, 252, 253
Singhbhum fold belt, 42, 75, 81, 82, **209–226**
Singhbhum group, 69, 70, 71, 79, 81, 209, 211, 213, **215**, 216, 217, 218, 219, 220, 221, 222, 223, 224, 225
Singhbhum shear zone (SSZ), 66, **71**, 210, 211, 212, 213, 214, 216, 217, 219, 220, 221, 225, 226
Singhbum granite, **68**, 69, 211
Sittampundi complex, 268, **272**, 275
Soda granite, **71**, 215, 216, 217, 219, 220, 221, 226
Sonapahar high grade area, **100**
SONA zone (Son-Narmada lineament), 41, 42, 61, 62, **178**, 179, 180, 181, 182, 183, 188, 189, **190**, 191, 200, 201, 202, 203
Son-Narmada lineament (SONA zone), *see* SONA zone
South Delhi fold belt (SDFB), 145, **149**, 150, 152, 160, 161
Southern granulite terrain (SGT), 42, 44, **46**, 57, 100, 136, 250, 263, 264, 265, 266, 272, 273, 276, 278, 281, 297
South Tibet detachment system (STDS), **118, 129**, 131
Stratigraphic status of Chhotanagpur granite gneiss complex, **81, 82**
Subduction zone, 1, 2, 12, 13, **14**, 17, 18, 19, 20, 21, **32**, 33, 36, 38, 53, 117, 118, 122, 133, 134, 147, 162, 191, 198, 222, 275, 276
Subhimalaya (Outer Himalaya), **119**
Sukinda thrust, 66, **231**
Sukma group, **58**, 59
Sung Valley carbonatite, **100**
Suspect terrains, **38–39**, 123
SWEAT model (SW US-East Antarctica Model), **247**
Sylhet trap, 99, **101**
Syntaxis
  Kashmir (Nanga Parbat), **118**
  Namche-Barva, **118**

T
Tamkhan granite, 180, **181**
Tamperkola granite, **71**

Tanwan group, **85**, 88
Tertiary fold belt, **2**
Tethyan (Tibetan) sedimentary zone, 122, 126, **129**, 130, 131
Tibetan block, 119, 121, **123**
  See also Asian block
Tirodi gneiss, **59, 60**, 62, 63, 178, 183, 184, 185, 186, 187, 189, 199, 202
Tomka-Daiteri (D) basin, **70**
Transform fault, 13, 14, **15**, 21, 25, 27, 28, 133, 162
Trans-Himalayan batholith, 118, **121, 123**, 124
Transition zone (TZ), **46, 231**, 232, 233, **244**, 247, 248, 255, 264, 267, 268, 274, 276
Trench, 12, 13, 14, 15, 16, **17**, 18, 19, 25, 26, 27, 31, 32, 33, 36, 37, 117, 124, 160, 161, 163, 165, 168, 171
Triple junctions, **24–28**, 159, 162, 163, 165, 166, 168
Trivandrum block, 264, **265**, 269, 270, 273, 276, 277, 283
Tso-Morari (crystalline) dome (TMC), **124**, 130

U
Ultra high pressure metamorphism (UHP), 124, **125**, 138
Ultra high temperature metamorphism (UHT), 78, 236, **237**, 245, 246, 269, 270, 281
Untala granite, 89, **90, 92**, 93, 153

V
Variscan orogeny, *see* Hercynian belt
Volcanic arc, 4, 6, 17, **18–20**, 22, 31, 32, 33, 117, 161, 164, 222

W
Western block, *see* Western Dharwar craton (WDC)
Western charnockite zone (WCZ), 232, **233**, 241, 244, 247, 248, 250, 253, 255
Western Dharwar craton (WDC), 45, **46–47**, 51, 264
Western Himalaya, **120**
Western khondalite zone (WKZ), **233**, 244, 248, 254, 255
Wilson cycle, **35–38**, 39, 159, 223

Y
Yelagat granite, **52**

Z
Zone
- Arc-continent collision, **117**
- Central charnockite-migmatite, **233**, 244, 248, 254, 255
- Central Indian tectonic, 58, 61, 63, 81, **177**, 178, 179, 183, 188, 189–191, 195, 200, 203, 219
- Continent-continent collision, 2, 36, 37, **117**, 118, 122, **125**, 130, 156, 157, 169, 199
- Delwara dislocation, **145**
- Eastern khondalite, 232, **233**, 244, 248, 254, 255
- Indus-Tsangpo suture, 118, **119**, 122, 123, 124, 133
- Shyok suture, 123, **124**
- Son-Narmada lineament, **41**, 42, 61, 148, 178, 179, 182, **190**, 202, 204
  See also SONA zone
- Western charnockite, 232, **233**, 241, 244, 247, 248, 250, 253, 255
- Western khondalite, **233**, 244, 248, 254, 255